Bodywork for Sensory Integration

Harmonizing Anatomical Challenges to Sensory Wellness

Susan Vaughan Kratz

Ten|16
PRESS

www.ten16press.com - Waukesha, WI

For information, please contact:

Ten | 16
PRESS

www.ten16press.com
Waukesha, WI

Art Director: Kaeley Dunteman
Cover Designer: Ashton Smith
Interor Designer: Ashton Smith & Kaeley Dunteman
Photography: Jenna Rortvedt @ Jaunt Photography
Imagery:
        Shutterstock
        Wiki Commons
        Wiki Media

DEDICATION

For my mother, Joan B. Vaughan, for preparing all those noontime meals.

Other books by Susan Vaughan Kratz

A Touch Better: Two Therapists' Journeys and the Lessons They Learned from Dr. John E. Upledger about CranioSacral Therapy

Co-authored with Gayle Mya Breman

A Touch Better describes how Dr. John E. Upledger's work of CranioSacral Therapy (CST) inspired a shared journey of two therapists who put his work into practice. This book offers a promise of guidance for newer students to embrace the work, as well as for the sage therapist who is interested in learning about how the Upledger Institute International came into being. Embedded are personal observations of "Dr. John's" philosophy and character, and his eventual rise to become one of Time magazine's Top 100 Healthcare Innovators of the 21st Century. The authors share details, including their learning curves, of how they developed their CST skills and how CST profoundly changed their lives. Therapists learning the work will gain helpful tips in applying CST in the manner Dr. Upledger intended. Case studies are offered, which highlight excellent clinical outcomes. This book includes a review of scientific research and evidence supporting the use of CST as an accepted healthcare treatment. Healthcare professionals and lay persons alike searching for a deeper understanding of CST and a way to articulate their experiences will find this book beneficial. There is something for everyone who wants to make our world 'A Touch Better.'

Bodywork for Babies

Employing a fusion of craniosacral therapy, lymphatic techniques, modified visceral-myofascial-interstitial mobilization, Kratz shares treatment routines honed through years of therapy encounters with babies. Written for any healthcare professional who can practice manual therapies, this book is also a gift to all parents who wish to understand their baby's development and behaviors better. When their baby simply is not 'growing out of an issue', there are answers waiting to be discovered in the body. The body knows what the problem is and will guide the trained therapist to the exact spot. The methods offered in this book have assisted in resolving digestive issues such as colic, reflux, constipation, and breastfeeding difficulties. Torticollis, tongue-tie, movement delays and reflex integration, misshapen heads, irritability and sleep difficulties are also routinely helped in significantly less time than traditional therapies. Self-regulation and sensory modulation are behaviors that communicate from the autonomic nervous system, and Bodywork for Babies has proven to be a direct treatment that helps the parasympathetic system become active, which turns on the "rest and digest" mode. Bodywork for Babies can be a wellness "spa-like treatment" as well as remedial intervention when functional problems don't resolve on their own.

Contents

Acknowledgements

This sharing of this accumulation of therapy experiences would not be possible if not for the visionary and courageous careers of many people. A. Jean Ayres, Ph.D., OTR; Karl and Berta Bobath; Dr. John E. Upledger, DO, OMM; Bruno Chikly, MD, DO, LMT; and Jean-Pierre Barral, DO, RPT, have their voices represented within these pages. In addition, dedicated teachers associated with these greats have expanded and mentored the original lessons to my and future generations of practitioners. Your lessons allowed therapy careers to be transported to realms often not dreamt of in undergraduate school.

At one point in my career, I wished to pursue a doctoral degree for scientific study of the observations I had of connections between craniosacral system findings and sensory integration dysfunction. Following the advice of Dr. John Upledger, I chose a different path and gained a diplomat-level certification in his methods of craniosacral therapy. He thought that doctoral training would educate the instincts and intuition right out of me, both necessary to practice good if not excellent craniosacral therapy. This book was a natural extension of my first book, "Bodywork for Babies", another collection of observations I gathered following this advice.

I want to thank all those who contributed to this book's final product. The devotion and shared vision for the contents of *Bodywork for Sensory Integration* from these individuals are exceptional. To Shannon Ishizaki at Orange Hat Publishing, who continues to recognize the need to get these books out to the public, and the publishing support team of Kaeley Dunteman, Ashton Smith, and Lauren Blue. I am forever grateful for all their sustained belief in and efforts toward getting us to the finish line. Lynne Kurtz was invaluable for supportive editing throughout the writing process. Thanks go to Linda Bradak, Lisa Upledger, and Susan Steiner for the treasured feedback as prospective clinical users of the content. My bottomless appreciation is extended to these ladies for their leadership and complete belief in supporting this project.

And finally, gratitude goes out to my family: Randy, Emily, and Daniel. Your cheerleading, patience, and tolerance for my writing days and career choice are felt to the depths of my soul. I love you all so very much.

Such affirmation from all these people is what such endeavors need to succeed.

Foreword

by Emily A. Francis

[A discharge report from our Developmental Pediatrician stated]:
Impression: A 4-year, 9-month child with a history of an autism spectrum disorder, motor coordination weakness, a feeding disorder, developmental delays, speech apraxia and who now has transitioned out of the ABA center into a traditional Pre-K program where child has been functioning extremely well. Child has made progress developmentally over the past 8 months, and there have been no behavioral issues at school. The teacher did not report any social skills deficiencies on the SRS-2. Most recent speech evaluation (on April 24, 2017) revealed that receptive and expressive language skills are now falling within the average range for chronological age. Therefore, (name removed for privacy) no longer meets DSM-5 criteria for an autism spectrum disorder diagnosis. Mild residual articulation weakness persists and should be monitored, along with academic progress with maturation.

I am the mother of a child with recovery from an Autism Spectrum Disorder (ASD). When I share our story of incredible healing, people tend to shudder and make the assumption that we were not diagnosed correctly if my child could have the diagnosis removed. We are taught to believe that once you are on the spectrum, you can never get off. I am here to challenge this belief and to offer an alternative to the stigma that specific diagnoses are only addressed through maintenance and without the option of possible recovery.

I had my child evaluated through three separate avenues, and all agreed on the same diagnosis. And after our intense two-year journey, they all stated they had never witnessed a complete healing of ASD in their practices. All agreed with our lead Developmental Pediatrician in removing the diagnosis completely. As a sidebar, we moved from the United States to a European country a few years later. To be overly conscientious, I had my child meet with a developmental diagnostician there, giving him no history of the prior diagnosis and our therapeutic journey. I wanted to see whether he could pick up on any spectrum disorders or other diagnosable issues. His evaluation revealed that there was no sign or trace of autism spectrum we knew so long ago. He was surprised when I shared our history with him.

When my child was three years-old and first diagnosed, we enrolled into an intensive Applied Behavioral Analysis (ABA) school. Programming was four hours every single day with 1:1 therapy year-round without holiday or summer breaks. My child's development, which was tracked by the developmental pediatrician every six months, only improved 3 months over a 6-month timeframe. That meant development was half the rate of neurotypical peers, even with a strict diet and extensive daily therapies in place.

After six months of solely ABA therapies, we added Craniosacral Therapy (CST) to the routine. This was the only change made during that timeframe. Schooling was still at the ABA program, eating the same diet and working with the same therapists. After our first CST session, my child went from speaking only two words at a time to a fully coherent 7-word sentence. Speech apraxia was re-tested through a speech pathologist, and retesting revealed it was no longer present. Developmental indicators improved by FOURTEEN MONTHS over that second six-month period! The following and final assessment another six months later revealed that my child no longer met any criteria for a DSM-5 diagnosis and was discharged from the pediatrician's services. My child was now matched developmentally to the chronological age.

In our region of the USA, a mainstream approach for most children with an ASD diagnosis included: a change in diet, standard occupational, physical, speech, and ABA therapies. The part I take the greatest issue with is applying a "one size fits all" treatment routine. This system is outdated and only helps to minimize what other symptoms may be presented from the child. There are so many alternatives to healing, and real improvements cannot be attained through a blanket approach. Somehow, as a collective, we have dismissed the fact that recovery or self-correcting abilities are even possible.

On a personal note, craniosacral therapy, and another alternative practice called NAET (Nambrudipads Allergy Elimination Technique), are the two therapies we continued long after we discontinued the ABA school and the diagnosis removal. The NAET eradicated the need for such a strict diet without behavioral regression. Also, in keeping with our recovery and protecting it with everything we've got, we moved to a country that has a full ban on GMO foods. It was my number one driving force in finding a home abroad that allowed us the freedom to eat foods that were not sprayed with heavy chemicals, herbicides, fungicides or pesticides. The country also has a full ban on the dangerous food dyes that wreak havoc on bodily systems, particularly on a compromised or sensitive system.

There are no specific pieces of training that we, as parents, are taught to look for when finding and working with any kind of therapist. This means usually we just see one through our public school system or maybe outside of school, but we don't know what skills to look for other than the letters after their names. It was at the tail end of our journey that I learned to work with an OT with specific "sensory integration training." I now wish therapists were exposed to a wider range of interventions that can be applied to each child. My first advice to parents looking for the best help would be to find a therapist who has a significant level of extra training to focus on your child specific needs. Sensory integration, self-correcting, and body balancing are essential techniques that OT, PT, and specialized massage therapists should become familiar with.

As a professional clinical massage and lymphatic drainage therapist myself, as well as a mother with a recovered child, I am often asked from members of the standard medical and therapeutic community: "Why don't people know about this as a treatment? Why aren't we learning this with Occupational Therapy or other hands-on therapies? Why aren't insurance companies covering the costs of this incredible work? Why are these therapies not shared throughout the mainstream practices?" This comes down to one real answer: lack of education and acceptance of alternative therapies as real treatments throughout the medical and financial communities. It is my firm belief that the pediatric therapy community should be able to apply hands-on therapies in their practices. I also believe that well-qualified and properly licensed manual therapists should be employed in hospitals for such treatments as well. Most notably, craniosacral therapy needs to be an accepted and covered service across the board for properly accredited therapists in any appropriate therapeutic setting.

Working in my professional capacity, I understand there are different specialty areas in which CST and other bodywork methods can address the body's sensory needs in unique ways. A book such as this could have helped me tremendously while on our journey to better understand our child's behavior from a sensory system perspective. With a solid understanding of the sensory systems now, I am able to recommend bodywork to parents of any aged children, as well as for themselves if it is needed! This book is a perfect guide to expand the practice and clinical competency of any pediatric therapist. What needs a child has that can't be expressed in words or behavior, can now have an interpretation framework from the body's communication.

I am the author of several books on body healing, including one entitled The Body Heals Itself. My focus is on adult bodies and the muscular system. I knew nothing about the sensory system when I took on the role of recovering my child. Through my work, I believe with all of my heart that our bodies have self-correcting and self-healing mechanisms within us. I also advocate that you can't do this alone. Use this book as a roadmap to help yourself become a better pediatric therapist, or become more enlightened as a parent into your child's needs. Use this book to learn about the sensory systems and how an extremely knowledgeable and highly-qualified therapist can support you on your family's health restoration journey. Learn about the body and expand every avenue for recovery and balance within a sensitive or compromised system. Had this book been available to me sooner, I would have had a much better understanding of my child's sensory system and made different choices.

Hands-on therapies have always had an incredible effect on the body, and a young body especially is so receptive to the correction that a knowledgeable therapist can lend. Susan Vaughan Kratz is one of those remarkable and knowledgeable therapists on the topic of sensory integration, and she skillfully blended bodywork methods within her specialty practice of helping children and their families. My hope is that with such an incredible book as this all practitioners who profess to address the neurological and behavioral needs of children will add these bodywork techniques into their practice as standard treatment. Our babies and children will thank you for the rest of their lives for what can be achieved with the knowledge that lies ahead. My story of my child's healing would never have come to be had I remained solely within the mainstream set of therapies and not sought treatments that exceed our standard operating systems.

Bodywork for Sensory Integration is one of the first books of its kind. It illustrates how balancing this part of their child's body can affect the overall dynamic of their growth and development. The course for possible healing as we enter the world of bodywork for children starts here with this book. In conclusion, I hope our story of health and healing is comforting and inspiring. Our story should not be one in a million, but one in many. If you are a parent reading this, please remember to follow your instincts above all else. If you feel that there is more to be found, never stop looking. To quote our pediatrician on the day that I showed her the removal of the diagnosis, she shook her head and smiled, and said "If you want to see a miracle, put the mother in charge." If you are a therapist reading this, I hope that you expand your knowledge base to add in these hands-on techniques that will not only change your own way of looking at the world around you but will continue to foster incredible amounts of healing and recovery beyond your wildest imagination for everyone you touch.

I highly recommend this book. Use it well and refer to it often.

With gratitude and tremendous respect for the work presented on these pages,
Emily A. Francis

Emily A. Francis: Has a BS in Exercise Science and Wellness with a minor in nutrition and A Master's degree in Physical Education with a Human Performance Concentration. She studied Clinical & Neuromuscular Massage Therapy from The Atlanta School of Massage, and completed all levels of training through the Dr. Vodder School in Manual Lymphatic Drainage and Combined Decongestive Therapy. Emily trained through Upledger Institute International for Craniosacral Therapy Level 1 & 2 and introduction to autism. She is a Certified Pediatric Massage Therapist through the Little Kidz Foundation and the author of the following books:
- Stretch Therapy
- The Body Heals Itself
- Whole Body Healing
- Healing Ourselves Whole
- The Taste of Joy
- Co-Author: A Leg Up! On Lymphedema
- Contributing author: Autism Essentials

Preface

In the small town where I grew up, older kids had the option to leave school and go home for lunch. Most days, I chose Mom's home-cooked meal over the gray mystery meat often served in the cafeteria. The radio was always on during those noon meals broadcasting the ABC news. The newscaster, Paul Harvey, consistently ended his broadcasts with a human-interest feature known as The Rest of the Story. I remember listening with rapt attention to his baritone voice telling accounts of random, forgotten, or surprising facts. Fascinating tales such as Abraham Lincoln never once slept in that famous bed in the White House, or how defective toilet paper became the original roll of paper towels, or how the heritage of chastity led to the creation of graham crackers. He would spin that yarn of a tale with such artistry as he led us through the golden particulars. Harvey ended each story with the tagline, "And now you know….(pause for effect)….the rest of the story."

I would walk back to school feeling a part of an exclusive society, and curious about how he found those lost facts. Reflecting on those years, I've come to realize Mr. Harvey's recordings affirmed my inquisitive nature and launched my trajectory developing the skill of deductive reasoning. I have carried and nurtured that trait throughout my life. Those stories shaped my digging beyond the obvious, seeking more profound knowledge. It always felt important to me to understand the whole picture.

I became practiced at being a sensory detective working as a pediatric occupational therapist, experimenting with many treatment methods. Sensory theories gave me many answers to understand the challenging behaviors of children who struggled with daily life. After learning bodywork methods like craniosacral therapy, I realized there was so much more the body wanted to reveal. Behavioral theories leaves a lot of the details of a child's existence out of our analysis. Tension and tone of organs, blood vessels, lymphatic movement, and the suspending weave of fascia play a significantly underappreciated role. A long-held phrase in my professional circles, "behavior is communication," had become the starting point of analyzing children's struggles through the lens of sensory processing. Manual therapies help reveal the physicality of behaviors as an expression of the autonomic nervous system. The physiological state of a child offers a novel backstory for us to learn from. From the lessons learned from over two decades of using bodywork methods in sensory integration therapies, a new phrase worthy of promotion is "Physiology is Behavior."

Bodywork for Sensory Integration allows practitioners to gain more of the rest of the story to expand understanding of neurodiversity and become informed of possible structural issues behind sensory challenges.

Introduction

"Your body is your subconscious mind"   *Candace Pert, Ph.D.*

Sensory wellness has become an evolutionary step in understanding the function of all the nervous systems. This book offers novel ways of considering sensory integration processing, from structures that serve neurological behaviors to the under-appreciated anatomy of fascia as a sensory organ. *Bodywork for Sensory Integration* has become a valuable resource for all who wish to gain access and inclusion in neurological wellness and acceptance for neurodiversity.

Analyzing behaviors of sensory difficulties, regardless of one's profession, has yet to consider the patent integrity or receptor field for each sensory organ. Existing theories come from a vantage point of central control within the brain. Sensory-behavioral approaches to treatment assumes that the cellular anatomy and sensory nerve pathway is healthy and intact. Behavioral analysis has yet to routinely incorporate manual palpation for tone or tension of sensory structures. Novel ideas found in this book give us ways to treat the autonomic nervous system directly at the organ level, promising efficient and effective ways to help children with regulation and modulation difficulties. Manual methods are fresh concepts when considering the needs of people with diagnoses of autism, attention deficit disorder, sensory defensiveness, and other neurobehavioral issues. Neurodiversity has an expanded voice with the insights in this book. Two decades of clinical observations of this new paradigm aim to bridge a knowledge gap learned through applying bodywork to people with sensory processing challenges.

Through the ages, it has been accepted that humans have five senses: touch, sight, hearing, smell, and taste. Senses feed information to the brain so we can perceive events and actions, form understandings, and execute a reaction to function in the world. As neuroscience evolved over the past half of a century, several additional sensory systems are now appreciated. As many as ten different sensory classifications have been identified. It is presumably more accurate to describe those five basic systems as senses of conscious awareness (in most people). The other senses are generally recognized as subconscious or non-consciously processed. We call them hidden senses. Proprioception, kinesthesia, vestibular, interoception, and enteroception constitute these other five sensations, which are processed with vast neuronal distribution and extensive numbers of receptors. In addition, layers of touch perception fall into either category of awareness, with each cellular layer possessing profound effects of reacting to the things that touch our skin.

The hidden senses manage a greater volume of background information, allowing us to navigate gravity, move against unseen structures and resistance, and maintain posture and balance. Humans do this rather autonomically, without much thought or language in discussing how mastering gravity was accomplished on any given day, or how the body stayed in contact with the floor because the bottoms of feet understood what the floor means.

Each sensory system has receptor organs that stimulate a corresponding nerve and pathway. However, new knowledge of connective tissue has challenged the premise that all sensation is managed through the routes of nerves. As we learn more about the physiology of fascia, it may become apparent that not all sensation travels nerve pathways as previously thought. Receptors surrounding visceral organs and woven into connective tissues may serve as feedback mechanisms to inform or help us feel and understand our insides. The whole-body organ of fascia has been theorized to have sensory qualities reflected by retractive and expansion abilities, which most likely react in the body's best interest of homeostasis. The cellular network of fascia acts like a fiber optic cable system, providing body background feedback to supplement the transduction of sensations traveling via nerve pathways. It sends signals at warp speed, faster than the speed of sensory nerve pathways. Fascia tension may also be a player in the communication of the autonomic nervous system.

Fascia anatomy and physiology deserve inclusion in scientific pursuits of exploring sensory input and processing. It may be a significant player in interoception and enteroception. Fascia's extension from tendons and ligaments, as well as the weave through muscle cells, may also be a player in proprioception and kinesthesia.

> **Fascia anatomy and physiology deserves inclusion into future research to advance understanding of sensory integration and processing function**

In undergraduate school, my neuroscience professor instilled the critical reasoning skill for recognizing and working with the "behaviors" of the nervous system, particularly the autonomics. I've accumulated therapy methods and skill sets based upon this foundational teaching throughout my career. All post-graduate training I sought had a theme of helping nervous systems recover or adapt to the struggles and challenges of life. Working with people with brain injuries caused by illness or trauma, I witnessed firsthand how therapy activities designed in our clinics assisted people's neurological healing and recovery. Generalizing neurological rehabilitation to Ayres' Fidelity of sensory integration methods was a natural progression in comparison.

At some point in the late 1980s, I first became aware of the work of Dr. John Upledger (creator of contemporary craniosacral therapy). I learned how his methods emerged while he worked with children with severe autism. He opinioned that autism was toxic encephalopathy, though the behavioral diagnosis had been (and still is) the prevailing practice. At the time professionally, I was focused on studying sensory integration theory and practice, applying the approach to my growing caseload of children with developmental problems, brain injuries, and autism. Though it would be several more years before I embarked upon an exciting journey of training through the Upledger Institute, I gradually became aware of this innovative approach's lasting impact on people's wellness.

My eventual lessons learned in Upledger's formal classes were life-altering and career-defining events. Once I started to apply craniosacral therapy (CST) clinically, I witnessed impressive responses in several pediatric clients. Sometimes improvements would occur quickly (with just a few treatments), and sometimes it took extended periods to reveal measurable changes. I compared and contrasted the CST approach to other methods, including myofascial release, infant massage, and soft tissue mobilization in orthopedics. CST became a "structural" explanation for many sensory-based behaviors we treated. Building confidence and competence for the layers of CST work, I gravitated to other complementary methods (lymphatic drainage, lympho-fascia mobilization, visceral manipulation, and somato-emotional methods). I began to generalize bodywork with different populations and eventually these methods became my practice's primary service. The demand from the community has kept the practice very full and busy. I began to see predictable patterns of responses to treatment. (See Part IV for specifics pertaining to subgroups of sensory function.)

Historically, sensory integration therapists have analyzed and measured behaviors through the adaptive response but have not necessarily considered anatomy beyond cursory awareness. There is little information correlating the structural compromise of organs and nerve pathways and little to no reference to the human circulatory system in the literature about sensory integration and processing function. However, vascular responses are the hallway of sympathetic nervous system dominance. Even less information exists on how to treat the vascular system for a therapeutic effect on the troublesome behaviors related to the modulation of sensory input. The tools of bodywork, combined with sensory integration and neurodevelopmental therapies, ushered in changes at a more rapid pace in many clients. Blending neurobehavioral approaches with historical structural medicine principles brought forth the observations included in *Bodywork for Sensory Integration*.

Human behaviors reflect how the brain manages, sorts, shifts, deciphers and responds to information. Information can be current sensory events, memories of sensory events, and perceptions of both. Over of my forty-year career, I have had ~45,000 patient encounters, and each one offering a lesson in a fundamental yet profound truth about the autonomic nervous system (ANS). The reality is that the ANS wears its heart on its sleeve, and one can palpate the state of the ANS anywhere in the body. The ANS has two words in its vocabulary: Tight and Loose. Those two messages are expressed through the tension and tone of a nerve, its innervation target, and neighbor organs in the vicinity.

Palpating body structure for their qualitative natures is how we communicate directly with the ANS. The hypothesized, though never proven, Type A and Type B personalities were traits used to describe hard wiring for any predispositions of heart disease; however, lessons from bodywork challenge the notion that personalities define physiology.

Instead, our observations have been that moods and emotions change following alterations in the ANS state during and following bodywork. Taut visceral organs and connective tissues forming vessels seem related to the behaviors associated with stress, anxiety, depression, modulation, and regulation struggles. And behaviors we described as well-regulated and optimal readiness to engage in the world and with people are absent of tension and free of too-high tone. As if neurochemistry changes with the change to the structural tone of the viscera. Routinely, family members announce, "We have a different child" after receiving treatment. *Bodywork for Sensory Integration* experiences has led to the new hypothesis that personalities may develop, or at least be influenced by "Type Tight" and "Type Loose" of organs regulated by the ANS. When we can assist the ANS towards the parasympathetic system, expressed as varying degrees of "looseness," we assist in resetting and perhaps even affecting the neurochemical expression. Direct treatments to the fascia surrounding organs and vasculature, including key neurovascular bundles, the bony container housing the brainstem, and cranial nerve exit sites, proved helpful to down-regulate sensory systems.

Health and wellness thrive in a body where all systems are integrated and communicate without distress, without mechanical compromise, and without electromagnetic resistance.

Structural strains and compressions can hinder sensory conduction and prolonged protective retraction can skew the perceptions of sensations.

Even when sensory challenges don't necessarily involve autonomic regulation but instead reflect coordination, balance, or motor planning difficulties, manual therapies have demonstrated a positive effect on postural awareness and adaptive responses. A sense of self-worth in mastering gravity and the physical worlds does contribute to overall mental and emotional health and wellness.

In additional to these structural medicine observations, the cellular terrain of the body has also cast new light onto microscopic influences of what we call sensory-based behaviors. Biomedical and naturopathic practitioners networking with our clinic routinely discovered sources of systemic inflammation or gut dysbiosis in our clients. Certain viruses detected were suggesting an etiology for neuritis and neuropathies. Harmful bacteria known to secrete neurotoxins outnumbering good bacteria, have been identified within our younger clientele and treated by these practitioners with observed behavioral improvements. The toxic load of heavy metals such as mercury or aluminum have manifested in some parallel neurobehavioral patterns as sensory struggles.

We have witnessed many naturopathic interventions provide [additional] knowledge for the etiology of sensory-based behaviors, and have observed organic recovery. Positive changes in sensory-based behaviors are consistently observed in the co-occurring detoxification process. Bodywork combined with naturopathic medicine has proven effective for many children seen in our clinic who present with extreme sensory and out-of-bounds behaviors (that can't be attributed to learning or family relationships).

These approaches help the body self-correct and heal the biological terrain. Bodywork then seemed to address the fallout of soft tissue inflammation (resulting in chronic tightening of fascia and other soft tissues). Restricted tissues increase resistance and hinder optimal fluid flow in all the circulatory systems: interstitial, lymphatic, blood, and cerebrospinal fluid. Inflamed visceral organs and circulatory systems tighten into sympathetic tone (vasoconstriction), which creates a state of protective retraction in organs and blood vessels. Methods such as lymphatic and interstitial mobilization help restore tissue flexibility towards improving fluid exchange. Parasympathetic states help clear the effects of inflammation as the detox pathways relax and expand and are more conducive to exchanging fluids and mobilizing inflammatory markers.

Biological effects on organs and the fascia surrounding them can affect the person's adaptive behaviors. When nerves heal and gut bacteria are balanced, behavior changes. Cells communicate through vibrations of fluid and receive feedback through resonance. They react to immediate changes in tension or terrain alteration. Gut and organ health reflect the optimal tone and motility. Tensions in the body reflect the health of the cellular and organ terrain.

Bodywork experiences identified the structural location where protective retraction can be chronically held within the body, with observed pre- and post-treatment behavioral changes. Protective retraction is a homeostatic function of the fascia organ collectively. Not only do blood vessels constrict during stress, but the singular widespread fascia organ does as well. Left in chronic tension, these organs and blood vessels drive the person to remain in some degree of sympathetic dominance, and their behaviors may reflect attempts at coping and seeking strategies for returning to homeostasis. With proper training and experience, manual therapists can palpate where sympathetic tone and protective retraction persist and directly treat those areas specifically to assist parasympathetic recovery. All other endeavors for the child will become easier if the ANS becomes regulated.

However, not all people presenting with sensory challenges have a biological issue or emotional dysregulation. Sometimes a pure structural problem can present itself, which may be more likely related to in-utero positioning, the birthing process. Bodywork observations have led me to believe it is a false assumption all babies fully decompress their body structures from in-utero positioning or the birthing process. It is a philosophical belief also held by chiropractors and osteopathic practitioners. Structural compression into central nervous system structures can directly affect keeping varying persistent sympathetic states of alarm or stress.

Bodywork for Sensory Integration first offers a historical perspective of one therapist›s experiences, plus the philosophical and theoretical basis for her interpretations. This book then proposes a fresh view of the anatomy of each sensory system to consider the functional ANS behaviors associated with each. From such insights, practitioners can raise their awareness of any of their pediatric client's underlying anatomical communication related to behaviors and performance. The third section in this book reviews the fusion of methods and techniques used to treat the scope of all the sensory systems. It is not an exhaustive collection of technique, but rather an overview. Other similar methods can be included and generalized within the scope and interest of the pediatric practitioner. Finally, the last part of the book addresses historical sensory processing subgroups that contribute to neurodiversity, with observations of specific bodywork methods that can help each sensory system.

Bodywork practice is an individualized interest embraced by many professions carrying a license to "touch people." No one profession owns bodywork methods, just as no professional group owns anatomical treatments. Effective bodyworkers will be defined by their skill sets and advanced training level rather than by a professional title. The savvy consumer of health and wellness services would be wise to interview practitioners of the types of exercise and their clinical outcomes. Word-of-mouth referrals are a primary driver in bodywork practices.

The practical thing about manual therapies is that they can be generalized to different populations of all ages and reasons. From health and wellness to medical remediation to facilitation of trauma care and release, manual therapies have proven to be a cost-effective, efficient way to achieve relaxation. The power of achieving and maintaining relaxation to improve quality of life for a person and their family should never be under estimated. The quality of skill sets for the practitioner is an artform, training institutes that provide peer training and mentoring raise the predictability that these non-invasive methods are appropriately applied with no risk of harm.

Sensory processing is optimized by the autonomic nervous system's ability to react, modulate, and respond appropriately to sensory input from many systems. As we use our innate senses to touch, we can expand our perception and gain assessment data by practicing bodywork. Compassionate and therapeutic touch methods can correct, stimulate, and soothe the autonomic nervous system. *Bodywork for Sensory Integration* shows you how to do this in practical ways. This book focuses on the typical developmental progression of sensory integration. The bodywork methods are for young and older children, though treatments can be generalized to adults. My experience with all bodywork methods is that people, even children, recognize the effects of their bodywork within one to three sessions. The risk is low in trying bodywork methods when performed by a well-trained practitioner.

Regardless if we reach a consensus that sensory processing should or should not be critiqued as something to fix or remediate, the bottom line is the person who owns the sensory systems determines the degree of feeling uncomfortable or stressed. The degree in which that person's family is struggling with the individual and the collective struggle with homeostasis and sensory recovery contributes to the seeking of therapy services. Quality of life measurements can help families determine if an intervention trial is worth taking. Therapists must continue to find ways that are cost effective, inclusive, and readily available to all communities.

Part I

Existing Pediatric Therapies and Methods of Bodywork

Part I
Existing Pediatric Therapies and Methods of Bodywork

This chapter synthesizes the resources that shaped *Bodywork for Sensory Integration*, though it is not exclusive to a limited set of approaches. Other methods can be included as other professionals join the ranks and blend unique training and experiences. An eclectic approach yielded more significant results than a singular approach in clinical application.

Noninvasive bodywork methods are a great equalizer among professions. Just as no single vocation owns exclusive knowledge of anatomy, no profession owns sovereign rights to body palpation or skill sets of manual therapy. There are, of course, judicial limits to scopes of practice, such as spinal mobilization and other invasive procedures, or as defined by licensing agencies. *Bodywork for Sensory Integration* blends traditional pediatric therapy approaches with methods that work with fascia, interstitial and lymphatic fluids, the craniosacral system, neurovascular pathways and bundles, and viscera organs. Personal interest in postgraduate training prepares the clinician for this area of work.

One cannot learn competent manual skills by reading a book or attending isolated weekend workshops. Beyond formal training and workshops, skill sets are developed through practicing time on the body. Institutes that offer certification in their methods help the standard of reliability of a particular brand. Regarding the components of *Bodywork for Sensory Integration*, skill sets are defined as someone being 'qualified' rather than a particular profession laying claim to a method standard. Maturation of skill sets propel practitioners toward higher competency levels. Receiving mentoring is valuable and highly recommended.

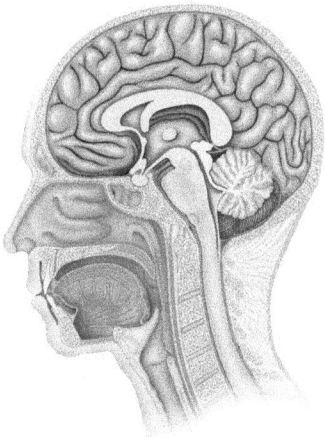

Sensory Integration is a neurological and biological process of registering, modulating, and discriminating input received through ten sensory systems. This process results from human behaviors, skill sets, and even meaning of life. Social-emotional, communication, physical capabilities, cognitive aptitudes, and self-care are the adaptive responses of Sensory Integration. Challenges or deficits can interfere with optimal learning, playing, and working, as well as socializing with people and engaging in appropriate behaviors. Much of what is processed needs to be inhibited so that selective attention to relevant stimuli can occur. Noticeable sensations are managed at the same time that background sensations are processed. As children grow, they typically refine and mature in appropriately registering, modulating, and discriminating sensory information to support the development of effective emotion regulation. Social and imaginative play skills and fine motor and gross motor skills reveal how well those background and noticeable sensations are filtered, attended to, or ignored.

Sensory modulation handles the onslaught of sensations as the child grows and interacts with their world. Modulation reflects the resiliency of the autonomic nervous system (ANS) to alert to a stimulus, respond appropriately, plan a reaction, or disregard the sensation. Modulation also refers to the child's ability to grade responses to match the demands and respond in ways that are neither over-reactive nor under-reactive to the situation. Modulation means the ANS can down-regulate from excitement and stress.

Sensory discrimination is the brain center's higher-order reaction to learning relationships between differing sensations. Discrimination refers to the child's ability to accurately perceive and utilize the sensation in a refined way to produce adaptive, functional behaviors. Discrimination of information leads to the development of preferences and avoidances. Both of these processes are needed for children to successfully develop the occupational performance skills that allow them to learn from, and interact with, the world in which they live.

Sensory Integration Treatment Concepts

Key Concept: The measurement of the brain making sense from sensory input is reflected by the array of adaptive responses to the complexity of the world. A balanced ANS requires not over- or under-reacting to sensory events, which assists the best adaptive responses. Sensory integration treatment applies methods that are child-directed, and activities designed to be the "just right challenge" for targeted sensory channels, scaffolded for greater success in mastering more complex skills.

The occupational therapy (OT) profession is credited for carrying out the conceptualized theory of SI and processing from the work of A. Jean Ayres, Ph.D., OTR, FAOTA. Dr. Ayres had several contemporaries who continued to advance the science towards a practical approach of "functional neurology" in OT interventions. Therapy equipment and activities were designed and constructed to entrain weaker sensory systems through the child's active participation. Equipment wasn't just fancy gyms, but rather therapy tools to activate sensory channels not processing or modulating well. Identifying and modifying environmental triggers with modifications is also a significant aspect of assisting someone with sensory processing differences, as well as education and advocacy to support people within their families, homes, and communities.

Occupational Therapy (OT) has a hundred-year history of addressing a client's psychological, neurological, physical, emotional, and social needs. The human being functions as a whole when the development of these traits is maximized. Professional training for OTs guided the evolution of practiced-based sensory and motor strategies to strengthen and assist the maturation of the child's nervous system when it did not occur naturally. When there is a weakness, or challenged difference, the OT considers all parts of function upon how a client occupies their time and how these components work for purpose and independence. Interventions have expanded in collaboration with other professionals on a child's therapy team. Raising awareness of the impact on human function of underlying sensory processing is now a master-level foundation in the education of becoming an occupational therapist. In addition, exploring a continuum of sensory health, neurodiversity, and the parameters of actual dysfunction has become a new narrative.

Like all other professions, OT has promoted research and scientific evidence to support practice methods. In conjunction to scientific evidence, clinical reasoning for inventive practices remains a vital aspect for forward-thinking therapists. Dr. Ayres was an innovative OT and neurobehavioral scientist, serving as a role model for future practitioners to raise the

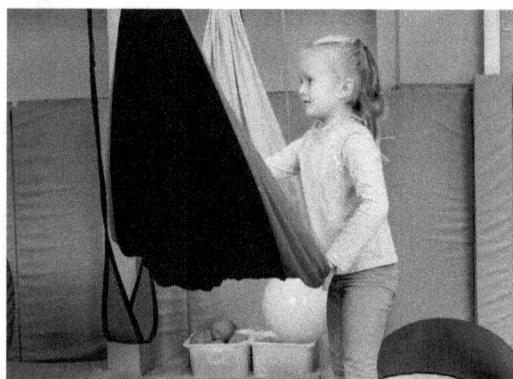

caliber of the profession. She dedicated her life to studying the impact of sensory information processing on a child's ability to learn and control their body, emotions, and behavior. Now referred to as Ayres Sensory Integration®, the fidelity of this approach includes the theory, assessment, patterns of sensory integration and praxis dysfunction, and intervention concepts.

Sensory integration theory proposes that information processing is a neurobiological activity that modulates and organizes information with a resulting expression of an adaptive (or maladaptive) response. The end products of sensory integration have been studied extensively in OT and applied to practice with parallel discoveries made in neuroscience. Ayres' work has launched the careers of thousands of OTs specializing in her methods, aiding their clinical reasoning and research using neurobehavioral analysis.

Functional Outcomes of Typical Sensory Integration

- Modulation to sensory input; emotional stability; autonomic nervous system balance
- Self- regulation and self-control (to sensory input or memories)
- Postural-Ocular control for balance and body stability
- Bilateral Coordination of both sides of body (arms, legs, eyes)
- Somato-sensory processing (meaning of touch and meaning of touch for dexterity skills)
- Praxis skills: motor planning, execution, organization, sequencing, constructing

Methods utilized in sensory integration practices may include, not limited to:

Child-directed methods

- Sensory integration gyms designed for a variety of child-directed sensory input
- Therapeutic activities that provide deep pressure input
- Therapeutic equipment that activates vestibular system
- Therapeutic exercises structured to activate kinesthesia & proprioception
- Therapeutically-designed sessions for sensory entrainment, scaffolded to promote increasing complexity of reflex integration & voluntary control of movement patterns and motor planning

Imposed input methods (augmenting Ayres' approach)

- Therapeutic Brushing
- Astronaut Training
- Therapeutic listening to engineered music
- Massage and Qigong massage
- Weighted vests
- Swinging (linear & rotatory)

Historically, sensory integration was grounded in therapeutically-designed activities that stimulate sensory input while guiding the child toward maximal adaptive responses. Interest and motivation from the child are paramount in this child-directed approach. The SI approach gives a functional neurology interpretation of behaviors and skills learned through engagement or avoided because of sensory aversion. Rarely was therapeutic touch involved, and if it was, it often involved a brushing technique or some similar imposed tactile input. The SI approach, combined with *Bodywork for Sensory Integration*, offers a structural explanation for aberrant behaviors and sensory-based personality traits.

Akin to the concept of the self-correcting mechanisms, the fidelity of Ayres' intervention was most authentic when the child actively engaged in the task and not just a passive participant . Though the techniques directly contact the child's body, *Bodywork for Sensory Integration* methods are also not things that are also child-directed in that the child's adaptive responses in the tissues and organs lead the session.

> ***Bodywork for Sensory Integration* assists in the occupation of regulating & modulating sensations**

Neurodevelopmental Therapy - Theory and Practice

Key Concept: Distal body parts (hands & fingers, feet & toes, mouth & tongue) move best if the proximal (core) of the body has achieved balanced symmetry & flexibility. The movement of all body parts is a balance of stability and mobility, controlled weight shifts, and rotation over a stable base of support. Tension or spasticity at far distal body parts reflects tight or unstable proximal parts.

Neurodevelopmental Therapy (NDT) is a formalized training approach for pediatric [and adult] neurological rehabilitation for physical, occupational, and speech-language therapists. Originating in Europe by Berta and Karl Bobath in the 1940s, NDT consists of theoretical constructs guiding therapeutic handling methods to treat cerebral palsy, brain injuries, and other neurological disorders. The handling techniques applied by the therapist entrain standard movement patterns during rehabilitation and therapy sessions. Advanced certification requires postgraduate training of the licensed therapist to prepare one for the breadth of NDT palpation and treatment strategies. The foundation of learning afforded by NDT training affords the therapist astute listening of the hands, moving subtly with the baby or child, guiding imposed movement where needed. The handling skills of NDT are excellent preparation for other forms of manual therapy also mentioned in *Bodywork for Sensory Integration*.

Palpation during NDT informs our hands which movements are unavailable for the patient, and then therapeutic touch contact in handling helps activate movement patterns. Therapeutic handling provides tactile and proprioceptive input to the client during intervention sessions. Handling helps the therapist evaluate and treat the tone balance between antagonist and agonist muscle groups; attempting to normalize posture, body alignment, and transitional movements. The NDT methods facilitate weight-shifting in various positions and establish core body stability to gain more fluid distal movement of extremities (arms & hands, legs & feet, eyes, and tongue/lips/jaw).

Assisting active control over movement patterns assists with the integration of primitive reflexes. Control of distal body parts (fingers, toes, tongue and mouth, even eye movements) is optimal, with core stability gained by control of proximal muscle groups and active control over primitive reflexes. Fine motor, eye tracking, oral-motor, and feeding skills are distal movements resulting from good proximal stability. Interventions following NDT principles help improve head and postural control, balance reactions, rib cage mobility, breath support for chewing and swallowing, and core.

A Different Take on Reflexes and Muscle Tone

Numerous sources define and sequence primitive reflexes. Periodically, 'new' reflex integration treatment methods emerge and share similar premises. Typical musculoskeletal reflexes are fundamental to activating and mastering motor skills. Abnormal muscle tone and retained reflexes interfere with attaining typical movement abilities. Handling methods of NDT are designed to facilitate movements to reduce spasticity and tone imbalance and minimize the effects of nonintegrated primitive reflexes due to a central nervous system injury (i.e., cerebral palsy).

The purpose inherent to each reflex and how the specific sensory stimuli activate that reflex may be woven into the cellular matrix of the fascial web. The experiences gained from years of applying bodywork from several populations offer a new hypothesis of a relationship between reflexes and fascia. Clinical observations of change suggest that tight connective tissues and meninges may be one cause of retained primitive reflexes. The mechanics of the fascia restrictions may be the stimulus that triggers the reflex activation, especially at the level of the dura mater.

Once thought to be a function of spinal and cerebral miscommunication or injury, primitive reflexes can reemerge when the child cannot meet the postural demands of physical activity. Tight fascial structures appear related and can impede the timing and placement of body parts. Sensory integration therapists have long worked with retained reflexes when addressing postural and sensory-motor control, as well as conditions of dyspraxia (where timing and rhythm skills are challenged).

Bodywork therapies are similar to handling methods in NDT, moving as the child moves, with the child's tissues leading the treatment process. Retained reflexes such as the asymmetrical tonic neck, labyrinthine reflexes, and the Galant reflex can interfere with coordination and motor planning skills. Clinical experiences demonstrate that bodywork, reducing structural tension assisted several children integrate these methods without much work on gym equipment, and in a relatively short period of time.

It is now theorized that fascia is a sensory organ which translates information via vibration. If this is true, nerve conduction theory is challenged regarding how the body obtains adequate speed of transduction and sensory feedback. The tension in the fascia can change the vibration (information) at warp speed through the fascial organ to the spinal cord and brain. When working with the geriatric population, we observe the re-emergence of what appears to be tonic neck reflexes with stiffness and loss of muscular flexibility in head turning (and increased tone in arms and legs). Primitive reflexes in adults have previously been interpreted as related to brain disease. Controlled studies have indicated that primitive reflexes occur in the normal elderly and are not correlated to cerebral atrophy or cognitive dysfunction.

If the fascia matrix serves in communicating the degree of expansion, monitoring system-wide tightness or flexibility, retained fascia tightness could be a sensory input that elicits a reflex. That would change the understanding of retained reflexes and the lack of integration through typical childhood movement experiences. This observation begs for more study into the mechanisms of the fascia structures as they may contribute to a greater understanding of reflexes and how they work.

Emerging Perspectives to Analyze Sensory Integration

Two [re]emerging branches of alternative interventions have helped shed light on the cellular framework and physiology that contributes to the organic whole of sensory processing. The biological terrain and the structural medicine (osteopathic) perspectives have found their way into traditional sensory integration clinics, with practices pushing the boundaries into innovative options.

Biological Terrain and Energy Medicine

Somewhere between 40-100 trillion cells make up the human body which is bathed in a nutrient-dense fluid matrix called the biological terrain or internal milieu. These interstitial fluids flow in, around, and through connective tissues suspending neighboring organs and blood or lymphatic vessels. The internal milieu can be expressed overtly by behavior and tissue integrity, especially ANS behaviors of state control and self-regulation. A healthy biological terrain (optimally adaptable) is one which balances pH, hydration, and nutrition levels. ATP is the energy that fuels the necessary actions and reactions in the cells. Metabolic wastes, toxins, and pathogens need proper filtering and disposal through detox pathways. Healthy cellular terrain implies that fascial and interstitial tissues holding and supporting fluids are free of structural compression, torsions, and adverse resistance. When structural compromise exists, fluids are less than optimally exchanged, which can raise the alarm of the sympathetic nervous system. The sympathetic nervous system state elicits protective retraction and vasoconstriction. System-wide cleansing occurs when tissues are in a parasympathetic state and vessels dilate with ease.

Sensory processing behaviors can reflect the unease (or dis-ease) of the terrain. Inflammation, irritated nerve endings, less than optimal nutrition or food absorption, and other entities can be considered at points of entry into the sensory pathways, and sites that may imprint initial sensory reaction. Cells can be in a state of trauma and hyper-protection in the presence of microbial assault, depletion of micronutrients, and inorganic toxins (which the body may not be able to eliminate through normal physiology). Analyzing the biological terrain through urine, saliva, fecal samples, and blood are functional medicine practices that can offer valuable information about underlying imbalances.

Genetic status and epigenetic factors can also affect cellular terrain and sensory processing. One of a few examples is a gene implicated in a lack of proprioceptive awareness (PIEZO2). Another gene getting much attention is the MTHFR gene, the instruction guide for helping the body process folate by producing an enzyme called methylenetetrahydrofolate reductase (MTHFR). This gene provides a blueprint map to create a protein that processes amino acids to use folate (Vitamin B9). Symptoms of the MTHFR gene mutation co-occur in attention deficit and autism spectrum disorders, autoimmune disorders, digestive disorders, and other chronic health conditions. The mutation is suspected in less-than-optimal body detoxification. When autoimmune dysfunction co-occurs, the immune system can be kept in hyper-overdrive, secreting cytokines, mast cells, and histamine (an excitatory neurotransmitter leading to inflammation). Histamine can be a player in self-regulatory behavioral problems. Food sensitivities, allergies, chemical intolerances may co-exist.

Toxic-load evaluation is not a standard pediatric medical practice, unless a pronounced expression existed, such as lead poisoning as an etiology of developmental delays or nutritional depletions. The world of naturopathic and functional medicine has awakened a global awareness of this issue. Energy medicine, utilized for decades in multi-national aerospace programs, as health monitors for astronauts and fighter pilots, is now available to the discerning healthcare consumer.

Outside the realm of allopathic medicine, bioresonance in energy medicine evolved from biomedical engineering and mechanical physiology to discern what micronutrients are present or depleted. Naturopathic practitioners have become purveyors in the practice of applied quantum physics through biological body scans. These methods emerged from the combined works of Albert Einstein, Nikola Tesla, and Dr. Royal Rife, all visionary pioneers in applied physics. Several different types of bio-resonance scanning equipment are available, but the science is essentially the same. Scanning systems read, communicate, and then stimulate cellular energy changes.

Directed frequencies treat distortion to raise or lower electrical potential after analyzing energetic fields of body regions and organs. This process has demonstrated a direct stimulant to the vagal nerve system and activates a homeostatic restoration function. Names used to describe scanning (and treatment) systems include quantum medicine or energy medicine. Incorporating ancient oriental knowledge, historically used herbal remedies embraced these theoretical physics principles.

Comparative studies of body scanning to traditional lab tests and scanning methods are in full swing. These systems are non-invasive and cost-effective methods to evaluate the body's health and wellness. The machines are based on spectral analysis of the cortex magnetic field found around all biological matter. Scanning uses magnetic inductors and special sensors that detect the frequency of cellular activity and measure the presence and locations of bacteria, parasites, viruses, and fungal entities. Each generation of scans has become more sophisticated. More vitality equates to a healthy cellular terrain. How the body experiences and handles the energetic function is now being studied for chronic pain conditions, autoimmune issues, headaches and migraines, chronic fatigue and fibromyalgia, addictions such as smoking, mood swings and depression, allergies, and other chronic issues. The scan's findings determine functional and nutritional treatment plans.

Though once considered pseudo-science, the science of bioresonance has been firmly established, earning an exclusive subject index in biomedical literature. Indexed under magnetic field therapy, non-linear scanning, or bioresonance body scanning, the reader can further investigate the emerging science. Bioresonance treatment has been shown in clinical studies to positively affect symptoms of depression, reduce allergies and asthma, and assist in smoking cessation. In a recent clinical study comparing treatments for people with chronic

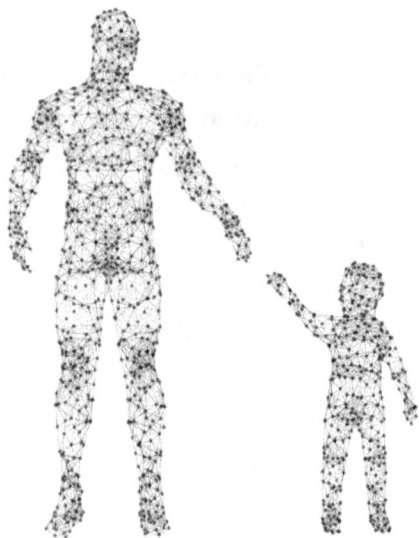

depression, outcomes of bioresonance therapy were significantly higher than the selective serotonin reuptake inhibitor medication. In our clinic, we witnessed children with levels of mercury, lead, and aluminum (confirmed via traditional medical testing) demonstrate organic recovery with improved autistic and neurobehavioral features following such treatment that safely chelated. These children were long-term clients, and such changes were not happening with other interventions. Children with severe self-abuse behaviors have had scans suggesting parasite load or viral meningitis, presuming behaviors were extreme reactions to unrelenting pain. Once aggressive anti-parasitic and anti-viral supplements were added, occasionally with dietary alterations, extremes in behaviors began to subside (perhaps as the brain healed). Bodywork palpation in these children consistently communicated tight and restrictive mobility of meningeal tissues under skull bones and down the length of the dural tube of the spine. As the source of inflammation was eradicated, the tissues became more soft and supple. These children then often began to seek out craniosacral therapy, verbally or nonverbally, asking us to hold their heads and "move the bones" for apparent relief. We have witnessed this so often that in cases of extreme fascia tightening (palpate in meninges, peritoneum, mesentery), I insist upon body scanning intervention to families to help us gain insight into possible sources of systemic inflammation. Much more clinical study needs to be promoted.

A group of parents introduced us this technology through their own research and opened our minds to this world. These are parents who have looked beyond behavioral diagnoses for more answers in their children's autism spectrum disorder, PANS/PANDAS[1], sensory processing dysfunction, behavioral issues, and other neurodevelopmental deficits. Observations raised the understanding that many children with sensory issues can also have systemic inflammation, from leaky gut to latent infections. Examples include parasites, Lyme spirochetes, mercury and aluminum, other toxic agents, and gut-disruptive bacteria such as H-pylori or E-coli. Most of these pathogens produce neurotoxins.

1 PANS (Pediatric Acute-onset Neuropsychiatric Syndrome) is acute behavioral deterioration with severe symptoms of: obsessive-compulsive, anxiety, sensory am plication, movement abnormalities, regression, tics, mood and sleep, and urinary symptoms. Triggers are believed to be one or more pathogens. PANDAS is described as the same condition, with the distinct feature of sudden onset, but with the distinct identification of streptococcal infection. Both essentially describe functional expression of meningitis or encephalopathy.

Structural Medicine (Osteopathic & Chiropractic Perspectives)

Structural medicine philosophy is based on the principles of the unity of the human body, and dates back to the days of the father of modern medicine, Hippocrates. An interrelationship of structure and function is exemplified. Generations of practitioners providing manipulative treatments saw truths in the human body's traits of holding trauma within body tissues, the body's emotional-somatic component to wellness and disease, and body's ability to regulate and heal itself.

Osteopathic physicians embraced the above principles as core practices developed by the founding father, Andrew Taylor Still, M.D., D.O. (1826-1917), the world's original osteopath. Before becoming a physician, he was a soldier with a Kansas battalion during the North American Civil War. Dr. Still cared for many of the fallen in his army unit. He later watched his wife and several of his children die of the orthodox medical practices of the day. Most medicines used mercury chloride, administered without standard dosages, often ending with the poisoning of the patient. Still went on to train as a traditional allopathic physician, but his personal experiences led him to study non-pharmaceutical ways to treat disease. Based upon meticulous anatomical dissection (without preservatives), he studied how body structure directs function. When a structure is compromised, then so is function. He was the founding father of the world's first osteopathic school of medicine in Kirksville, Missouri. (Modern-day osteopathic medicine shows a trend of fading from mandatory training in manual methods, despite their long history. Conversely and on a global scale, allied health and manual practice professions are seeking training in manipulative medicine methods to augment clinical practice.)

Chiropractic care embraced similar views as osteopathic principles, except for the belief that medicines and surgery could be avoided with preventative spinal adjustments. A contemporary of Dr. Still, David Daniel Palmer (1845-1913), was the founder of the world's first college of chiropractic methods in Davenport, Iowa. Despite paralleling visions, their rumored personal animosity and conflicts helped define the differences between osteopathy and chiropractic science. Dr. Still's osteopathic approach focused on the fluid and energetic movement of life itself (blood, air, energy) moving through the body's structure as the health baseline. Palmer's approach differed by focusing more specifically on the alignment of the spine to free nerve compressions and relieve stress at the innervation site (organ, muscles, vasculature). There were infamous debates between these two men over whose approach was "best" and who had ownership over philosophies.

William G. Sutherland, D.O. (1873-1954) was a student of Dr. Still and became a significant figure in American osteopathic medicine. He studied and identified rhythmical changes in the size and shape of the bony container of the skull. His experiences and experiments in compressing his own cranial bones resulted in (often extreme) behavioral changes. He used a manually-operated clamp to compress specific bones, and his wife observed and recorded any behavioral alterations. Sutherland's behaviors normalized when the clamps were released (allowing for self-correction of bone compression). He theorized that the behaviors were altered, in part, by damming of cerebrospinal fluid. His clinical practice embraced these observations, leading to a unique and revolutionary intervention called Cranial Osteopathy. His methods continue to be taught and practiced worldwide, mainly by osteopathic physicians and dentists.

With more discoveries, Sutherland's work spawned generations of practitioners to further the applied practice of cranial osteopathy. John E. Upledger, D.O., O.M.M. was a graduate of Still's osteopathic college and studied Sutherland's methods extensively. At Michigan State University, Upledger advanced the science and proved the existence of the craniosacral rhythm and physiological movement of cranial bones. He developed the modern-day version of craniosacral therapy in his endeavors to understand the function of the pumping system of cerebrospinal fluids within the confines of the meningeal tissues. Upledger's view was that bones were influenced by soft tissues and the exchange of fluids within and underneath the fascia container of the craniosacral system. Upledger recognized the trend of his fellow osteopaths moving away from manual interventions and was visionary in expanding training to other professions licensed to touch people. In order to meet the growing need of people seeking out forms of wellness, he believed that any professional with a strong education in anatomy, physiology, and pathology could practice craniosacral therapy.

Many branches of the same tree of interventions now exist due to these visionary giants of structural medicine. Each branch has its share of purists and loyal followers. Discerning practitioners trained to compare and contrast tend to embrace what works for their clients, fusing the methods in which they have been trained. What osteopathy and chiropractic care did not profess in their original constructs was the anatomy and function of fascia. Connective tissues were routinely indiscriminately removed and tossed on the floor as waste in cadaver study as it was considered an impediment in anatomy labs in revealing the "real" structures. Connective tissue was unknown, unloved, or under-appreciated until the twenty-first century. Supported by the Ida Rolf Institute and other structural medicine organizations, the first International Fascia Research Congress[2] was held at Harvard Medical School in 2007. This conference was dedicated to promoting the science of fascia in all its forms and functions. Rolfing is one of the first formal approaches in bodywork to address the structural integration of body form, recognizing that fascia permeates the entire body. Though historically known as somewhat invasive, rough, and occasionally painful, Rolfing has evolved in its approach.

2 Fascia Research Congress https://fasciaresearchsociety.org

Notable pioneers in fascia research include Bruno Bordoni, Donald Ingber, Jean Claude Guimberteau, Helene Langevin, Tom Myers, James Oschman, and Carla Stecco. They, and others, have expanded our understanding of the fascia network. Chiropractic care addressing misaligned vertebrae that may cause nerve interference and osteopathy addressing structural hindrance of movement and exchange of fluids have contributed to our knowledge of body dysfunction. Fascia science may prove that both structural medicine theories are correct. More on this subject later in this chapter.

The sensory integration approach is based on decades of analyzing behavior and function and making opinions about neurology through well-researched testing methods. However, the approach misses insight into the "behaviors of tissues and organs" in analyzing a child's nervous system. (Observations from *Bodywork for Sensory Integration* suggest there are *fascia behaviors* to consider).

We need observations of the body's structural integrity and contributions to adaptive behaviors. We must refrain from analyzing behaviors observed from solely across the room. Palpation is required to feel organ tones and tensions to further the science of how practitioners can help with sensory challenges. In the routine sensory assessment, the practitioner watches how the child performs tasks or interviews parents and analyzes results from extensively researched questionnaires. Historically, palpation of the body is not included in a sensory processing evaluation.

Conversely, Neurodevelopmental Treatment (NDT) and other similar methods guide therapists in handling skills that give a foundation for palpation (of discerning motor or movement disturbances). NDT practice is a good gateway training for bodywork methods. Sensory integration practitioners assume the central nervous system is where processing issues occur, somewhere within the brain structure and between brain components. There are assumptions that information enters the central nervous system through "obvious" sensory pathways from the peripheral nervous system. New science about fascia now considers that the layers of body tissue and individual cellular structures may play a function in background sensory feedback.

It is a novel concept that the body's structure may not be 'ready' to send or process the massive amounts of information needed. It is an osteopathic and chiropractic viewpoint of finding retained structural impact of birthing or actual trauma or organ stress persisting in body tissues. It is also an alternative belief that retained structural compromise can influence behavior. It is well known that maternal stress can be a precipitating factor in children with sensory integration challenges. No research exists on the tensegrity stress of the cellular matrix of connective tissue that may impact sensory processing function. The osteopathic viewpoint has proven the relevance of palpating and directly treating the autonomic nervous system, maximizing transduction of sensory input from structural and mechanical stress removal, trauma-informed care, and other terminology aligning with Ayres' original work in neurobehavioral development. Combining the two approaches is the focus of this book.

Sensory Health and Wellness

Studying a host of perspectives would be helpful in moving the science of sensory processing and integration towards optimizing human potential and quality of life. In no order of importance, examples include:

- Nutrition – availability of micronutrients to feed sensory systems
- The health of the cellular terrain determines sensory receptor vigor, and ultimately behaviors
- Brain, organs, and vessels free of toxins is a state of sensory wellness
- Brain that drains and cleanses itself leads to healthy sensory systems
- Brain and spinal cord free of structural compressions or torsions leads to optimal sensory systems
- Gut lining health (Enteric nervous system health) is the basis of most neurochemistry
- Neuropathies and neuritis as a function of sensory integration challenges
- Meningeal restrictions as a function of sensory integration and primitive reflex challenges
- Neurodiversity as a concept of inclusion and acceptance

Bodywork Methods that Augment Sensory Integration Practices

- **Craniosacral Therapy** (Upledger methods)
- **Lymphatic Drainage** Massage of the Lymph-Fascia Complex (Chikly methods)
- **Modified Acupressure and Reflexology** (accumulated from multiple sources)
- **Visceral Manipulation** (Barral method)
- **Viscero-Fascia Mobilization of Enteric Nervous System** (Treating peritoneum, mesentery, connective tissue walls of vascular-fascia complex)
- **Generalization of all above** with focus upon interstitial tissue spaces

Craniosacral Therapy (CST)

Key Concept: The craniosacral system, fascia and fluids housing brain and spinal cord, are core structures of the fascia matrix. Methods directly treat the meninges at specific anatomical points. The body communicates and works well if this core fascia structure is free from mechanical compromise and if the flow of cerebrospinal fluid is free from constriction or impingement.

One form of craniosacral therapy, mentioned earlier, was developed by John E. Upledger, DO, OMM. Emerging from his clinical research in the 1970-80s, methods were further refined in his long and storied clinical practice as an osteopath and surgeon. His research proved the existence of the craniosacral system and the craniosacral rhythm of the central nervous system. Skilled in traditional methods of osseous adjustments, Upledger evolved the specificity of treating soft tissues and fascia structures to promote self-correction of the fluids and tissues of the craniosacral system. Key hand placements directly treat the tissue networks to gain expansive correction of surrounding structures, leading to the central nervous system (the spinal column, the brain, and the projecting cranial and peripheral nerve pathways).

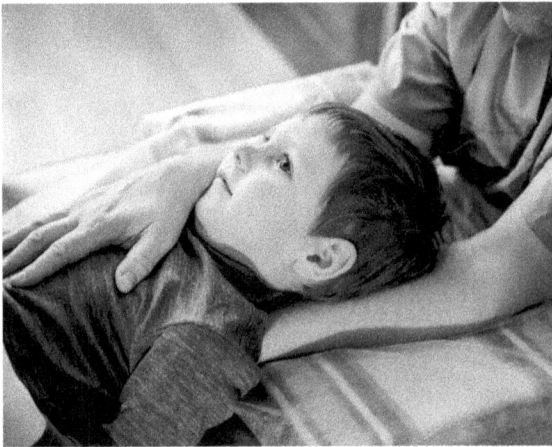

A beginner's protocol is the first foundation of learning, designed for client safety when applied correctly and with a reliable application method between practitioners. A certification process reinforces reliability. Non-invasive touch contact and allowing adequate time for self-correcting tissue changes is key. A therapist must not rush with this method. Advanced skill level evolves by honing the art of 'arcing' towards primary sites of treatment need and differentiating anatomical structures palpated. Trauma and emotional history are observed to be something that can spontaneously arise during skilled treatment. Upledger studied this effect and shaped guidance for therapists to assist the client through a "somato-emotional release" process. Craniosacral therapy has far-reaching effects and is easily blended with other modalities.

Years of bodywork practice have revealed that mechanical forces (from events like birthing) do not always decompress or resolve. Tensions in the meninges can contribute to maintaining a higher sympathetic tone in the central nervous system affecting physiology, coping behaviors, and interpersonal relationships. Sometimes it takes years for a structural problem to manifest symptoms fully. Craniosacral therapy is gaining evidence

Brain
Skull
Cerebellum
Cerebellar tonsils

Spinal cord

Cerebrospinal Fluid

Vertebral body

Sacrum

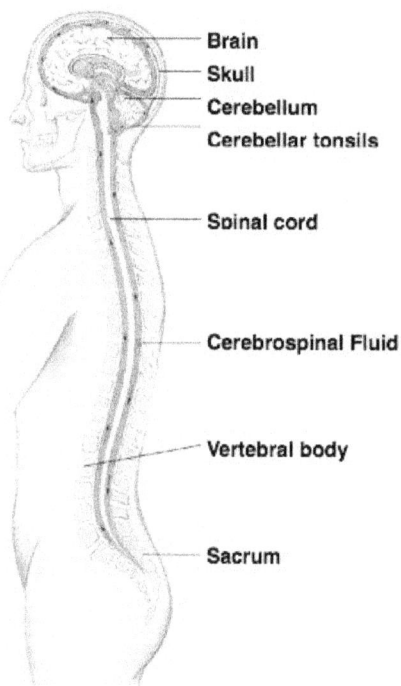

to support its use in various populations and conditions. Traumatic injury, stress (physical or emotional), toxins in the body (environmental pollutants, drugs, synthetics), prenatal problems, and a difficult birth can all compromise the craniosacral system and areas of connective tissues. Connective tissues appear to play a role in shock absorption.

Behavioral assessments cannot determine structural compromise of the craniosacral system. There is yet no body imaging that can determine meningeal tightness, though the restriction of space or movement of cerebrospinal fluid could be studied by space size and influence of fluid motion. Upledger expanded on the historical osteopathic assessments of palpating tensions through this core connective tissue system and developed methods that challenged the concept of quick adjustments to an approach more aligned with a body's self-correction. A body self-corrects when tissues are held long enough for fluid exchange. CST training helps develop subjective skills in evaluating the flexibility of the entire fascia system. Craniosacral therapy enhances and complements other interventions and has repeatedly demonstrated assistance in self-corrective neurological function and sensory-based behaviors.

Treatment consists of light touch contact at specific fascia-osseous sites, and typically, the therapist's hands rarely appear to move, if at all. The dura mater attached to the bony container of the skull has been demonstrated to be "moved" by lift, measured in microns. Hand placement at specific osseous sites helps "lift" the bony plates to gain minute space for fluid exchange between bone and dura mater. (The former statement has been proven in a cadaver study, and the latter has yet to be proven.)

Treatment can be a whole hour or more, or shared time during other therapeutic endeavors, depending upon the child, the treatment environment, and the needs. As in any therapy, no prediction is made on how the nervous system will express a shift in the autonomic nervous system or body self-correction. The non-invasiveness of this method makes it an ultra-conservative and risk-free option (when applied as taught). The therapists utilize the bones more like handles to hold the tissues in a gentle, non-invasive, and sustained stretch.

Fusing the treatment philosophies of CST with the fluid models of lymphatic and glymphatic flow patterns, the assumption is that cranial bones are gently lifted by sustained

Cranial suture

Cranial bone

arachnoid layer

periosteal dura mater

sinus cavity

meningeal dura mater

pia mater

pia mater

brain

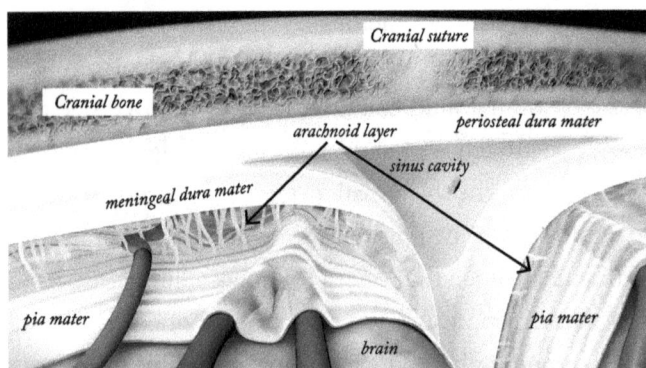

stretch gain space between meningeal layers [measured by microns]. Decompressing tissue stress facilitates exchange of fluids in underlying arteries, veins, and glymphatic fields. Emphasis is on gentle and sustained hold of cranial bones. When considering the structural aspects, the fluid movement is likely one factor in the self-correction of the system.

In twenty-two years of applying this work in practice, the author has yet to see a child who does not demonstrate some degree of positive change within a session or two. The improvements can be drastic and impressive, or more subtle. The child's tolerance for hands-on therapy must be considered within their limits. Symptoms caused by factors unrelated to the craniosacral system (such as gut inflammation or family dysfunction) are not likely to respond by manual therapy alone. However, the response to treatment is usually noteworthy when problems are due to structural restrictions of the meningeal system and causing resistance to cerebrospinal fluid and lymphatic flow.

Advanced levels of CST train students to go beyond the level of the dura mater into the deeper meningeal layers and brain and spinal structures. This advanced work has been shown to effectively treat post-concussion symptoms and chronic inflammatory conditions affecting the central nervous system.

Theoretical constructs of CST provides a structural explanation for sensory processing difficulties

Lymphatic Drainage Massage and the Lymph-Fascia Complex

Key Concept: Manual lymphatic drainage is a precise massage method that evacuates stagnant lymph fluid away from tissues, moved toward the heart via vein networks, and eventually the liver. Restrictions in soft fascia and connective tissue can hamper lymph movement and circulatory systems.

The human lymphatic system is a network of hundreds of large lymph nodes, countless smaller chains and vast micro-vessels carrying typically clear lymph fluid. This system continuously circulates its fluid to remove bodily waste and transport white blood cells in the daily inhibition of infection. Any obstruction in soft tissues can cause lymphatic fluid to stagnate. Bruno Chikly, MD, DO, is considered the modern physician-scientist who expanded our knowledge of the extensiveness of the lymphatic system's anatomy map and specific lymph flow directions. The lymphatic anatomy is a whole-body water shed terrain with drainage pathways arranged in specific mapping.

This immune system highway includes bone marrow, spleen, thymus, specific lymph nodes, and systemic lymphatic vessels. These thin tubes branch like blood vessels, traveling alongside arteries and veins into all body tissues. Lymph fluid eventually dumps into venous blood and eliminated via the liver. Stagnant livers can indicate a stagnant and backed-up lymphatic system. Lymph-collecting vessels have a layered wall of endothelial cells, smooth muscle, and collagen fibers that contract rhythmically to pump lymph flow. The term Lymph-Fascia Complex implies an intimate relationship between the fluids within the fascia that flow into the lymphatic and the cells that construct both systems.

The vagus nerve system mediates lymphatic movements, and fluids are pumped globally by the heart and locally by smooth muscles of the lymph vessels. Local glial cells and other interoceptors probably monitor lymph volume at fluid exchange sites. Tiny muscular units set along the length of lymphatic vessels, called lymphangions, assist fluid movement without increasing blood filtration or lymph node collapse from too much pressure. Lymphatic fluids may not drain as efficiently when the body is chronically stressed, carrying a toxic burden, or depleted in energy or nutrients.

Where swelling and edema exist, structural pressure backs up the lymphatic pathway, elevating sympathetic tone and even pain. The sooner edema is evacuated, the

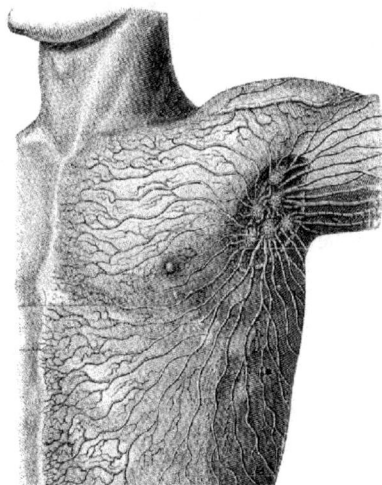

Atlas drawing of French anatomist Marie Sappey's mercurial injection of human superficial lymphatic system (1874).

sooner the sympathetic tone in the area can release itself back to a relaxed state. Furthermore, clinical experience suggests that stagnant lymph fluid can spill into surrounding tissue, creating the basis for tissue adhesions (that may take months or years to manifest fully). With lymphatic palpation and evacuation of lymph content, the practitioner can gain a more thorough assessment of this fluid exchange within tissues and make clinical judgments as to the state of health of tissues.

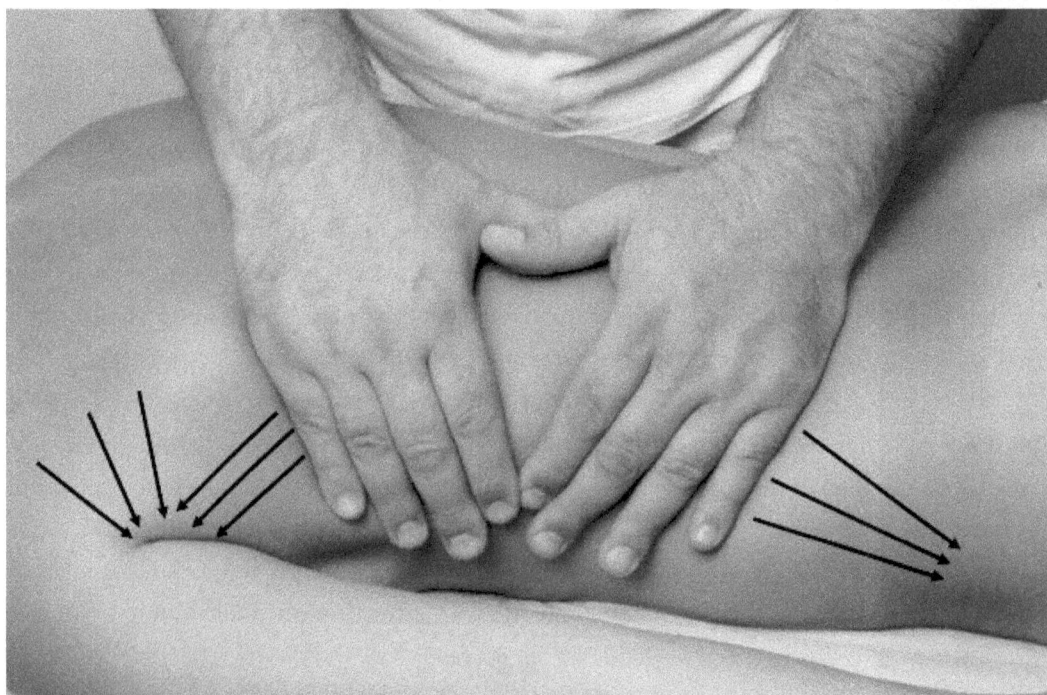

Lympho-fascia, and lymphatic drainage massage, is used for body detoxification for the evacuation of stagnant lymph fluid holding toxins, metabolic wastes, and the effects of inflammation. The skill set of lymphatic drainage with attention to the surrounding fascia is a vastly different level of touch input from the therapist. With just enough touch pressure to "move water" under the skin, this method (like CST) requires an intention of precision for therapeutic effect. Lymphatic massage, and the fusion of it with other non-invasive methods, is the polar opposite of deep tissue massage.

There is strong evidence that lymphatic massage positively affects the autonomic nervous system. Chikly has also developed an advanced brain and spinal cord curriculum to help access nuclei and ganglion centers to assist in limbic and ANS downregulation. Our clinical observations is lymphatic drainage has help ease and even resolve behaviors suggestive of touch defensiveness, anxiety and irritability, and even toe walking. This work is woven into *Bodywork for Sensory Integration* and used on a daily basis.

Modified Acupressure & Reflexology

Key concept: Acupressure and reflexology by history demands application of physical pressure on trigger points or pressure points positioned along nerve meridians. Variations have existed through the centuries, evolving by changing viewpoints and perspectives.

Acupressure is believed to have originated in China more than 5000 years ago, predating acupuncture by approximately 2000 years. Acupressure and acupuncture are based on the same principles of acupoint activation at key endings of nerve meridians. Acupuncture in the United States requires a master's level education and licensure, whereas acupressure can be learned by anyone and practiced by anyone holding a license to touch people.

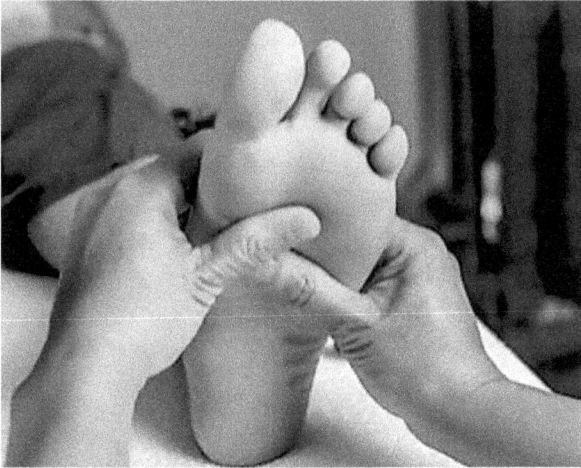

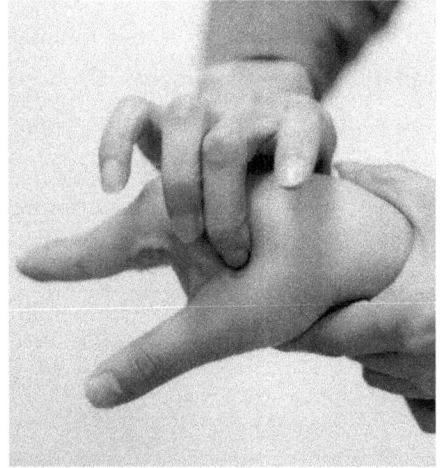

Both approaches are based on the theories of Qi (chi) (literally translated to "breath") and imply an energetic life force that flows through living tissues. Compared to other bodywork methods, the "breath" or Qi of life can also be discerned and palpated by perceptive hands. In some circles, it is now believed that Qi is the movement of interstitial fluids through connective tissue and the status of the flow of blood, lymph, and cerebrospinal fluids. From a structural viewpoint, anything that tightens and restricts increases resistance and slows fluid exchange. Reduced fluid exchange raises pressure, and the sympathetic nervous system is activated, even at the slightest level of tension. Compared to what we know about the parasympathetic state, optimal Qi parallels optimal parasympathetic tone throughout the body.

Acupressure resembles another form of touch intervention known as reflexology, though the latter is a protocol of applied pressure to predetermined reflex zones. Acupressure, by history, demands the application of physical pressure on trigger points or pressure points positioned along those nerve meridians. Meridians are nerve channels through the body that maintain the flow of Qi and thus steady homeostasis. Each meridian is connected to various organs and tissues for autonomic regulation. Bodyworkers learning the terrain of tensions through the client's body can palpate where Qi might be stagnant or blocked and use acupressure to help correct the flow.

The whole body is a cluster of acupressure points. Stimulation through applied pressure has been shown to elicit a function change. Tender points usually present nociceptor and polymodal pain receptors, and pressure at these sites can yield pain relief. Different tools have been used over the centuries in the practice of acupressure: stones and shaped gadgets, elbows, feet, fingers, and thumbs, elastic bands with spot buttons, and mustard seeds on adhesion strips.

Several variations of acupressure exist based on origin and the combination of methods and techniques. The physician William FitzGerald is considered the founder of a method called Zone Therapy as the foundation of the national profession of reflexology. Other variations of this approach have been traced to ancient civilizations of the Roman Empire, India, and Native Americans. Other contemporary methods include Zone Analgesia by physician Shelby Riley and Reflexology by Eunice Ingham, a physical therapist. Ingham brought undisputable contributions to the practice of reflexology by discovering that body parts are reflected in a nerve distribution over the bottoms of feet and, and with alternating pressure, can stimulate a positive response.

The history of reflexology has origins back to ancient Egypt as evidenced by hieroglyphic inscriptions found in the tomb of a physician named Ankmahor (2330 BC), suggestive of describing healthcare. The inscription has been translated as the patient stating "Don't hurt me" and the practitioner replies with "I shall act so you praise me." Source: Wikimedia.

Bodywork for Sensory Integration employs modified acupressure nearly every session at sites of specifically held tensions, but not necessarily following specific nerve meridians. If more thorough knowledge is of interest to the reader, formal training in acupressure is recommended.

Observations from the experiences of applying acupressure revealed that many trigger points are, or near, engorged lymphatic vessels or nodes. Consistent muscle spasm relaxation is achieved by gently combining acupressure with lymphatic drainage specific to the site. Acupressure combined with localized lymphatic drainage effectively treats growing pains, often found in the anterior shins. Children can be just as prone to structural stress as adults, ranging from retained birthing strains to bumps and injuries acquired in the job of growing up and playing.

Visceral Manipulation and Modified Visceral Techniques

Key Concept: The suspension placement of thoracic and abdominal organs, held by ligaments and fascia layers, can be misaligned or compressed. Structural compromise can negatively affect natural physiological freedom of movement of organs and surrounding vasculature, as well as mechanics of the spine. Using a high degree of specificity to locate viscerofascia, mobilization of impingements is the objective. (Modifications align with co-applying craniosacral therapy, rather than from a perspective of mobilizing organs; implying self-correction of the organ occurs).

Another branch of innovative manual therapies, Visceral Manipulation (VM), was developed by Jean-Pierre Barral, DO, MRO(F), RPT, in collaboration with Alain Croibier, DO. Organ-specific methods assess and treat fascial and surrounding structures to enhance the motility and mobility of internal organs but also positively affect the whole body's movement. Techniques are precise, using soft, noninvasive manual forces to encourage the normal movement of the organ within its cavity and ensure suspension fascia and ligaments are flexible for optimal physiological action. The focus is taught from a physical therapy-osteopathic perspective (gently and without organ invasion). However, clinical practice of this for younger clients (even some adults) holding chronic stress and enteric system discordance proved to be too invasive and alarming, especially for those with ANS regulation issues. Babies' systems taught us how to modify these techniques by treating spaces between organs to avoid alarming their tiny systems.

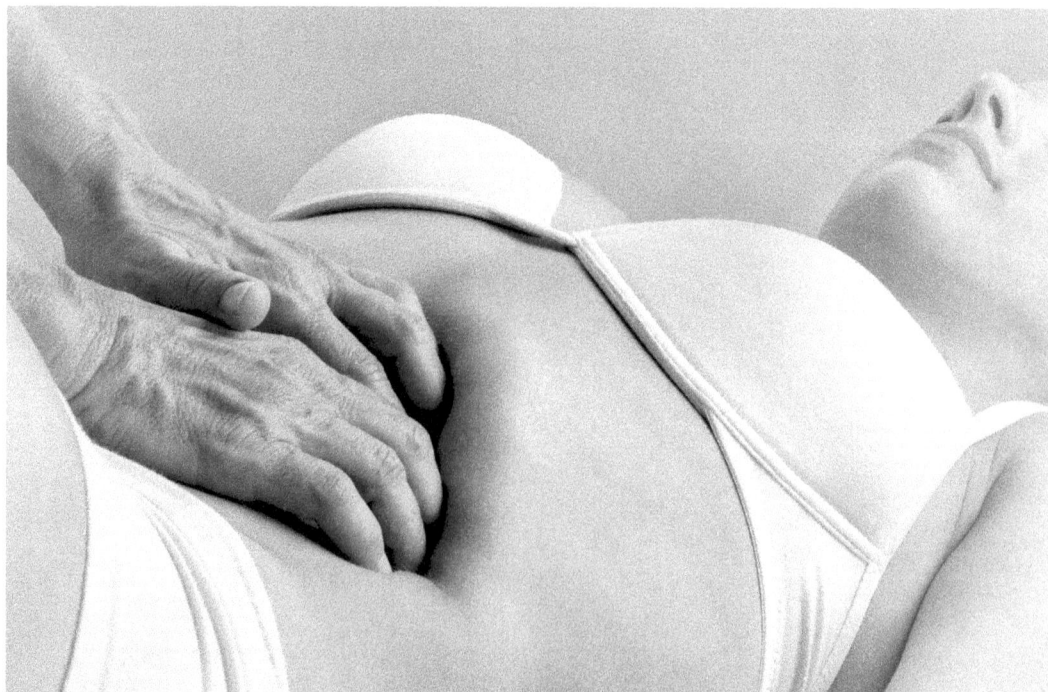

Generalizing these observations to the sensory integration population, modified visceral techniques have been a significant part of gaining direct effects on sensory modulation and self-regulation. The sympathetic state creates a higher tone around organ walls; conversely, the parasympathetic state reduces tone, making organs softer and more flexible. The body's protective retraction is often found in the visceral organs.

Treating vascular concentrations and nerve plexuses have been included in these modifications. Direct treatment of the autonomic and enteric nervous system in the visceral-fascia layers, vagal nerve pathways, nerve plexuses, and the mass of mesenteric vasculature can lead to self-regulation and modulation.

Clients' reactions guided the modifications by helping us to tune into the tension and "behaviors" of organs, more precisely the tension of surrounding connective tissues. Through these experiences applying VM to the population of sensory integration practice, many of whom also have digestive disturbances, we stumbled upon what we believe is direct manual therapy to the enteric (gut) brain structures and ANS regulation. We observed first-hand how the visceral environment directly influences human behaviors. The enteric nervous system can be communicated through tone and tension and can be adversely affected in function by structural restrictions. The gut-brain can be palpated and understood more clearly through manual communication of the mesh of nerves woven through the peritoneum and mesentery tissues. The extensive nerve mesh that wraps under and through the peritoneum and mesentery layers projecting nerve fibers into the organs' walls can communicate the ANS status.

A contraindication for Barral's visceral techniques is the use on anyone under the age of eight months. However, many younger babies coming to our clinic had a different idea and let us know that they, too, benefitted from specific organ work. Babies struggling with colic, digestive challenges of chyme and fecal movement, self-regulation, and self-soothing taught us how to locate the 'just right' spaces within the peritoneum. Babies taught us how to relieve organ tension, loosen digestive sphincters, down-regulate ANS nerve plexuses' innervation sites, and resolve those behavioral and physiological problems. Babies guided our listening hands to multiple and consistent locations during our learning curve. Treating 'vicinities' rather than specific ligamentous or organ structures is the primary difference between the two approaches. Modified techniques evolved as the need to "treat with tissues" instead of «treat the tissues» was the lesson. Increased awareness of the mesentery and peritoneum visceral fascia complex and the collective continuance of fascial suspension around mesentery blood and lymphatic vessels and the adjacent nerves has significantly contributed to the evolution of this part of Bodywork.

Barral's general listening techniques had been instrumental in building trust when Baby's pattern of tissue distortions brings us to twists in the tummy. Common findings included: remnants of umbilical cord distortions and compressions, spots where normal intestinal migration in the growth process may have met resistance, and pulls at digestive sphincters interrupting reflexive peristalsis. The recent discovery of a deep front line of fascia (foundational suspension of organs and viscera within the soft skeleton) has revealed that adverse tension through this structure can affect distal and core parts alike.

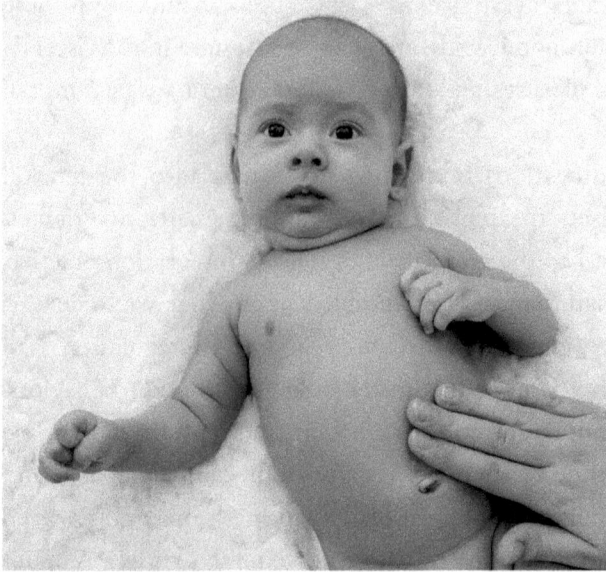

We are treating viscera to fill in a considerable knowledge gap about how to assist the autonomic nervous system. **Treating surrounding layers of connective tissues has taught our hands how to assist in releasing protective retraction before any other method is utilized**. The result of treatment is balanced organ tone, tension, and freedom in mobility; all directly affecting the autonomic nervous system and ultimately mood and behavior. Generalizing what babies taught us helped develop *Bodywork for Sensory Integration* for children with self-regulatory and modulation difficulties. However, a great deal more is needed to be discovered and studied.

Myofascial Release

Key Concept: Historically, these manual techniques were directed towards the tough fascial membranes that wrap, connect, and support the skeletal muscles. Postural symmetry, range of motion increase, and pain management have been primary objectives for treatment. Myofascial release continues to evolve in style and touch methods with the growing knowledge of interstitial tissues and the behaviors of fascia.

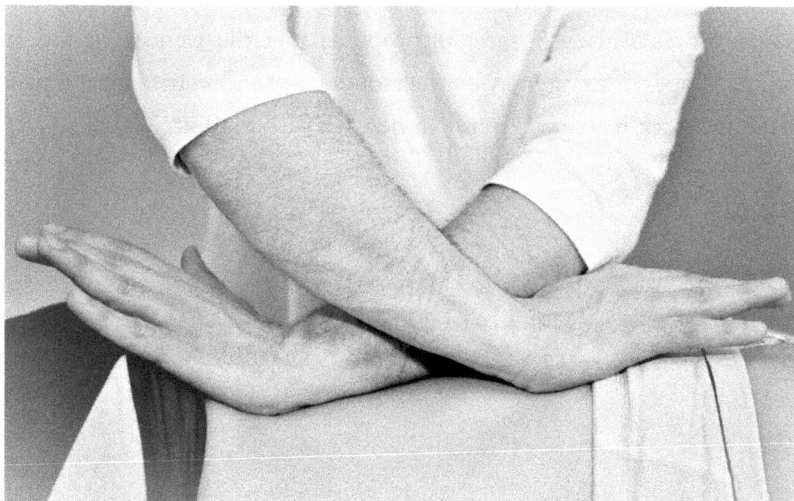

Myofascial release (MFR) is related to the historical paths of Rolfing as developed by Ida Rolf. Robert Ward is credited for coining the phrase the cellular interplay between fascia and muscle structures. Ward and John Barnes are considered the two primary founders of MFR, and Barnes created a series of evaluations and treatment techniques to address chronic pain and movement dysfunction. Pain, postural, and abnormal movements have historically been the focus for applying MFR (from a massage or physical therapy perspective). Like craniosacral therapy, MFR has been embraced by many professions. Focused manual pressure and stretching in myofascial release free restricted movement in a particular area, indirectly reducing pain and increasing range of motion. Then another area is sought for treatment. The therapist is guided by the body's responses, dictating how much force to use, the direction of the stretch, and the duration of holding a stretch. Small areas of muscle are simultaneously stretched.

Stacking the 3-dimensions of fascia – vertical, transverse, and rotational planes of structure, is the body's tensegrity. Existing evidence supports MFR contributing to pain relief. The mechanism for how this works continues to be studied, and the International Fascia Research Congress has contributed significant discoveries in recent years. Though it has been proven that the length of fascia tissue does not increase with manual methods, the positive changes might involve a rehydration and self-correcting tissue expansion through the mechanical process of applying treatment.

In *Bodywork for Sensory Integration*, MFR is a complimentary, though not a primary focus. However, there are common instances where MFR is the 'just right' modality to treat a restricted muscle such as the psoas or sternocleidomastoid muscle with precise application. These sites can be related to the body's chronic stress related to sensory modulation challenges.

It has been said within the practice of MFR that fascia takes about 90+ seconds to respond to gentle, sustained stretch. Comparing this view to experiences working with protective retraction, *Bodywork for Sensory Integration* proposes that the waiting period is the body's trust monitoring. Tissue softening may be the reduction of protective retraction and ultimate parasympathetic recovery, relaxing systemic tone with vasodilation.

Generalizing Methods Towards Interstitial Layers

Key Concept: There is a discernible difference from the previously mentioned techniques when focused treatment is between interstitium and fascia. Evolving with the emerging science of interstitial micro-anatomy and physiology and the improvised use of all these methods, it feels that we are working in the "spaces" between structures in concert with the targeted structures. These spaces are where we have found protective retraction and sympathetic tone, the effects of systemic inflammation, or retained mechanical compressions. The spaces are where organs feel compressed or adhered into their neighbors, hindering optimal body tensegrity.

A group of scientists led by Petros Benias garnered worldwide attention when they recently discovered a 'new organ of the human body' within the abdominal cavity. In studying visceral organs, they tried a novel method of freezing biopsy tissue (rather than using the standard fixative method, which essentially dehydrates tissue). By utilizing this alternative way to prepare connective tissue slides for microscopic study, they preserved peritoneal tissues that surrounded the stomach in their natural state. They discovered that the cell layers of this fascial tissue were inflated with fluid. (Traditionally, fascia appears as flat lines in connective tissue samples.) These scientists described this fluid-filled tissue as "human interstitium." In subsequent discoveries, they realized these fluid-filled compartments are found extensively throughout the body, beneath the skin, within the gut lining, lungs, blood vessels, and muscles, and joined together to form a network of strong yet flexible protein fibers.

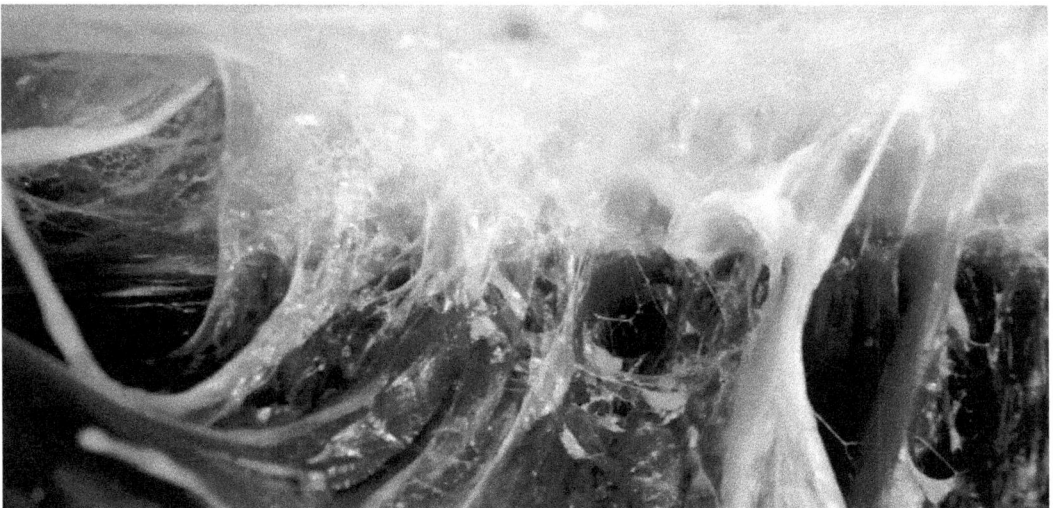

Benias' team described what, from their perspective, was a novel tissue, though they caught flack for claiming "fascia" to be a new organ. Nevertheless, they did make a significant discovery by their novel way of examining fascia with a remarkable comparison between living and dead tissue. Living tissue holds fluids, whereas dead tissue is desiccated. They, in essence, discovered a profound visual description of the cellular makeup and behaviors of fascia in its living, hydrated form. The cellular structure of these tissues is like a 'dynamic layer of bubble wrap' surrounding structures and organs, where fluid fills the bubbles and drains in and around cellular spaces. The fluid in these sacs may be a primary source of lymph fluids that travel through the lymphatic system and a source of blood plasma. This discovery suggests that much of fascia tissue is fluid-based, possibly giving rise to a fourth circulatory system that transports nutrients and even disease pathogens. The reader should know of the interchangeable terms fascia, interstitium, and connective tissues. The emerging anatomy discoveries will eventually require some work in uniform terminology.

Historically, manual therapists have assumed deep invasive or actively mobilizing pressure was needed to reach deep structures. This new tissue theory, in contrast, may explain why profound and lasting changes are seen in bodywork with gentle, noninvasive, sustained touch contact. The mechanism for improvements of function may result from mobilizing fluids of such exquisite tissues that afford them adequate time to rehydrate and self-correct in cellular expansion. When described as a method that 'holds space and waits for tissue response,' craniosacral therapy may be a therapeutic waiting for the tissues to expand the cellular terrain while holding the margins of the restrictions contained in the organ or the surrounding tissues. The concepts of 'listening and following tissues' and applying sustained noninvasive stretch may be functions of following the flow of waters and assisting a widening of structural spaces. We believe this to be the cause of the release of protective retraction.

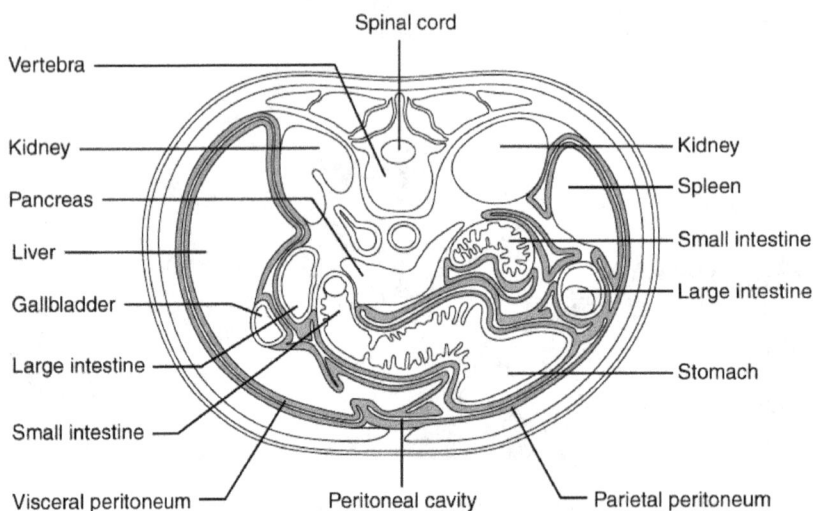

Interstitium is found connective tissues throughout the body. Peritoneum, mesentery, and tissues suspending organs and vessels are woven with interstitial cells where fluid-filled cells keep the tissues hydrated and buoyant. Photo source OpenStax College, Wikimedia: Creative Commons

Behaviors improve in children when bodywork has explicitly focused on the interstitial layers.

Lessons Learned from Bodywork

Thousands of babies and children, all active participants in their bodywork intervention, have confirmed the theoretical constructs that we can assist their bodies towards self-correction parasympathetic responses. Changes in behaviors suggest these efforts affect adaptability and resiliency. Valuable lessons were gained through the generalized application of blending bodywork methods to the population of autism, attention-deficit disorders, sensory processing and modulation dysfunction, sensory-based feeding difficulties, and other issues commonly seen in pediatric practices.

New Definition of Fascia

"The fascia is any tissue that contains features capable of responding to mechanical stimuli."

The fascial continuum is the result of the evolution of the perfect synergy among different tissues, liquids, and solids; capable of supporting, dividing, penetrating, feeding, and connecting all districts of the body

Types: endoneurium, perineurium, epineurium, fascia surrounding striated and smooth voluntary & involuntary muscle fibers, tendon, ligaments, cartilage, aponeurosis, endomysium, epimysium, perimysium, bones, meninges, and all viscera organs derived from mesoderm layer.

The continuum constantly transmits and receives mechano-metabolic information that can influence the shape and function of the whole body. Afferent and efferent impulses come from fascia.

Bordoni, et. al. 2020

Decades of clinical application have contributed to embracing these theories:
- **Fascia is a sensory organ** - that is reactive and retractive
- **Biotensegrity** - dynamic symmetry of the soft skeleton for posture & balance
- **Behaviors of the Autonomic Nervous System** – revealed in organs and vasculature
    - **protective retraction** is the language of organs and vasculature
- **Continuum of the ANS tone** – can be palpated anywhere

Fascia is a Sensory Organ

The continuous, whole-body fascia network is one of the most novel considerations in *Bodywork for Sensory Integration*. It is mind-numbing when thinking of the remarkable discoveries in anatomical studies, that took science until the 21st century to consider fascia's value. Down through the ages, anatomists and biology teachers were so conditioned to tell students that fascia was useless, instructing them to scrape away the shining film from the "really important" stuff. Until medical illustrators became conditioned and skilled at drawing fascia, when studying anatomy books, we need to train our minds to visually add in the fascia to the images because most of it is not included in the artwork.

Fascia anatomy consists of protein fibers of elastin and collagen plus a ground substance, which makes it flexible, elastic, expandable, and adaptable to position, posture, and movement. Collagen mixed with elastin and a liquid called hyaluronan is found in different ratios, creating different types of fasciae. Like a nylon stocking or a mitten, fascia can hold the shape of what it surrounds but is not strong enough to support its weight. Like the construction of citrus fruit, fascia is a web of connective tissue formed in bands or stretchy sheets that wraps around all the internal parts. Bands can be firm structure that anchor bone to bone, or can be the soft fluid-layers in which muscle cells (myocytes) develop, creating an inseparable weave of muscle and fascia layering and wrapping. Fascia surrounds and weaves through the brain and spinal cord. Fascia's traits and functions are why bodyworkers believe we can reach the craniosacral system meninges through *Bodywork* by communicating mechanically through those fiber networks in the periphery.

Fascia, interstitium, and connective tissues are often lumped together in reference sources. As the name implies, connective tissues often bind structures together, suspend organs, cushions and fill space where needed. Fascia serves the mechanical func-

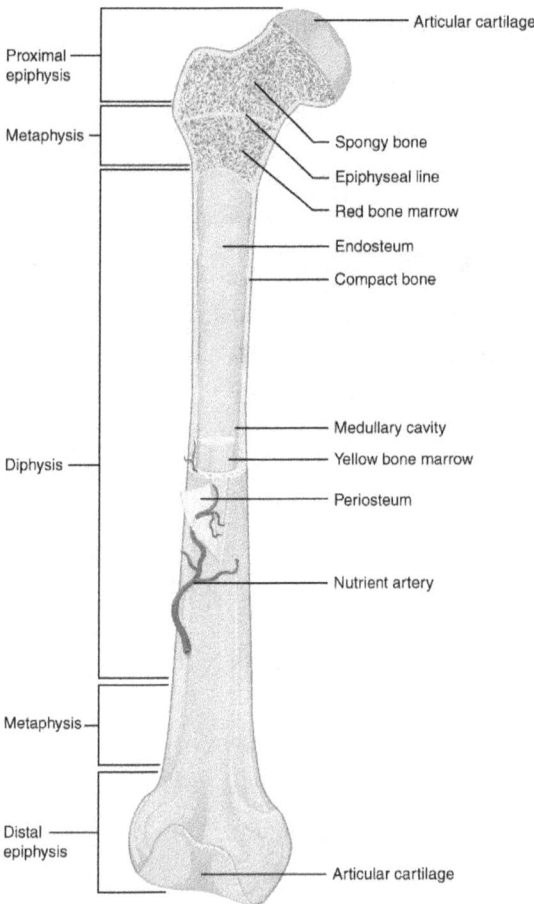

Proximal epiphysis
Metaphysis
Diphysis
Metaphysis
Distal epiphysis

Articular cartilage
Spongy bone
Epiphyseal line
Red bone marrow
Endosteum
Compact bone
Medullary cavity
Yellow bone marrow
Periosteum
Nutrient artery
Articular cartilage

PhotoImage source: OpenStax College, Wikimedia Creative Commons

tion of holding the body together to maintain form and shape but also houses life-giving fluids and transports wastes through the network of these fibers. The fluids of the interstitial layers may also be another circulatory system.

There are six general types of connective tissues:

- loose tissue [abdominal organs]
- dense tissue [tendons and ligaments]
- cartilage
- adipose tissue
- blood vessel-forming tissue
- bone

The complex cellular network of fascia is now believed to be a sensory organ and serves and reflects the function of a 'shock absorption' mechanism. Histologic studies have discovered nerves embedded within the fascial tissues. **Ten times more sensory receptors are embedded in the fascia weave than in muscle tissue.** Consider the analogy of a dry cleaner bag wrapping through muscles and around nerves and organs. Fascia is dynamic tissue with both afferent and retracting qualities. No matter what body part the practitioner palpates or engages, a tactile interface occurs with the fascia web enveloping that structure. Imagine how your tactile cues communicate with those layers when performing bodywork. And how might those fascia fibers be reacting to the level of intrusion your hands are making. Fascia is a trust sensor.

The measured velocity of nerve conduction cannot account for the swiftness at which humans demonstrate immediate and rapid postural adjustment or coordination refinements. It turns out, nerve pathways do not travel that fast. The fascia system now is given credit for postural and balance control from the sensory cues within local regions, though the signals still travel to the efferent nerves to the spinal cord, to the brain, and back down the motor pathways. The cellular connectedness of fascia forms a messaging system between distant tissues, organs, muscular structures, and sensory cells - explaining the quantum speed of sensory-motor reaction and response times. Proximity is key to communication between cells. The speed of nerve signaling from muscle to brain is not fast enough for an appropriately-timed reaction, say when you are standing in an elevator, and there is that degree of the floor moving as the elevator ascends in the shaft of the building. With a quick enough response, you adjust, often without thinking about it. Alfred Pischinger (1899-1982) was a noted histology scientist who first described a "neural" control of the extracellular matrix.

As quoted by the Fascia Congress, fascia has been referred to as an energy superconductor by quantum physicists. We know that every cell in the human body has receptors for neuropeptides which tell cells how to react. In response to thoughts and feelings, the chemical release of neuropeptides happens simultaneously throughout the body in the immune, nervous, endocrine, and gastrointestinal systems. Dr. Candice Pert (scientist and author of the book *Molecules of Emotion*) called these "nodal points on a vast superhighway of internal information exchange taking place on a molecular level," arguing that cells are not the only sensory communicators, nor the seat of consciousness.

James Oschman teaches that cell communication happens via vibrations, frequencies, and nerve conduction throughout the connective tissue or extracellular matrix, which he refers to as "the living matrix." Vibration and signaling between cells connected within the matrix contribute to the protective nature of fascia as a modulator of all it touches, and that touches it.

It is known that nerves transmit signals over 120 meters/second or 275 miles/hour. The ion transfer mechanism of nerve impulses needs to be faster to account for the massive amount of information the brain and body process. It is not possible. The current nerve conduction model is the sole source of cellular communication for the trillions of cells having more than 100,000 reactions per second. The fiber-optic network serves as background sensory processing function bathing all cells, organs, and structures with information in the form of vibration, energy, sound, nutrition, oxygen, biochemical and hormones, and even light.

Furthermore, long fibers have been recently found and identified as interstitial cells throughout the web of fascia believed to also hold a 'communication function.' These fiber cells are known as telocytes. They are incredibly long and thin and appear to reside everywhere that interstitium tissue exists. In daily life, the sensory aspects of connective tissue may be an underlying determinant of movement quality, mood, alertness, and general well-being.

Fascia comprises the walls of arteries and veins, lymphatic vessels, capillaries, and venules. It is a system wide communication system for stability, movement, form, and adjustment. This fact alone begs for further investigation into the behaviors of fascia within the scope of sensory integration theory.

Due to the sheer quantity and richness of nerve fibers, the fascia matrix is now considered the largest sensory organ of the human body. The liquid-crystalline make-up of the connective tissue fibers functions as a whole-body source of communication. Fascia could significantly contribute to total body awareness and kinesthetic processing of position and movement. The function of fascia has significant implications as we treat sensory integration disorders. Does it send sensory input directly to the cortex, the cerebellum, or other structures that process and disseminate body information? Although it can expand and contract, we do not fully understand this relationship to sensory input or as an adaptive sensory integration process outside the central nervous system.

Consider that the child's fascia cells certainly know your hands are present during *Bodywork*, to further this paradigm shift beyond thinking fascia does more than hold an organ's shape and binds structures together. To a bodywork therapist, the body is no longer considered a system of separate parts but a woven structure of cellular connections that communicate through a high-speech symphony of micro-optic signaling. To the sensory integration therapist, considering the body under a completely different paradigm can transform how you treat your clients with sensory challenges.

Biotensegrity

Nearly all of bodywork methods address the systemic organ of fascia. Referred to as the "soft skeleton," fascia is the structural system that gives the body its shape and form, maintained even when standing and moving. The structural suspension and the sensory qualities create an integrity of reciprocal tensions. Tensegrity, a term from the world of architectural design, is a hybrid union of the words "tension" and "integrity." This design principle applies to the building of a structure that can stabilize itself when parts are assembled in a way where compressive forces are held against non-compressive forces to sustain whole-system balance.

Biotensegrity, a term coined by Dr. Stephen Levin, applies tensegrity principles to biological structures (such as muscles, bones, fascia, ligaments, tendons, and even elastic cell membranes. These entities are made strong by the unison of tensioned and compressed parts but with just the 'right amount' of mechanical forces. Too much and the structure becomes imbalanced and eventually weakens. Not enough, and the structure loses integrity. The combination of fascia structural suspension and the sensory qualities for rapid efferent and afferent messaging serves this whole. Fascia holds structural shape while allowing organs and tissues like muscles to expand and contract while maintaining the upright form of tensegrity. Our fascia matrix moves with the musculoskeletal system between symmetry and asymmetry yet can maintain its form and structure to move. If the fascia is restricted, then muscle, bony joint excursions, and organ motility are restricted to similar degrees.

> In *Bodywork for Sensory Integration,* we are not moving or lengthening fascia. We are not massaging, unless directly applying lymphatic drainage evacuation. Instead, we are communicating with fascia and the tissue does its own thing in the treatment process. (See Responsive Hold pp. 64-67)

Peripheral fascia networks may absorb shocks to protect the cerebrospinal dynamics. A criteria function of fascia is the housing of interstitial fluids, now believed to give rise to lymphatic fluid. The interstitial cells of fascia have been described as 'fluid-filled' bubble wrap. Thus, fascia and interstitium play a considerable role in the body's homeostasis, such as maintaining hydration levels and fluid balance. Fascia is the structure that most likely keeps water and fluids from pooling down into our legs, contributing to optimal vessel tone for fluid suspension and fluid exchange. Interstitium flushes out toxins at an extremely high rate as a function of the immune system and detoxification pathways.

As already stated, it has been proven that fascia is not lengthened through stretch, but there is evidence that manual stretch to fascia may assist rehydrating tissue. Tissue responses during treatment may be the self-corrective mechanism more clearly defined. Furthermore, we are facilitating the exchange of fluids as a result. All body systems make up Biotensegrity. Fascia transmits sensory messaging as a backup nervous system. Studying the human form in the upright stance is a stunning example of structural tensegrity dynamic and instantaneous self-equilibrating maintenance of form, movement, and posture. When actively engaged in physical activities, the bones, muscles, and ligaments form joints and adjust the body's form through an orchestra of localized moments of compression and tension to accomplish load resistance and balance. All are governed by the body's drive to keep the head protected.

The solid and the fluids of fascia shape this model of biotensegrity of the human body. Bodily fluids modify the shape of structures. Bodily fluids are silent witnesses to mechanotransducive information, allowing adaptation and life and transporting biochemical and hormonal signals. As the solid fascia tissue divides, supports, and connects the different parts of the body, the fluids of fascia tissue feed and transport messages as intersystem communication. (That implies sensory communication through cellular signaling by the mechanics of flow, contributing to the maintenance of body tensegrity).

Considering the lymph–fascia complex, we should include arteries and veins with walls that move fluids as another aspect of interoception and the role these underappreciated structures have in autonomic regulation for sensory modulation and self-control. The fluids are probably expanding when we feel tissue reactions (under the gentle, noninvasive touch of treatment). Fluid moving either against or without tissue restriction and resistance yields immediate "sensory" feedback in the cellular terrain. This structural expression is the continuum of the ANS. Therefore, movement of bodily fluids should be considered a source of sensory information within the body. Interoceptors consider the fullness or the movements of an organ and vessels. Does fluid stagnation prompt us to move? Pain from occluded circulation when legs are crossed would surmise as such.

The tensegrity of the human body is not unlike the tensegrity of citrus fruit. Clients should be handled more like an orange: juicy bits that are soft and pliable, held together by a communicating and structurally supporting membrane. Blending interstitial tissue work with craniosacral therapy, myofascial release, visceral methods, and lymphatic drainage has given clinical access to the foundation of *Bodywork for Sensory Integration*. Benias and his team's discovery strengthens the argument that the clients should be approached as vessels containing water rather than the inanimate, aggregate mass of flesh.

Compression   Tensegrity of suspended fluids   Traction

An example of the Biotensegrity suspension of fluid membranes. The soft skeleton of this orange slice extends from the core to the underside of the peel and mechanical influence of the whole can be made through contact and melding with the peel. Visualize the possibility that fascia and interstitial makeup of the human body might be replicated in such a manner.

Behaviors of the Autonomic Nervous System

The behaviors of tensegrity parallel the study of neurodevelopmental and sensory integration interventions. Regulation of postural control, ocular motor, vestibular balance, and bilateral coordination can be considered through the science of biotensegrity. The human body relies on the background tensions (or the lack thereof) of the fascia continuum to support posture and balance during movements and possibly as a background sensation for discrete and quick adjustments to gravitational and physiological demands. Behaviors of the ANS are communicated in tissues, organs, and blood vessels, reflected by the presence or absence of protective retraction.

Protective retraction is a core concept and essence of understanding and performing *Bodywork for Sensory Integration*. Feeling and identifying locations of protective retraction through palpation is the beginning of helping balance or reset the autonomic nervous system. This is one of the basic truths learned through the thousands of patient encounters previously mentioned. Palpating the retraction of structures raises therapeutic awareness of the cellular layers. Tone changes are caused by reactions to sensory, cellular, or mechanical events. *Bodywork* aims to [re]gain something like a dynamic poise[3] between tight and relaxed tone.

To further expand the concept protective retraction, consider a similar phenomenon observed in the world of marine biology. Animals with simple nervous systems recoil for protection at the onset of a threat (real or perceived). A colony of sea anemone is a perfect example of a simpler nervous system that retracts its tentacles, for protection as well as body homeostasis. This is not unlike human beings returning to the safety of their homes at the end

3 (The term "dynamic poise" is credited to Tad Wanveer through his advanced level trainings in CST for the central nervous system.)

of the day. We turn into ourselves with various senses of threat, danger, unwell, fatigue and overload, and physiological adjustments (after eating a big meal).

Images of sea anemones: a garden of waratah anemones in various stages or absence of protective retraction. (John Turnball 27742276225). Wiki Commons.

Turtles and tortoises, too, utilize protective retraction in ways that are helpful understanding a more complex nervous system. It's easy to see the full-body effect of extremities and head/neck of the animal as it retreats into the comfortable safety of its shell. Most Cryptodira (tortoises and most turtles) are divided into two different groups by how they retract their necks, either by pulling back sideways or straight back into their shells. Four legs can also fully or partially retract. These creatures retract for the same reasons other animals possess defensive postures, either from threat or for autonomic homeostasis. By the way, only turtles and tortoises have two shells. Sea turtles, in comparison, have only the one dorsal shell.

Protective Retraction – the language of organs and vasculature

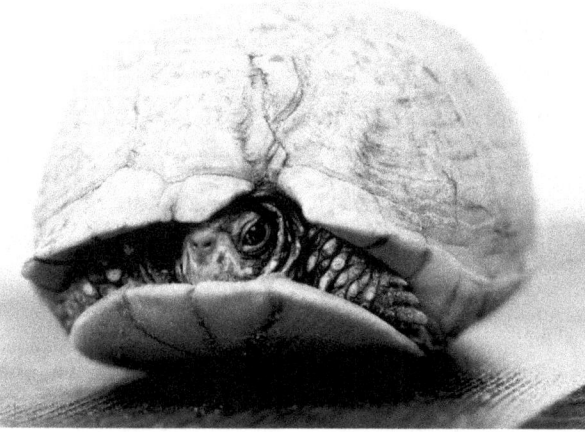

Protective Retraction (full) from protopathic threat or stressors. Homeostatic recovery occurs within for as long as needed.

The continuum of Protective Retraction (partial) responses of the sympathetic responses and the adaptive responses that emerge within the parasympathetic responses.

Absence of Protective Retraction. Ready to engage, think, learn from discriminative responses to sensory input.

The human autonomic nervous system reactions in a similar way, though more complex cortical structures [hopefully] learn to modulate and intercede with threats. This enables a person to think through events and gain emotional stability and resiliency for autonomic recovery. Our higher brains thus learn the difference between real and non-real threats. Newborns have not yet developed this part of their brain's executive cortex. Baby's body staggers through various stages of pulling in for protection and expanding and emerging into relaxation to become ready to engage. There are so many ways to describe this, but we all understand this concept of 'pulling in' when overwhelmed or when the sympathetic firing occurs. We begin learning how to manage protective reactions in infancy. *Bodywork for Sensory Integration* lessons were learned by treating thousands of babies and observing their responses.

Continuum of the ANS tone

Protective retraction explains the structural response to stress (emotional, chemical or physical, and mechanical). Vasoconstriction will be the first sympathetic response from the three nervous systems (central, peripheral, and enteric). Protopathic (meaning protection) sensors detect pain and temperature and one of the first physiologic responses is vasoconstriction. Vasoconstriction can be measured by galvanic skin testing, but *Bodywork for Sensory Integration* proposes that "soft, listening hands" can also detect such protective retraction. Protopathic sensors set off the need to retract. *Bodywork for Sensory Integration* can help determine the continuum of tension (vasoconstriction) of the child's internal structures.

The therapist must look beyond just the outward expression in body posture (Moro reaction or tight fetal positioning). Protective retractions will reveal themselves wherever the ANS nerve project and innervate. Nerve plexuses are begging for treatment when bodies are stuck in sympathetic tone and tension. With *Bodywork* we palpate deeper for internal protective reactions where the centers of the ANS are located.

> **From skin to the pia mater, with vasculature in between, *Bodywork* assists the release of Protective Retractions**

Starting in infancy, how a child reacts to sensations promotes a reaction from those around her. Mood, behavior, and personality develop, in part, around the structural state and resiliency of visceral organs and vasculature to easily recover from sympathetic signaling. Interpersonal relationships are built around the status of the autonomic nervous system which determines whether sensations are perceived as 'ok' or a 'threat'. The way we react to the world begins with how our structures are prepared and ready to engage efficiently with the world. Structural tightness is going to put us in a more protective and cautious state. Conversely, a relaxed, flexible, and balanced system (nerves, muscles, organs) is the manifestation of the parasympathetic state. Where resilience thrives is in a quick recovery to protective retraction (sympathetic) reactions within viscera and circulatory systems.

Bodywork for Sensory Integration offers us a palpable understanding of the state, the balance, and the adaptability of tensions within the autonomic nervous system, as well as helping to determine what might be hampering it. If body physiology is hindered or restricted for prolonged times, and when a child is expending too much energy in the sympathetic mode and could benefit from a 'reset' a parasympathetic tone imprinting.

> **Protective Retraction is a core concept and essence of performing *Bodywork for Sensory Integration***

Vascular vessels are extensions of the fascia field, all having sensory and retractive qualities. Not only is vasculature poised in a regional fascia matrix, but three layers of endothelial connective tissue also construct most vessel walls. Considering the mechanism of vasoconstriction, which is the crucial feature of sympathetic firing, and sensory modulation behaviors, it is a safe assumption that fascia plays a role in sensory modulation. *Bodywork for Sensory Integration* guides the therapist to treat large vascular fields: the scalp, over the skin surface, buried in the thoracic organs, and woven through abdominal mesentery. Motility at these sites indicates the absence or presence of structural resistance that affects peristalsis.

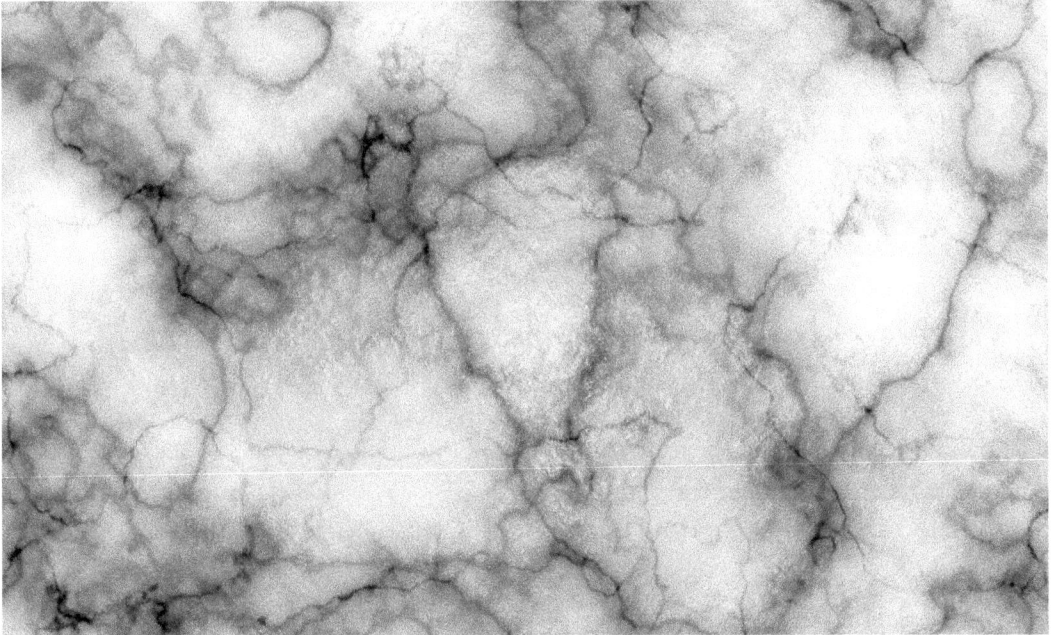

Human blood vessels suspended by fascia tissue. Source: Shutterstock

The hallmark of autonomic nervous system physiology is vasoconstriction in sympathetic (SNS) and vasodilation with parasympathetic (PNS) recovery. Both ends of this continuum raise or lower the tone of muscles, organs, skin, and the fascia matrix itself. Vast fields of vasculature are created and suspended by epithelial fascia. Even vessel walls are layers of fascia. Treating blood vessels suspended in fascia fields give therapist direct access to the ANS, and the ability to treat with specificity. Learning the sliding scale of the ANS tone continuum arose from hundreds of palpations of living bodies, making clients your best learning tool resource.

Enteric and Vagal nervous systems

Two peritoneum layers line the abdominal cavity and covering the organs. The peritoneum is just another extension of the fascia continuation throughout the body. These layers of tissue separate the organs from the contents of the retroperitoneal space and comprise a layer of mesothelial cells that resides next to the organs. SNS and PNS tone can be palpated in the visceral organs, presumably with signaling arises from all sources of innervation. Gut-brain tension palpated in the discrete peritoneum and mesentery have informed us of sensory modulation, mood, and state control. In turn, *Bodywork for Sensory Integration* assists self-regulation by treating visceral organ fascia and the concentrated field of blood vessels woven through the mesentery. The ANS and the Enteric state influences how the child will respond to sensory reactions and perceptions. No sensory approach has guided us until now, but *Bodywork for Sensory Integration* has shaped how to palpate the ANS directly and assist homeostasis flexibility.

> ***Bodywork for Sensory Integration*** **palpates the ANS and Enteric system directly and assists homeostasis recovery by resetting parasympathetic firing through entire systems simultaneously:**
> **vasculature, lymphatics, organs, mesentery and peritoneum**

Other Manual or Therapeutic Touch Methods

Though not included in the historical aspects of this book, individual pediatric bodywork practices can consist of other methods. Some examples include Total motion release, Reiki, Qigong, Strain-Counterstrain, Feldenkrais method, Pilates, and endeavors specific to the osteopathic and chiropractic fields. These different methods also reflect an increasing and collective awareness of listening and working with soft tissues and understanding the fluid nature of working with fascia and the energies inherent to these fluid tissues instead of applying strict mass-directed protocols.

The Continuum of the Autonomic Nervous System

Sympathetic	(reciprocal tensions can be palpated)	Parasympathetic
	Anatomical	
Thoraco-lumbar spine (pre-ganglion nerves pass through cervical spine)	Outflow from CNS	Brainstem and sacral plexus
Closer to brain/spinal cord	Location of ganglia	Closer to organ/gland
Throughout the body	Distribution	More limited
	Functional	
Excitatory - generally	Homeostatic Effect	Inhibitory- generally
Mobilizes energy	General metabolism	Conserves, restores, excretes
Raises awareness and attention to sensory events or threats	Effect on Sensory Systems	Restores attention and awareness back to base levels also
Cardiac muscle - circulation		
Increases	Heart rate and blood pressure	Decreases
Increases	Force of ventricle contraction	Little effect
Vasoconstriction	Effect on blood vessels	Vasodilation
Skeletal muscles		
Increases rate of glycogen to glucose conversion for muscle activation & power	Muscle activation	No action
Smooth muscles		
Constricts distal vessels	Blood vessels	Little effect*
Pupils dilated	Iris of eye	Pupils constricted
Inhibits bulge; vision stays distant	Eye (ciliary) muscles	Bulges eye for close vision
Bronchi to lungs	Dilated under tension	Relaxed into constriction
Inhibited; causing tense organs	G.I. tract	Stimulated but relaxed organs
Contract (usually)	Sphincters	Relaxed; absence of tension
Shuts down peristalsis		Activates peristalsis
Inhibited; stops flow	Urinary bladder	Contracted (to void)
Contract sphincter (bowels move less easily)	Rectum	Relaxes sphincter (bowels move with ease)
Glands		
Stimulates (viscous)	Salivary glands	Stimulates (watery)
Constricts; dry eyes	Lacrimal glands	Stimulates tear production
Inhibits secretion (usually)	Digestive glands	Stimulates secretion
Stimulates	Sweat glands	No effect
Stimulates adrenaline	Adrenal medulla	No effect

*Some blood vessels in facial and pelvic regions dilate under parasympathetic impulse

Responsive Hold (RH) - Primary Technique in *Bodywork for Sensory Integration*

The child's skin will read the "intention" behind the therapist's touch, regardless of method. Non-invasive touch with altruistic intentions has the best chance of not alarming the sympathetic system, and building trust. That trust should never be violated or taken for granted. It is sacred.

In determining where on the body to begin, Upledger teaches arcing as a way of scanning energetically for entropy and tissue stagnation. Barral teaches a listening method of hands feeling for an epicenter of a pull or energetic field. Informal learning of older osteopathic methods taught how to feel for lines-of-pull through fascia fields by direct palpation. In CST, we are taught to first work on older kids and adults before attempting to work on babies. That is sage advice, but my practice at the time was primarily children under the age of eight. Once I learned just how valuable CST, and my clients' needs were presented to me daily, I was thrust into the land of working with teeny tiny body structures. Though I already had competence in handling babies with my NDT training, they were an incredible learning source for fusing treatment methods. In the two decades of blending all these methods with NDT handling skills, I found I needed to adapt a few things.

After years of analyzing the concepts of blending and melding, a handling description became clearer. I call it Responsive Hold (RH) and I attempt to describe my perspective of what may be occurring where therapeutic hands meet tissues.

A technique like no other, RH guides therapeutic hands when utilizing any of the methods in *Bodywork for Sensory Integration*. RH begins as soon as you lay your hands upon a child's body with static touch. RH is used for both evaluation as well as treatment. The hands will not move on the body, unlike when the situation calls for myofascial release or lymphatic massage. The length of time holding static touch depends upon the child's unique system, the degree of ANS stress, and past experiences with touch contact. Let their body scan *your* hands and *your* intention (through cellular communication of vibrations and warmth). If and when the parasympathetic system is activated, your hands may undoubtedly perceive movement of structures as vasodilation, fascia expansion, and relaxation commences. **The waiting time is how we discern the difference between tensions of protective retraction and visceral guarding, from any actual mechanical compressions and structural compromise.**

We need to get through the first stage of these ANS behaviors, and we do so by waiting long enough for the ANS to decide if it trusts our hands. Otherwise, you may be wrestling with vasoconstriction and a rejection of your presence within their space. This child did not come to see you for you to wrestle with their vascular walls. There are several indications that any protective retraction has subsided: softening of tissue tension, increased warmth, deeper breathing, digestive sounds, shaking or pulsation in tissues as energy is released, and an overt relaxation of the body. The softening action is believed to be self-correction of the interstitial layers as well as vasodilation. In babies, very little time is needed for release, unwind, and self-correct into parasympathetic mode. More time will likely be needed the older the client. (Thus, the modification of my approach and touch was actually taught to me by my very young clients and not on the adult bodies I learned the techniques on. Upledger suggested this, though, so that novel practitioners did not alarm or harm young clients.)

Continue to hold non-moving hands on the treatment site. Consider all layers of tissues and organs beneath the surface of the skin, including interstitial layers and fluid vessels. Reach the connections of fascia tissue to gross structures. The softening or spreading of tissue relaxation will likely feel like your hands are being moved by the body. They most likely are, even though outwardly it may appear your hands remain still. Infinitesimal movement occurs within the spaces between tissues and structures, reacting to the energy from the intention behind the touch contact. Only enough tissue expansion for a droplet of fluid to seep between cells is needed to facilitate release of compression or tension. If hands leave these margins too soon, the structures can re-constrict and can stop the correction process. Remember, the therapeutic touch method unique to Upledger's CST of blending and melding with tissues, is an allowance of the body to self-correct and expand from its protective retraction state. Other methods that don't lead with this intention risk alarming the system back or deeper into a sympathetic withdrawal.

Responsive Hold is a tactile interplay with structures that respond similarly to handling rising bread dough.

In offering an analogy to train your hands, RH can feel like holding a ball of actively rising bread dough. One can feel the edges of the dough expand ever so subtly if patient enough to wait out the process of the rise. As your palpation skills mature, this sensation will begin to feel like large movements as the extremely discrete sensory perception of your fingertips develops. To the outside observer, it will appear that nothing is being done to, or for, the person on the treatment table other than your hands laid on their body. The dough stays in contact with your skin, and the skin stays actively in contact with the dough. All the while, the dough continues to rise. Your hands can alter the shape of the dough through the use of gentle contact or from firm compression. Too firm and you can stop the rise. Equally, you holding the ball of dough can assist its expansion by merely intending to aid the process of cellular expansion. Your fingers will actively help the dough stretch itself. (Sounds far-fetched, right?) The gentler the intention of stretch, the more the dough ball expands within the hands holding it (because your hands are free from adding resistance to the dough ball.) The more firm and invasive your hold, the more the dough ball is likely to stop expanding.

Now, pretend you are that ball of dough being held in the air by someone with therapeutic intensions. Imagine how you would want to be handled and treated. This is where your intention starts, with that level of empathy. The child's body will perceive this through the vibration of your level of touch.

Listening to the tissues with hands and fingertips is critical. However, light *moving* touch (directly on bare skin) has often been problematic for those with sensory defensiveness. RH is a different kind of touch. It is not a moving touch, but a static touch of your hands reminding the skin of the safe touch felt by the pressures within the womb, hopefully, the original experience of safe touch. RH is a way for the therapist to create the right touch for the child by reading their cues and listening to autonomic tone and tension at the skin level. Continually monitor your level of touch input throughout the process of using RH and the child's response to it.

Bodywork for Sensory Integration begins and ends by palpating the tone of the ANS. Treatment begins as soon as hands are placed upon the body. For tissues to release tensions and return to a parasympathetic state, the therapist must hold the space for tissue expansion and be physically, mentally, and emotionally present. RH can induce the most profound changes to the release of protective retraction of the whole body. If this is not happening, perhaps the therapist is impatient or needs more confidence in skills. Perhaps the therapist needs treatment to help regulate ANS responses. If the client's body and ANS communication is not thoroughly listened to, you could create the opposite effect upon self-regulation.

Responsive Hold treats the space between structures, the interstitium that surrounds the gross organs and forms you are accustomed to thinking, the cellular layers around those structures, and the fascial network that weaves around and through those structures. Holding space can be done with muscles, organs, tendons, and craniosacral structures. Considering the new views of anatomy, the interstitial space may be where protective retraction and tissue restrictions are found.

Listening hands, not moving hands use RH to become informed about where the next place of structural imbalance, restriction, or compression is held. Moving hands do not perceive as much as in tissues as quiet hands do because moving hands will take your awareness to your proprioception and tactile awareness. As any treatment process proceeds, tissues may then call for the therapist's hand to move into a method that helps mobilize any compressed structure, but we do not start at the stage. Moving your hands at the wrong time will likely increase protective guarding. RH is employed in both direct and indirect methods. Again, RH is not the same as moving touch, as in massage or myofascial stretch methods. The sense of movement comes from the expansion structures, tissues, or organs treated. Static touch is what the hands do to begin contact, but RH is the reactions of your hands to the changes in the tissue. **It is imperative to understand and appreciate this difference.** Think of your hands touching the top of the water like a lily pad (static touch), and the movement of water currents below is what moves the lily pad.

Listening (static) Hands can learn more about body tissue than Moving Hands

The intention of the touch within RH is to develop the art of listening with one's hands. The experienced therapist working at fusing methods can move between techniques of CST, versions of myofascial release and lines-of-pull treatment, or lymphatic drainage. For the sake of the child's ANS, do not move too quickly from the initial holding. It is amazing to witness how much a body can self-correct if we hold all the connective tissues and communicating cells long enough. It is incredible what can be felt with your fingertips when your hands do not move. So much more can palpate and be perceived when your [loud] kinesthetic touch perception does not drown out the subtle and [almost silent] sensation of tissue expanding and fluids exchanging. Set your intention that using RH as an indirect treatment with a soft, gentle touch will yield sensory information to your fingertips about the state of structures lying beneath the surface.

Evidence-Based Practices and Practice-Based Evidence

Evidence-based practice (EBP) has become a new standard to review, analyze, and confirm empirical study supporting a therapeutic claim. EBP is a model developed to guide practitioners in deductive reasoning when choosing methods to apply in their practices. By definition, EBP includes clinical experiences and expertise, along with patient expectations and preferences. Comparing and contrasting practice experiences with empirical research (if it exists) strengthens the value of a particular treatment and the profession that promotes it.

The intent of EBP has sadly become corrupted in some circles by unchecked intrinsic bias towards particular therapies, demanding "proof" by empirically designed studies. The EBP model was never intended to prove if a method was to be used or abandoned, nor was it intended to squash innovative practices. Sadly, that is a reality many practitioners are facing. Sharing anecdotal information between peers, matched with altruistic motives, seems to have lost favor.

Any intervention with a history associated with repeated documented positive outcomes is also evidence. The medical record and collected anecdotal evidence is still evidence. EBP does take into account the level of experience and expertise of the practitioner, the practicality for any practitioner to gain such skill training, and the values, context, and availability for the target population. Integrating all arms of EBP is known as clinical reasoning, guiding practitioners in decisions of care for their clients.

For many forms of manual therapies, patient-reported outcomes hold a great deal of weight. It is, after all, the consumer of a service the decides personal value to an intervention.

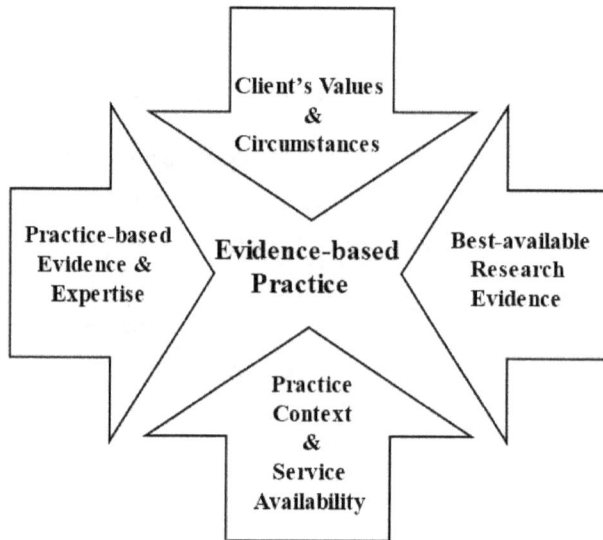

Evidence-based practice integrates information from four elements

Manual therapy for musculoskeletal addressing pain and movement dysfunction has a mountain of supporting evidence. Based upon a most recent systematic review (conducted in the United Kingdom) of thousands of published studies adhering to empirical standards, all forms of manual therapy used by various professions were given the most robust ranking. (III against a rating scale from the American College of Physicians and American Pain Society). For individual studies, the rankings drop to II or I. Nonetheless, manual treatments are well-supported by evidence to be applied in a correct manner. Few studies, however, addressed the effect of manual therapies on neurologic and neurobehavioral conditions.

Most studies of CST remain practice-based evidence, though a growing number of high-quality studies have now been conducted. CST has been shown to reduce the frequency and intensity of headaches, both migraine and cervicogenic in nature. A few studies have shown CST is safe and effective for reducing intensity and disability from neck pain lasting several months. One study also established a sham control method.

A recent meta-analysis of CST for pain suggested significant and robust effects lasting up to six months. A few studies have shown that visceral manipulation methods have had positive effects on low back pain, diastasis recti, and effectiveness for pseudo bowel obstruction. Both CST and VM have demonstrated positive effects upon urinary incontinency and post-concussion syndrome.

The wealth of science supports traditional views of manual therapy for musculoskeletal (stretch, myofascial release, joint mobilization), edema management (lymphatic drainage), and stress and wellness (adult and infant massage, therapeutic touch). These methods share one thing in common: protocols of how techniques are applied to a client's body. Innovative methods like CST and VM, however, are harder to measure patient-reported value and effects. Though basic protocols are applied for safety at entry-level practice, the best work in CST and VM is usually done when the body tissues dictate the location and what needs to be done. Thus, every treatment session proceeds in a unique pattern of delivery. It is implausible to measure nonverbal directives from tissue other than through biomarkers of the ANS.

In addition, all bodywork methods have the potential to tap into emotional memories of trauma and injury held in the tissues. Resolution of deeply held emotional effects of trauma is a crucial aspect of CST and its global effects, but this fact also creates a formidable uncontrollable variable. Several studies from different professional viewpoints demonstrated a positive influence on the autonomic nervous system depending on the stimulation site and type of methods. It has been shown that a more significant parasympathetic response emerged when techniques were applied to cervical and lumbar regions, and a sympathetic response could be stimulated with techniques performed in the thoracic region. Regardless, it has been demonstrated routinely that various forms of manual therapy at the upper cervical/cranial base region promotes a parasympathetic reaction, thereby directly affecting the ANS.

In several smaller studies, CST also demonstrated a favorable effect on the autonomic nervous activity as measured by heart rate variability, skin conductance, and respiratory rates. Psychotherapists employing psychotherapy and CST found that clients access emotions and traumatic experiences more significantly than with psychotherapy alone.

For the pediatric population, a few studies showed moderate evidence of statistical improvements from CST and VM for crying and digestive discomfort due to colic. It has been demonstrated that CST is safe to apply to healthy preterm infants, showing no adverse changes in general motor patterns (an indicator of neurological stress). Osteopathic treatment reduced significantly the number of days of hospitalization and is cost-effective on a large cohort of preterm infants.

The best way to study bodywork methods such as CST has yet to be developed. Grounded theory research may also be a preferred method to study CST by elaborating and analyzing existing constructs. The existence of a craniosacral system has been proven, and the physiological rhythm, as described by Upledger, can be visualized during neurosurgery if the dura mater is exposed. Over four decades of scientific and clinical evidence supports CST, strongly satisfying two elements of EBP guidelines.

Addressing concerns about the placebo effect in children with manual therapies, the twenty-plus years of study on the topic derives from studies of ADHD, depression, and migraines with the clinical use of medications. There is little data on the placebo effect, whether learning and conditioning play a role in the placebo effect, or if placebo even exists in younger children. Only a few studies investigated placebo mechanisms in children younger than age six.

Interventions with carefully designed constructs that provide peer-tested certifications can raise the level of application reliability between practitioners. Upledger (CST), Chikly (LD), and Barral (VM) institutes all provide certification processes that elevate reliability and higher standards, though none require certification to practice any of the methods. These institutes ensure that a professional license exists before acceptance into [advanced level] classes. The pediatric realms of SI and NDT also maintain certification tracts in training, with the SI programs recently undergoing rigorous updating.

Two decades of clinical reasoning helped create the contents of this book. I have practiced sensory integration, neurodevelopmental, and neurological rehabilitation for forty years. The lessons learned observe how the nervous system responds to interventions and expresses injury and recovery. Tens of thousands of observations from patient encounters shaped the opinions within the pages of this book. My own medical records (including repeated endoscopy and echocardiogram) proved that both a hiatal hernia and mitral valve prolapse fully resolved with craniosacral therapy. Even the cardiologist remarked that he had never witnessed a spontaneous correction of a mitral valve prolapse and was very interested in what had been done to correct it. Changes to my reactions and management of stressors improved from those moments. CST also was responsible for "lifting" the left temporal and parietal bones (from a head injury thirty years prior). Those results brought reduction to life-long auditory defensiveness, with improved attention and concentration despite background noise. There is a collective appreciation, noted by both me and my family members, of mood stabilization and more flexibility in stressful events involving noise. My own patient-reported changes after treatment have lasted for over two decades.

Furthermore, during treatment for Lyme disease with co-infections, my body experienced swollen and stagnant lymphatic systems and extreme pain from systemic neuritis and neuropathies. (Caused by Herxheimer reactions from cellular die-off). During Lyme flares and subsequent die-off from treatment, touch defensiveness suddenly became a reality for me. Wearing deep-pressure clothing and dry brushing was a daily need for months. Lymphatic drainage massage proved beneficial in reducing the pain under the skin by facilitating the body detoxification process. The cellular world under the skin healed after the debris field was evacuated. The appreciation for agony of touch defensiveness gained yet another paradigm shift for me and I gained a new theory based upon these experiences. These methods were then generalized and offered to clients with touch issues, and we observed similar positive results.

Craniosacral therapy is yet another sensory-based, non-invasive, and non-alarming touch method that may also be a form of sensory entrainment, perhaps to interoceptors. Making contact activates touch receptors in the skin, vessels, and organs. Skin is the distal layer of the projecting fascia complex. The skin has an indirect connection to the meningeal layers through the communication network of cellular structure as well as direct connections through the trains of fascia layers. Adapting to touch often takes time, often longer than usually thought of, when sensory modulation is an issue. The best way to objectively study protective retraction is through ANS biomarker methods (heart and respiratory rate variability, skin conduction, and oxygen saturation).

It is firmly established that therapeutic touch, massage methods, and other manual therapies can produce a relaxation response (parasympathetic activation). For people with self-regulation and modulation struggles, this effect is often what they need most urgently. In traditional sensory integration treatment, waiting weeks and often months to observe behavioral changes from behavioral and sensory methods has always been a point of contention because many variables can contribute to changes. However, bodywork can produce immediate or quick changes; the effects can be more readily attributed to manual therapies. Indeed, more research is needed in this area.

Manual therapies with an undercurrent of therapeutic intention are also relationship-based. The basic method to connect with another human being is through touch. Human connections are not language dependent. If sensory cues are misfiring, then the intention of touch can be perceived as somewhat threatening. Trust, and a therapeutic relationship must be formed at the onset. It has been observed that the more sensory challenges a child has, the more they may tolerate CST, but only if the process is child-led and the pacing is under the control of this child. Though if, in the rare case, a child does not tolerate or understand the premise, several CST techniques can be taught to parents for application at home. This book is based upon encounters made during regular clinic practice for populations ranging from brain injury, autism, attention-deficit disorder, seizures, and sensory processing dysfunction, as well as typical infants and children seeking health and wellness treatment.

Further reading on Therapeutic Methods

Barnes, John F. "Myofascial release." *Complementary Therapies in Rehabilitation: Evidence for Efficacy in Therapy, Prevention, and Wellness* 59 (2004).Barral

Barral, Jean-Pierre, and Alain Croibier. *Visceral Vascular Manipulations*. Elsevier Health Sciences, 2011.

Beyer, Lothar, Stephan Vinzelberg, and Dana Loudovici-Krug. "Evidence-based medicine in manual medicine/manual therapy: A summary review." *Manuelle Medizin* 60, no. 4 (2022): 203-223.

Blackburn, Jack, and Cynthia Price. "Implications of presence in manual therapy." Journal of Bodywork and Movement Therapies 11, no. 1 (2007): 68-77.

Bleeker, Deborah. Acupressure Made Simple: Easily Treat Yourself for Common Ailments. Draycott Publishing. (2019).

Bordoni, Bruno, and Thomas Myers. "A review of the theoretical fascial models: biotensegrity, fascintegrity, and myofascial chains." *Cureus* 12, no. 2 (2020).

Bressler, Harry B. *Zone Therapy*. Health Research Books, 1996.

Bronfort, Gert, Mitch Haas, Roni Evans, Brent Leininger, and Jay Triano. "Effectiveness of manual therapies: the UK evidence report." *Chiropractic & osteopathy* 18, no. 1 (2010): 1-33.

Chikly, Bruno. *Silent waves: theory and practice of lymph drainage therapy: an osteopathic lymphatic technique*. IHH Pub., 2004.

Chikly, Bruno, Jörgen Quaghebeur, and Walter Witryol. "A controlled comparison between manual lymphatic mapping of plantar lymph flow and standard physiologic maps using lymph drainage therapy." *Quality in Primary Care* 23, no. 1 (2015): 46-50.

Dang, Deborah, Sandra L. Dearholt, Kim Bissett, Judith Ascenzi, and Madeleine Whalen. *Johns Hopkins evidence-based practice for nurses and healthcare professionals: Model and guidelines*. Sigma Theta Tau, 2021.

Davis, Carol M., ed. *Complementary therapies in rehabilitation: evidence for efficacy in therapy, prevention, and wellness*. Slack Incorporated, 2009.

Duncan, Ruth. *Myofascial release*. Human Kinetics, 2021.

Frigard, L. Ted. "Still vs. Palmer: A Remembrance of the Famous Debate." *Dynamic Chiropractic* – January 27, 2003, Vol. 21, Issue 03.

Hull, Ruth. *Complete Guide to Reflexology*. Lotus Publishing, 2020.

Geri, Tommaso, Antonello Viceconti, Marco Minacci, Marco Testa, and Giacomo Rossettini. "Manual therapy: exploiting the role of human touch." *Musculoskeletal Science and Practice* 44 (2019): 102044.

Guimberteau, Jean-Claude. "Human Living Microanatomy." *Fascia: The Tensional Network of the Human Body-E-Book: The science and clinical applications in manual and movement therapy* (2021): 239.

Ingber, Donald E. "Tensegrity as the architecture of life." In *Proceedings of IASS Annual Symposia*, vol. 2018, no. 27, pp. 1-4. International Association for Shell and Spatial Structures (IASS), 2018.

Jones, Tracey A. "Rolfing." *Physical Medicine and Rehabilitation Clinics* 15, no. 4 (2004): 799-809.

Krebs, Charles T. "LEAP®: The Learning Enhancement Acupressure Program: Correcting Learning and Memory Problems with Acupressure and Kinesiology." (2007).

Kunz, Kevin, and Barbara Kunz. "Understanding the science and art of reflexology." *Alternative and Complementary Therapies* 1, no. 3 (1995): 183-186.

Langevin, Helene M. "Fascia mobility, proprioception, and myofascial pain." *Life* 11, no. 7 (2021): 668.

Myers, Tom. "Fascia as an Auto-Regulatory System." *Structural Integration* (2014): 38.

Myers, Thomas, and James Earls. *Fascial Release for Structural Balance, Revised Edition: Putting the Theory of Anatomy Trains into Practice*. North Atlantic Books, 2017.

Oschman, James L. "Fascia as a body-wide communication system." In *Fascia: The Tensional Network of Human Body-E-Book. The science and clinical applications in manual and movement therapy*, pp. 103-105. Elsevier Edinburgh, 2012.

Parramore, Karin. The Essential Step-by-Step Guide to Acupressure with Aromatherapy: Relief for 64 Common Health Conditions. Pub: Robert Rose. (2016).

Pirri, Carmelo, Caterina Fede, Lucia Petrelli, Diego Guidolin, Chenglei Fan, Raffaele De

Caro, and Carla Stecco. "An anatomical comparison of the fasciae of the thigh: A macroscopic, microscopic and ultrasound imaging study." *Journal of Anatomy* 238, no. 4 (2021): 999-1009.

Stecco, Carla, Veronica Macchi, Andrea Porzionato, Fabrice Duparc, and Raffaele De Caro. "The fascia: the forgotten structure." *The Fascia: the Forgotten Structure* (2011): 127-138.

Taveras, Jennifer Chellis. *12 Acupressure Points for Pediatric Sleep Improvement and Wellness Support.* Xlibris Corporation, 2014.

Upledger, John E., and Jon D. Vredevoogd. *Craniosacral therapy.* Vol. 236. Seattle: Eastland press, 1983.

Upledger, John E. "Craniosacral therapy Part I: Its origins and development." *Subtle Energies & Energy Medicine Journal Archives* 6, no. 1 (1995).

Upledger, John E. "CranioSacral Therapy Part III: In The Future." *Subtle Energies & Energy Medicine Journal Archives* 6, no. 3 (1995).

Upledger, John E. (Ed.) *Working Wonders: Changing Lives with CranioSacral Therapy.* North Atlantic Books, 2005.

Wales, Anne, and W. G. Sutherland. "Teachings in the Science of Osteopathy." (1990).

Part II

Structural Medicine Concepts Related to Each Sensory System

Part II

Structural Medicine Concepts Related to Each Sensory System

Bodywork for Sensory Integration assists each sensory system at the entry point of vast receptor fields. Structural medicine principles are applied to the theory of sensory information processing. Sensory integration theory analyzed the end-result of information reaching the central nervous system. The cellular world down in the body has much more information to reveal.

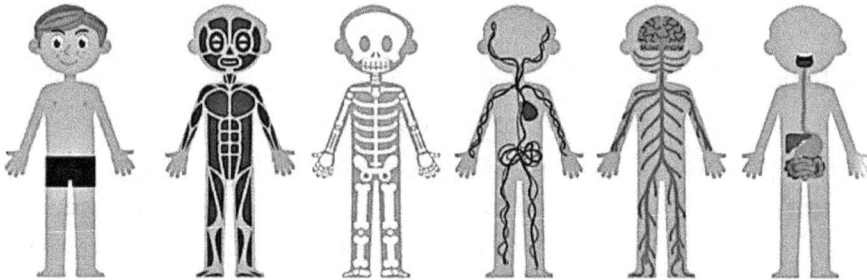

Sensory integration BEGINS at the level of sensory receptors of the cranial nerves, the extensive somatosensory systems, the connective tissues surrounding organs and vessels, and the sensory capacity of the whole-body fascia matrix before information even reaches the brain

Sense

"Sense" ….

nerve channel through which a body can observe, feel, or experience the outside world and itself
—*Cambridge Dictionary*

Most of the world believes humans only have five senses. We've been taught this from childhood. Dr. Ayres opened our minds in the 1970s to the function of additional "hidden" senses such as proprioception and vestibular processing. She enlightened the world on how humans utilize these senses to complete most skills that are expressed through traits and behaviors. Science has since increased the number of sensory channels to between fourteen and twenty distinct body systems.

This knowledge depends upon how one defines a sense. There are certain senses that only some people seem to possess while others may not. Proprioception is the sense of pressure and position and Kinesthesia is the sense of moving without visual perception. Equilibrium is balance and control over those sensations. Thermoception, the detection of temperature, is different than the sense of touch. Chronoception is considered a sense of the passage of time (how long have we been outside in the cold? How long has it been since I last ate food). In the animal kingdom, there exists electroception, the perception of electrical currents (in sharks and dolphins), and magnetoreception, the detection of magnetic fields (in bats and some birds). For practical purposes, we can simplify the human senses down to just three property types of function:

- Mechanical (touch, hearing vestibular, kinesthesia, and proprioception)
- Chemical (taste, smell, and internal senses)
- Light (vision)

The nervous system has two main reactions to all sensory input: **alert or alarm** the system, or **calm and regulate** the system. These two functions serve to first protect the body for survival from a threat (real or perceived). The sensations giving such an alarm are known as protopathic and happen when receptors of pain, heat or cold, or sense of danger or injury through intensity, are stimulated. These sensations alarm the sympathetic nervous system causing physiological ANS responses. Other senses contribute to the parasympathetic side of things once the system has recovered and homeostasis has calmed and regulated. In the parasympathetic mode, learning and discrimination between subtle differences in sensations occurs. This gives rise to meanings such as soft, fuzzy, gentle, melodic, sour, sweet, tickle, rocking, and many other descriptives of sensations and our perception.

Therapists practicing sensory and developmental methods for children have focused on the end products of how sensory channels manage information and formulate responses. Anatomy is a big part of learning sensory integration theory and practice, but more attention should be given to the structural aspects of the sensory receptors and pathways. We have applied therapies to receptor sites (i.e., therapeutic brushing and vestibular input via swings) but have only assumed the receptors are optimal. We need to consider if all sensory receptors are in their best mechanical and structural state as being "ready" to process the sensations from the world.

The twelve cranial nerves are very much involved in sensory processing. These nerves deserve greater study for their vulnerability to structural compromise. The cranium is quite vulnerable to osseous and fascia compressions from birth or retained tissue restrictions from acquired compromise to the craniosacral system (meninges covering the brain and spinal cord with cerebrospinal fluid production rhythm). Bodywork practices have explored this question by applying craniosacral therapy and cranial osteopathy. Let's explore each sensory system in kind to ponder the question of whether each system is optimally ready to receive information.

Olfaction-Sense of Smell

Bodywork objectives: Ensure freedom of retained or acquired structural compression into the frontal-ethmoid site and into the cribriform plate; reduce mechanical strains of olfactory tract (CN I) and bulb that can elicit sympathetic reactions to incoming input of odors and scents; build resiliency of autonomic reactions and recovery to smells.

Developmental Trends and Behavioral Considerations

Olfaction is a primal sensory pathway that serves the purposes of locating food sources for basic survival and discrimination and judging odors as noxious or pleasurable. Olfaction gives life meaning as it gathers information about aromas from food and the entire environment. Humans develop preferences based on tolerance and pleasure, and aversions are set when odors are offensive at mealtime and self-cares. Generally, in food preferences, we as children first gravitate to foods that smell good (and don't offend our noses).

Sensory processing issues can arise if birthing compression persists in the osseous and fascia structures housing the olfactory nerves. Commercially exploited products can assault the nose or cause injury to the mucous membranes of the nasal cavity. Perfumes, deodorizers, cleaning products, or offending scents from herbicides and pesticides can interfere with the sense of olfaction. And as the world came to recognize during the COVID-19 pandemic, pathogens can damage the microvasculature to nerve bundles causing loss of sensory function. Personal accounts abound of the profound impact intact smell and taste have on one's sense of well-being and emotional/mental health.

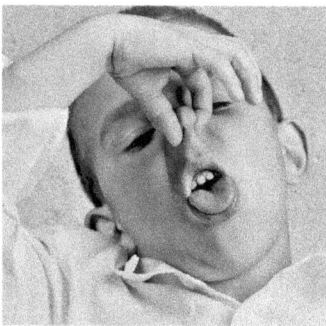

Sliding scale of sensory perception, adjustment, and adaptive response

The sense of smell is a product of chemical reception through the neuron activity of over 400 olfactory receptor cells. These chemoreceptors are activated by the molecular substances released from the aroma source. It is now believed that at least five different types of cells serve to perceive different chemical [odor] categories. This nerve pathway, which travels from the bridge of the nose deep into the middle of the brain between the eyes, can set off fight-or-flight instantly or, conversely, calm the brain with soothing scents. In addition, scientific evidence now suggests that a large majority of what we perceive as odor is a close intercommunication to how something tastes. On the other hand, decreased reception making someone unaware, can be dangerous for not detecting smells (and tastes).

Olfaction is a sensory system with a direct pathway to the limbic system. Regulating the limbic system and restoring physiology homeostasis helps the higher brain centers and executive functioning to be activated. Modulation helps regulate to either calm down or raise awareness appropriately of various smells encountered throughout one's day. Modulation behaviors allow the child to bond, alert to stimuli, engage or converse with others, or avoid and isolate from the environment and associating aromas. A typical and healthy olfaction system can habituate (tune out incoming input) with exposure over a period of time. A person with a typically responding nose can cease to smell good and foul odors but then, over time or after leaving the room and returning, can reactivate the perception of those odors.

On the other extreme, but equally related to health and wellness, is when the sense of smell is hyper-reactive when offending odor perception cannot be dialed down or turned off. This difficulty in habituating or modulation odors is a common issue in various levels of sensory processing dysfunction. Inhibiting or habituating this primal sense is important for the ease of attention and concentration in learning and work environments. Behaviors suggesting a challenge are discomfort or meltdown in a crowded, loud, and smelly school cafeteria. Other mal-adaptations may include: self-restricted food tolerances, limited diets, only wanting to eat at home, and eating in a particular room (away from the kitchen and source of many odors). For picky eaters, the place to begin expanding food choices and tolerances is often through developing smell tolerance.

Olfaction modulation is a chief feature of self-regulatory behaviors.

Adaptive Responses…

…allows the olfactory system to maintain emotional and autonomic equilibrium with the odor concentrations in the environment, and supports an appropriate level of attention, reaction, or ignorance to odors encountered in daily activities.

- Orient and acclimate to sources of scents
- Alert to, modulate alarm, and recover from noxious odors

- Differentiate odors, develop preferred odors (with tastes)
- Extinguish response to odor in time, with attention and concentration to other sensory stimuli and activities, or co-exist within homeostasis to engage
- Mood alterations with odors found to be pleasant, soothing, stimulating hunger, and memories
- Be able to eat a meal fully without bothered by smells of food or environment

> **The olfactory pathway deserves assessment and treatment due to its important role in sensory modulation and regulation.**
>
> **Osseous and fascia tissues can be compressed at ethmoid, frontal, and maxilla/vomer complex as well as the nerve endings within the foramina of the cribriform**

Anatomical Considerations

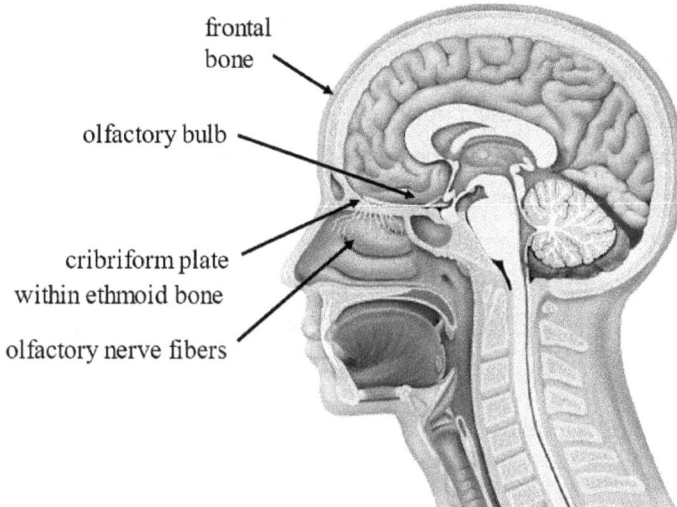

Olfactory Nerve, Bulb, and Cribriform Plate

The nose is a projection of skin and cartilage off the cranial structures of the ethmoid and frontal bones, with a fascial relationship of tissues connecting with maxilla/vomer/palatine complex. The nose itself is attached to paired nasal bones, is the primary entry into the respiratory system, and protects the nasal cavity. The cavity is walled with mucous membranes that humidify while warming and filtering inhaled air.

At birth, a newborn's paired plates of the frontal bone are connected by a central metopic suture (typically fuses between ages 3-9 months). The fascia construct of the nose extends from the union (the glabella) of this site of fascia. Under-developed at birth, the ethmoid bone is situated at midline under the frontal ridge, anterior to the sphenoid, posterior to the lacrimals, and superior to the maxilla. The bones of the face surrounding the nose are also under-developed osseous plates suspended in the anterior to the fascia container of the cranium. The ethmoid eventually will articulate with eleven facial bones when growth is complete. The location of the metopic fontanelle at midline is a shock absorbing site from birthing. The anterior surface of the sphenoid approximates the back side of the frontal bone. Two sites of compression related to olfactory pathway concern are common here.

The olfactory nerve (CN I) is the shortest of the cranial nerves, begins at the cribriform plate of the ethmoid bone, and then terminates at the olfactory bulb. This bulb sits just above the ethmoid bone and below the frontal lobe of the brain. In the nostrils within specialized epithelium, six to ten million bipolar olfactory neurons distribute along a surface of two to three centimeters. These neurons merge to become the olfactory nerve, it is the only cranial nerve that directly attaches to the cerebrum without traveling to the thalamus. The tract is unmyelinated for fast transmission. That means there are little to no other neurons from other brain centers in proximity to inhibit and slow the transmission of odor. Unmyelinated nerves are vulnerable to chemical and toxic injury from lack of the glial-cell myelin's protection. To reiterate, this fact makes the sense of smell a very important aspect of sensory modulation and state control

The olfactory bulbs have fascia intimacy with the falx membranes, other meningeal extensions, and cranial blood and lymph vessels. Fluid pressure of brain has been shown to self-correct via the glymphatic system through little holes where nerve fibers project into the cribriform plate. Intense pain when we get hit in the nose or eat excessively cold food is an intimate threat to the central nervous system. A "brain freeze" comes from a lack of flesh to absorb the shock of extreme cold touching the roof of the mouth (maxilla). Trauma to any or all of the associated nasal bones, can cause either hyper- or hypo-responsiveness of olfaction, depending upon the extent and depth of mechanical stress.

Adverse compression related to the birthing process to the forehead or face can persist beyond infancy. It is a false assumption that such compression always self-correct with remolding of the head. Compression into the frontal cortex can send mechanical strain posteriorly towards the limbic system. A perpetual alarm system can fire with such strains.

Assisting Function by Treating Structures—Olfaction

Clinical experiences from many clients who reported loss of smell have demonstrated compression of the frontal bone into the meningeal

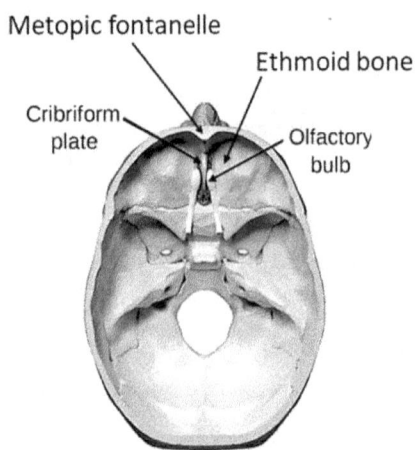

Fascia vicinity of metopic fontanelle to the cribriform plate, ethmoid bone, & sphenofrontal suture. Database Center for Life Science: Wikimedia 80383240

tissues surrounding the olfactory pathway. Ethmoid can sustain compression posteriorly into sphenoid having a direct effect upon the structures surrounding the olfactory nerve and endings. Sutural fascia, functioning as the shock absorber, can retain this compression. Ethmoid approximating the sphenoid, and the nasal bones approximating the ethmoid are the targeted treatment sites. Adverse fascia restrictions in and around the cribriform plate of the ethmoid bone, can occlude the olfactory nerve and interfere with sensory signaling (as seen in head trauma or concussions). Craniosacral therapy (sometimes in isolation and sometimes with other restorative measures) has been known to assist lost olfaction in cases of previous face trauma from falls or being struck by a projectile. CST can also reduce hyper-sensitivities to odor and other behaviors of sympathetic nervous system activation.

Bodywork Treatments that help olfaction

- Frontal bone decompression (See p. 227)
- Ethmoid-frontal mobilization (See p. 229)
- Sphenoid balancing (of fascia suspension) and decompression (See p. 228)
- Maxilla-vomer mobilization and decompression (See p. 236)
- Release soft tissue restriction through cribriform plate (See p. 229)

Gustatory – Sense of Taste

Bodywork objectives: Ensure efficient tongue movement to activate & modulate taste reactions with pressure into hard palate to control gag response. Release retained or acquired compression of occiput at cranial base and cranial nerve foramen. Though five cranial nerves are involved with taste sensation (CN – V, VII, IX, X, XII), two innervate for tongue movement (CN- V and CN-XII). Outlet foramen at the cranial base and near TMJ get focused attention. Ensure deep front line of fascia is free of adverse structural pulls between tongue midline, projecting through visceral cavities, & attachments at coccyx & big toes.

Developmental Trends and Behavioral Considerations

Gustation is a functional harmony of oral movement and sensory registration. Tongue movement activates and modulates tasting through moving fluids and chewed food, mixed with saliva, over the surface of the tongue to activate the papillae (taste buds). In learning how to eat, a developmental progression of tongue movements follows somewhat of a predictable sequence. From primitive to more mature action, six tongue movement patterns emerge:

- a reflexive suckling
- simple protrusion
- volitional sucking
- munching
- tip elevation
- lateralization

Lip mobility and jaw position also is needed for oral closure, providing a frame of stability for which the tongue moves against. A balanced resting tongue position without adverse retraction indicates a state of readiness for easy tongue movement. Adequate response time and coordination needs to match any food and liquid stimulus. The kinesthetic sensations of the tongue and throat coordinate timing with glottis closing for breathing without aspiration of fluid or food. Soft tissue restrictions or stress within the sympathetic state retracts the tongue for protecting the airway. Chronic structural tension can hinder the fluidity of sucking, chewing, swallowing with breathing. In addition, tissues and structures below the jaw and extending into the anterior and lateral neck can also contribute to throat tightening and tongue restrictions (structurally or behaviorally).

A child's behavioral reaction to tasting food and eating may be a direct relation to how well they move their tongue. Or how well olfaction is modulated with ability to press tongue to palate. Or how well the breathing mechanism function and glottis coordinate during the eating processing. Sometimes these struggles relate to a history of an ankyloglossia (tongue-

Sliding scale of sensory perception, adjustment, and adaptive response.

tie). Still, more often, this problem is one part of a global protective retraction of the tongue itself or somewhere else in the body. The autonomic nervous system can dictate the tension of the digestive tract, projecting towards the entry point which is the mouth and tongue. Tightness in the throat is a behavioral indicator of elevated stress and sympathetic tone. An adult can be aware of how tight the throat can become with such stressors as giving a speech, flying on a plane, or asking the boss for a raise. A child can experience similar stress but not have the words to overtly describe. One bodywork revelation that can indicate this state is the total mobility of structures around the hyoid bone. Ease in both swallowing and pitch of voice can also reflect such tension.

The tongue can be held in defensive positions for a myriad of causes. Tongue positioning, posterior in the mouth or elevating in the middle of the muscular tongue to block mouth entry, can indicate a sympathetic response to either taste or lack of movement control of the tongue. Aversive reactions to tastes set off a sympathetic response, possibly a gag or tongue retraction. If the tongue can't move freely, it can't quickly calm the hyper-reaction to a taste. Uncontrolled movement of food over the tongue's surface can threaten a gag reflex. A person may gag if the taste stimulus is not tolerable. Also, the gag reflex is stimulated when something is too far back in the oral cavity.

Adaptive Responses...

...develop over time from sensorimotor experiences of the tongue as it moves fluids and chewed food over the surface. Tongue movement helps develop control over the autonomic (sympathetic) protective responses, allowing discriminative (parasympathetic) flavor detection. An important adaption is control over protective behaviors in order to accept a well-balanced diet, as well as gain the ability to take medications or supplements should the need arise. Tastes expands quality of life through tolerating and then enjoying a multitude of tastes one will experience in life.

- Modulate reactions to taste through active tongue pressure into hard palate (activating proprioception); inhibiting any gag reflex.
- Activate tasting, first by reflexive tongue movements, and later with active control of mobility skills (lateralization, elevation, retraction, and protraction).
- Accept flavors, then swallows with ease.
- Remains calm and not alarmed by variations or intensities of taste; or quickly recover.
- Tongue moves fluid and food to appropriate place within the mouth and stimulates safe chewing and swallowing.
- Halts swallowing, inhibits gag and breathing, during chewing process
- Coordinates chewing with tongue movements
- Facial expressions signal taste registered, preferences, and social engagement

Efficient tongue movements both activate and modulate taste

Anatomical Considerations

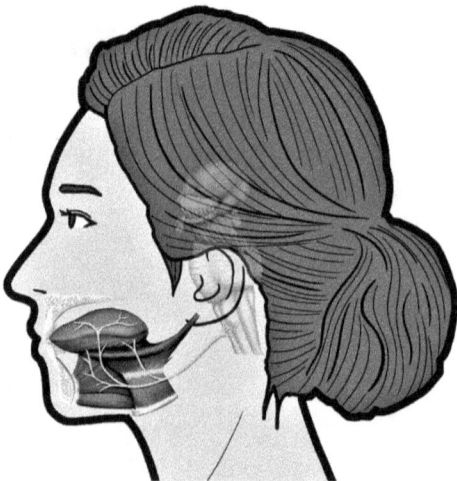

Hypoglossal (CN XII) Nerve

Three cranial nerves (CN) are involved in the **sensory** aspect of taste:

- **Facial** (CN-VII): fibers to the front two-thirds of the tongue
- **Glossopharyngeal** (CN-IX): fibers to the posterior tongue and pharynx
- **Vagus** (CN-X): fibers into the back of tongue and upper esophagus / epiglottis; elicits a gag when needed

Two cranial nerves are involved in general **sensory and movement** of tongue and mouth:

- **Hypoglossal** (CN-XII): the main efferent nerve for tongue musculature

- **Trigeminal** (CN-V): the mandibular branch general sensation of touch, pressure, temperature, and pain and can be considered a chief "sensory integration" overseer of total oral movement and position sense; innervates the floor of the oral cavity, monitors for protection and elicits a gag when needed

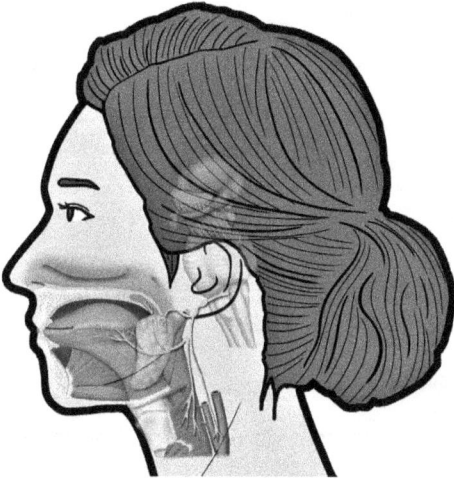

Glossopharyngeal (CN IX) Nerve

The osseous exit sites of these cranial nerves are of concern in (all) *Bodywork*:

- **Jugular foramen** (CN-IX, CN-X plus other nerves and blood vessels); located lateral to foramen magnum at the cranial base
- **Hypoglossal canal** (serves CN-XII) is a foramen "hidden" in the medial portion of the occipital condyles close to the foramen magnum at the cranial base
- **Temporal bone foramen** (CN-V) where the lower two branches of the trigeminal nerve pass-through
- **Stylomastoid foramen** where the facial nerve (CN-VII) exits the skull, posterior to styloid process of the temporal bone. After exiting the skull, the facial nerve turns superiorly; runs anterior to the outer ear.

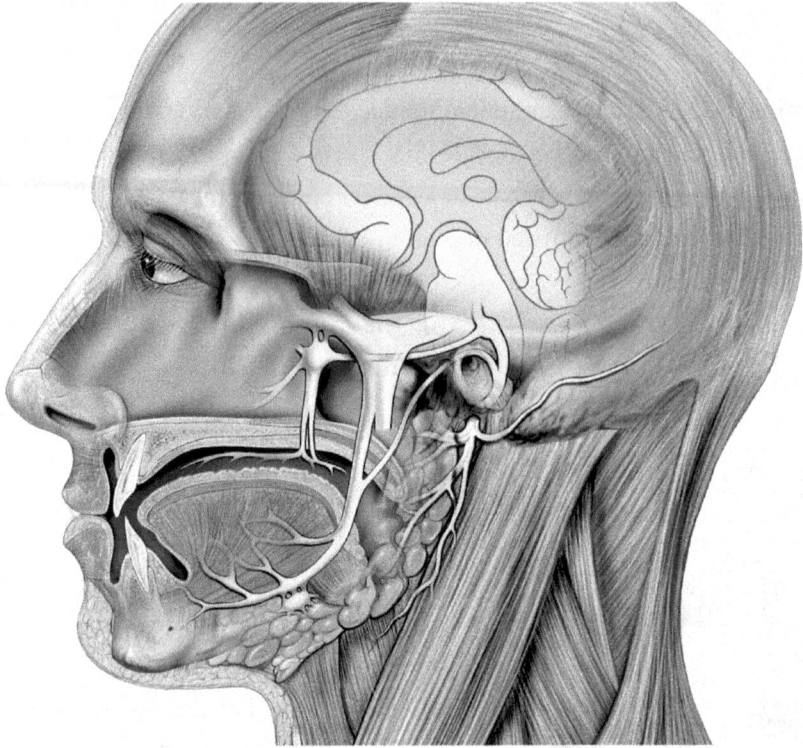

Trigeminal Nerve (CN-V) Patrick J. Lynch Creative Commons licenses via Wikimedia Commons

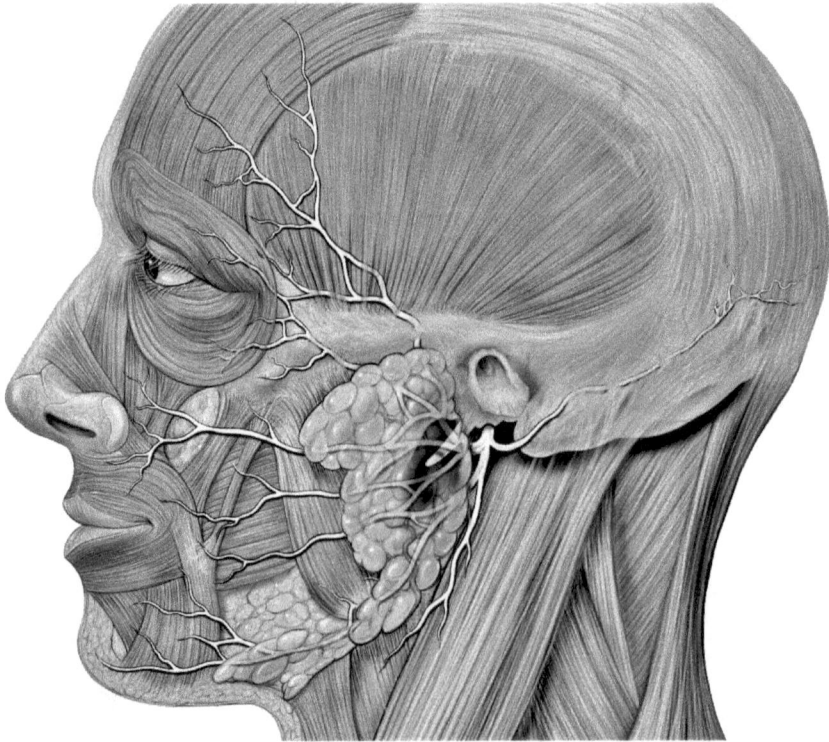

Facial Nerve (CN-VII) Patrick J. Lynch Creative Commons licenses via Wikimedia Commons

Extrinsic muscles of the tongue:

- **Genioglossus**: arises from inner curve of mandibular (symphysis); inserts into hyoid body, along entire length of tongue. It protrudes, depresses, and draws the tip of tongue back and down; innervated by CN-XII; allows tongue to expel things from oral cavity as a safety response
- **Hyoglossus**: arises from hyoid and inserts into side of tongue; depresses and retracts tongue; innervated by
- CN-XII. These movements are needed to put pressure on the hard palate and create the milking action on the nipple during feeding.
- **Styloglossus**: originates at styloid process and temporal bone and inserts into side of tongue; retracts and elevates tongue; innervated by CN-XII. attaches to the temporal bone and allows lateral sides of tongue to form a trough for swallowing (a bilateral coordination skill)
- **Palatoglossus**: arises from the palatine aponeurosis and inserts across the tongue; only tongue muscle innervated by CN-X (vagus nerve); attaches via fascial fiber to the palatine bone and elevates the posterior tongue during swallowing. A key regulator of the parasympathetic state.

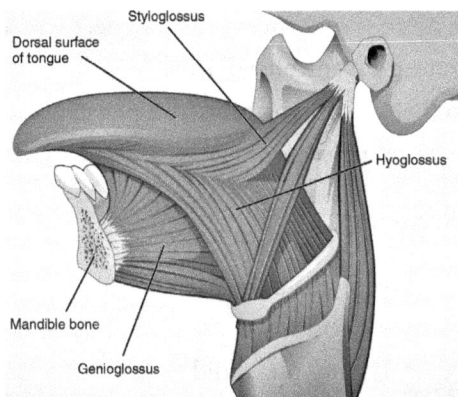

Muscles connecting tongue and hyoid. The hyoid is suspended by numerous muscular structures and great amounts of investing fascia. Upledger's CST and Barral's VM methods are two of a few approaches to even consider these associated structures. Besides important emotional regulation function, the muscles superior to the hyoid form the entire floor of the tongue. The deep front line of fascia is also influencing the structural tensegrity of the anterior neck and throat flexibility. This anatomy is underappreciated when considering best practice for ankyloglossia (tongue tie).

- Geniohyoid
- Hyoglossus
- Stylohyoid

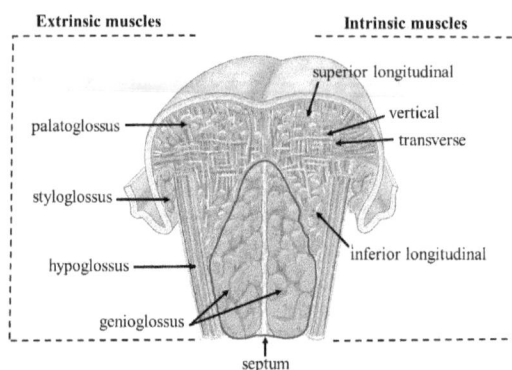

Intrinsic muscles of tongue: Four paired intrinsic muscles are housed within the actual tongue. These muscles alter the shape of the middle of the tongue and are more involved with articulation of speech, though they also provide discrete coordination for food movement within the oral cavity. There are four paired intrinsic muscles of the tongue and they are named by the direction in which they travel:

- Superior longitudinal: the muscle just under the surface of the tongue shortens and widens the bowl as well as dorsiflexes it to curl upward.
- Inferior longitudinal: more lateral within the tongue, these also shorten and widen the bowl and curls it downward (ventroflexion).
- Transverse: more into the middle structures, this one elongates and narrows the tongue.
- Vertical: striated above the transverse muscle, this muscle flattens the tongue

Mandible position and movement

Closes the jaw:
- Masseter
- Temporalis
- Medial pterygoid

Opens the jaw:
- Digastric
- Lateral pterygoid

Sensory Function of Taste

Taste is possible by taste buds (papillae) on the tongue's surface. There are many receptors in each taste papillae, and each can detect every category of taste (sweet, salty, sour, savory, and bitter). Despite a century of belief of a taste region (a map) receiving separate taste signals, it has been firmly demonstrated that every tastebud area of the tongue can detect all flavors. Besides differentiating flavors, taste buds also distinguish between safe, noxious, or rancid aspects of food. Interestingly, taste cells are not only found on the tongue but also farther back in the mouth, throat, soft palate, pharynx, and upper esophagus.

The activation of taste buds trigger saliva production and prompts reflexive swallowing. Also, to note that olfaction triggers saliva production as well, the first digestive juice secreted along the tract. A protective gag response or regurgitation can be elicited if the substance is judged unpleasant or toxic. The tongue also receives information about temperature and texture, (via the trigeminal nerve). Discriminatory sensory information via other cranial nerves travel to the thalamus and sensory cortex, then projects taste messages to many different brain centers.

The gustatory system operates in concert with the olfactory system and is modulated by the autonomics of the trigeminal nerve pathway and multiple proprioceptive pathways. Taste reactions are primal, at first for reflexive protection, and then mature via sensory experiences to become a primary discriminating system (the parasympathetic state). Taste responses occur when taste buds are activated by food being moved across the tongue's surface, both by tongue movement and saliva flowage. Deep pressure modulates noxious sensory reactions. The tongue's ability to exert and control pressure is one way modulation of taste occurs. The tongue must be free to move in all directions, without fascial restraint, to modulate the sense of taste efficiently, keeping the autonomic responses in check; otherwise, the risk of food intolerance and fussy eating could be in a child's future.

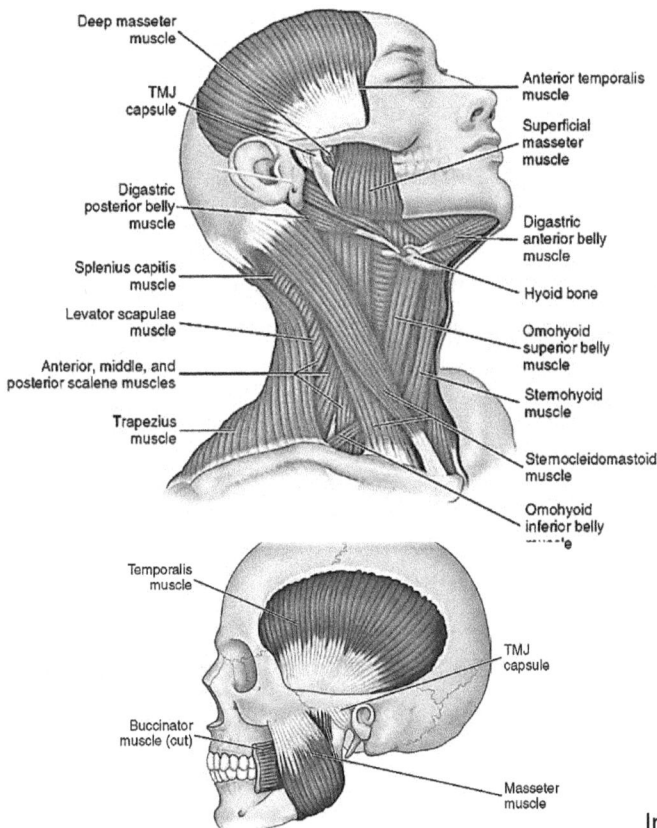

Images: Shutterstock

Motor Function Allows for Taste

In parallel fashion to the sensory structures, the musculature of the tongue is vastly complicated and integrated with many structures of the skull, the face, and the neck. A group of extrinsic muscles attach the tongue to the mandible and hyoid bone and form the tongue's floor, attached to the inner curve of the mandible. The attachments at the hyoid provides a myofascial brace for stability for tongue, pharyngeal, and laryngeal function. The hyoid's function in all tongue function is under-appreciated. Physiological functions that the hyoid bone supports include breathing, swallowing, and speech production. The hyoid also has a direct relationship and manifestation of the sympathetic nervous system and emotional centers because of its proximity to the parasympathetic and vagal nerve fibers.

In addition, sympathetic nerve fibers are also located in the anterior neck and upper thorax. The tension of the hyoid bone is a crucial indicator of protective retraction and tightness transferring to the tongue and oral structures from or related to the deep front line of the fascia. If protective retraction is evident at the hyoid attachments, it is easy to assume the tongue lacks complete freedom of movement. It is not a stretch to understand why any sympathetic tone anywhere in the body can result in constricted and restricted throat, tongue, and hyoid mobility.

Assisting Function by Treating Structures

A child's history with infant [breast]feeding can yield helpful information about the sensory integration and motor skills of the tongue and mouth. Cranial nerves of the tongue house nuclei in the pons (brain stem) are particularly vulnerable to superior compression of the head through the cervical spine from the birthing process. The base of the skull has multiple important foramina that allow the passing of vital tissues, primarily blood vessels and nerves. The two jugular foramina exist at the base of the skull lateral to the foramen magnum. Importantly, the internal jugular veins, which drain blood from the brain and intracranial tissues, make their way out of the cranium, terminate at the subclavian veins, and ultimately join the brachiocephalic vein. Additionally, cranial nerves IX, X, and XI pass through this foramen, sending stimuli to multiple extracranial targets.

The location of the primary taste nucleus in the brainstem, with its proximity to the upper cervical spine and the cranial base, is a focused area for *Bodywork for Sensory Integration* to ensure nerve pathways are uncompressed and ganglia is not hyper-activated. Compression at the brain stem and/or upper cervical spine can cause retraction of the tongue and other oral structures.

Any degree of compression of the trigeminal or facial nerve can give rise to a retracted tongue along with hyper-reactivity to tastes, temperatures, and textures. This can create behaviors of fussiness and pickiness in eating. The freedom and range of motion of mouth opening are directly related to the position the mandible is positioned on the face and cranium and can also interfere with the ease of tongue mobility. Posterior and superior compression of the mandible has been a common finding in babies who struggle with breastfeeding. When structural compression of the mandible remains unresolved, it is reasonable to assume the tongue is positioned or pulled along with the jaw into tongue protraction and protective elevation. The primary muscles and their investing fascia activate jaw positioning (masseter, temporalis, and pterygoids), which can be shortened or compressed by the birth process.

The focus of *Bodywork Sensory Integration* as related to taste is primarily at the cranial base, the entire neck (including anterior, posterior, and lateral aspects), the cranial nerve outlets with pathways to innervation sites, (extrinsic and intrinsic tongue muscles), and all tissues attached or associated with the hyoid bone. Compression at the brain stem or upper cervical spine can perpetuate the sympathetic state, causing retraction of the tongue and other oral structures. Tension held at and around the hyoid reflects the autonomic state as well as the availability of three-dimensional mobility of the tongue. One of the most overlooked structures pertaining to manual therapies for any age or condition is structures found on the anterior neck, especially the degree of investing fascia of the dozens of structures associated with the hyoid. There are dozens of muscular, ligament, and soft tissue attachments in direct and indirect relationship with the hyoid. The key point here is the degree of investing fascia and the superior end of the deep front line fascia band.

The focus of *Bodywork for Sensory Integration* is to assist with the movement functions related to tasting. Techniques help assure freedom of efficient tongue movement to manipulate food in the oral cavity, lateralize the tongue to sensory cues of food bolus, and inhibit gag reflexes as tongue stability is held to set off swallow response. Much of treatment focuses on attachment sites of the extrinsic muscles.

> **Cranial base & cervical spine compression, mandible restrictions, deep front line of fascia, and hyoid attachment restrictions are but a few structures can hinder tongue range of motion. Exercises do not necessarily increase expansion of these soft tissues.**

Bodywork Treatments that help TASTE

- Cranial base region - release compression hypoglossal & vagal nerve outlet (p. 211-213)
- Occiput condyles mobilization (p. 212)
- Temporal-mandibular joint decompressed inferiorly and anteriorly to release outlet sites for facial and trigeminal nerves and reduce tension in investing fascia (See pp. 232-233)
- Soft tissue mobilization around and attached to hyoid (See p. 209)
- Deep front line of fascia mobilization (See p. 214)
- Maxilla-vomer complex of hard palate (See p. 236)
- Palatine mobilization (See p. 236)

DO NOT WORK on hyoid structures without formal training

See also: Oral motor and feeding aversions (See p. 251)

Vision and Ocular Motor System - Sense of Sight

Bodywork objectives: Release retained or acquired structural compressions of cranial nerve pathways (CN-II, III, IV, V, VI, VII) through osseous-fascia tissue that could imbalance both sensory and motor nerve conduction. Promote autonomic regulation of pupils for light regulation by reducing compression through optic nerve through bony paths. Balance lines of tension in all extraocular muscles and suspending fascia tissues. Reduce eye strain & enhance skills of binocularity.

The visual system is a complex network of intercommunicating information processed with other sensory channels. Vision is a combination of sensory input and bilateral coordination of movement from both eyes. The whole of vision functioning has two key elements: Visual Acuity (simple act of seeing) & Ocular Motor (eye-teaming for navigating whole-body interactions with the world). Sensations of seeing are reactions to the rods and cones (sensory receptors) in the retina receiving color and light stimulation. The two eyes independently send information posteriorly to the brain via the optic pathways, where the signals cross at the optic chiasm on the path to the occipital lobe. The visual images blend to create depth perception and spatial orientation after extraocular muscles focus and adjust images. Muscles also track moving images and react to head & neck movements for balance & posture.

Developmental Trends and Behavioral Considerations

The eye develops embryologically as a direct extension of brain tissue, with dura mater wrapping the dorsal side of each eyeball. The bony orbit is shaped by several bones to protect each eye. A newborn's eyes respond first to light and then to crude shapes with strong contrast (usually black and white images), but their eye movements are still not yet activated. When a baby is moved, head movements help activate eye movements. Vision is often considered the primary sensation because it drives a baby's desire to move towards objects the gaze fixes. (Some argue that the vestibular system is primary as head movements activate the oculomotor system).

The optic nerve (CN-II) allows for "seeing," but eye movements create depth perception by working with other postural and balance mechanisms. The brain depends on eye-teaming (binocularity) to achieve focus and then interpret a particular image. The brain fuses separate images from each retina, activated by light, after which split images cross at the optic chiasm and travel to the visual cortex of the occipital lobe. The retinas perceive light and images in both central and peripheral fields. Peripheral vision is not very precise, is a portion of sight from one eye, and is a function of alerting to objects (which can startle the nervous system). Central vision perceives more light and images from both eyes, which allows a person to see shapes, colors, and details more clearly and sharply. The macula area of the retina provides this sensory function.

The brain can extinguish the attention of one eye if it cannot generate a stable image. Sometimes this is an obvious problem with a "lazy eye" that postures out of alignment. It is also commonly not obvious but can contribute to poor eye-hand coordination, reading, attention, and concentration, leading to background stressors. Teaming difficulties are commonly associated with attention, dyslexia, headaches, and other learning challenges.

The ocular motor system is maximized when the cranial nerves (CN III, IV, V, and VI), and the extra-ocular muscles are free of restrictions for osseous compressions. Four basic eye movement patterns contribute to efficient binocularity: gaze stabilization (fixations), visual tracking (pursuits), vergence movements, and gaze shifting (saccades).

- Gaze stabilization and fixations: Both eyes remain steady on an object, which is fundamental for developing binocular skills. Fixation brings the object's edges sharply into focus but requires postural stability of the neck and head. Gaze stabilization is a more complex motor skill where the eyes fixate on a stable object while the head moves from the neck.
- Visual tracking and pursuits: The eyes follow a moving target. Eyes track in linear and oblique planes with equal timing and smoothness. Advanced skills can combine planes for circular eye motions within the orbits. Spatial judgments are learned from such eye movements in development. These skills require equal tension and contractile activity from the teaming extra-ocular muscles.
- Gaze shifting and Saccades: Eyes move as a coordinated team, accurately jumping from one target to another with quick refocusing. A shift of gaze occurs with orchestrated finesse between eye movements and fixation. These are fundamental skill sets for reading, scanning the room, and reading people's faces, coordinating between central and peripheral vision.
- Vergence movements: The eyes team must focus on the action as objects approach the face and then move away from the face. Eyes moving medially towards the bridge of the nose is called convergence, and conversely, divergence as eyes return laterally back to midline position. These two motions contribute to binocularity. Timing, endurance, and accuracy are key features to smooth vergence. When adversely affected, abilities can be adversely affected.

The eyes are a direct extension of the neurocranium with vast communication from nerves and muscles. Seldom explored is the layers of connective tissue also present the surrounds, suspends, and can retain restrictions around the eyes

Adaptive Responses to Vision and Ocular Motor......

...is a complexity of communication between multiple systems simultaneously for maximal function. Stationary eyes can see, but vision function is the product of "teaming" of both eyes, background postural control from neck structures and vestibular input.

- Pupil reactivity to light levels for self-regulation and state control
- Convergence of both eyes to gain eye contact; first focus is food source or parent's face
- Tracking a moving object—jerky at first, then more coordinated
- Gaze stabilization when the head or body is being moved
- Shift gaze from one target to the next with ease in focus
- Crossing eyes is normal (over-convergence) during this time of ocular maturation

Anatomical Considerations

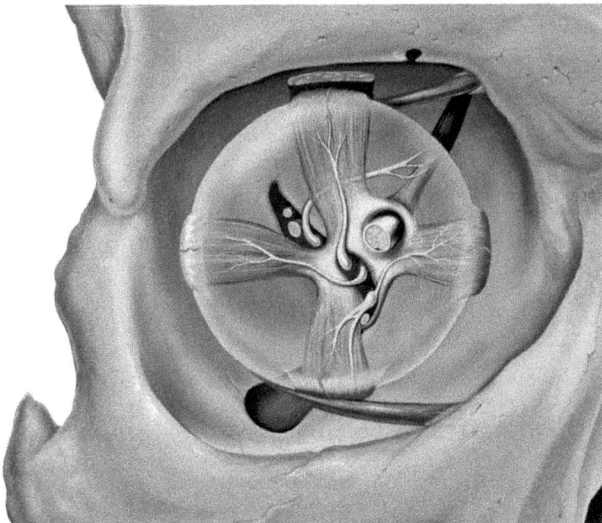

Photo: Modified. Patrick Lynch. Wiki Creative Commons License 2.

Anterior view of the right orbit, emphasizing nerve exits and paths (in relation to extraocular muscle attachments). The circular shadow of the sclera is conveyed, (connective tissue forming eye). Without sclera, there is no eyeball, which contains the vitreous gel. The fascia extends its weave with the extraocular muscles fusing to the sclera. The tensegrity of the eye is formed and dependent upon this fascia mainframe. We can palpate and treat eyeball tension tensegrity, and thus assist muscles towards equalized tension and readiness.

The bony walls of the orbits are created by a mosaic of four facial and three cranial bones:

- Optic canal: formed by the body and the lesser wing of sphenoid
- Superior wall & ridge: formed by the **frontal** bone and lesser wing of **sphenoid**; sphenoid segments remain more pliable until fusion is fully mature.
- Lateral wall: **zygomatic** bone
- Lateral posterior wall: greater wing of **sphenoid**
- Inferior-medial wall: **maxilla** along with a palatine bone projection
- Medial wall: primarily the **ethmoid** with contributions from **maxilla** & **lacrimal** bones

Thus, this group of bones need to be considered when testing orbit tensegrity.

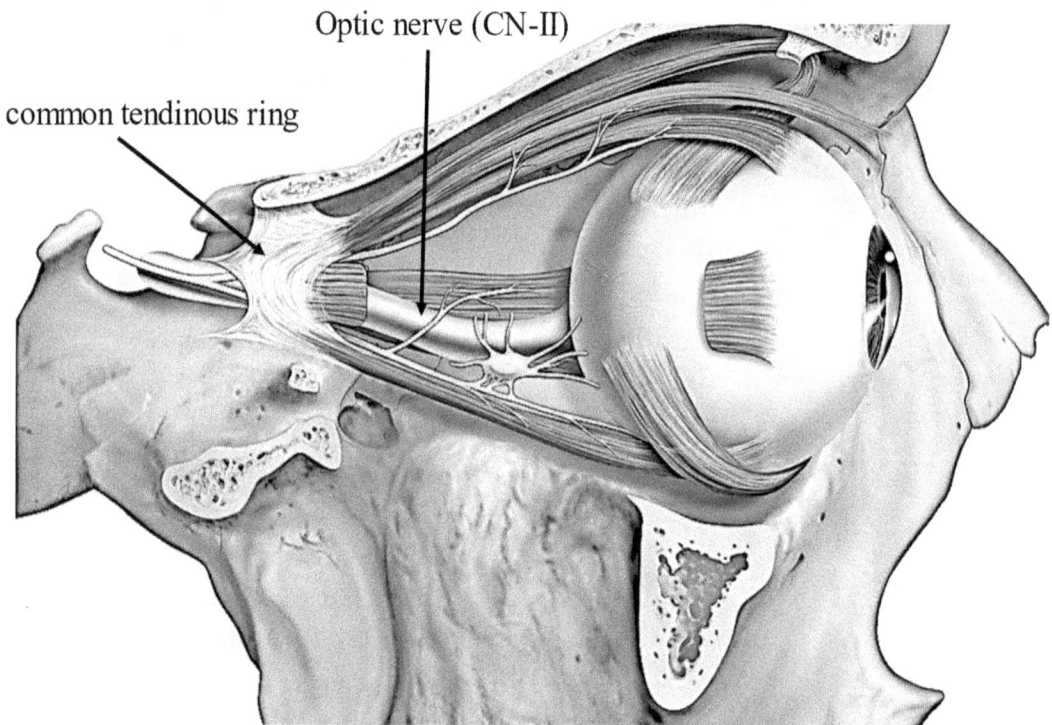

Optic nerve (CN-II)

common tendinous ring

Optic nerve illustrated by Patrick Lynch. Creative Commons licenses

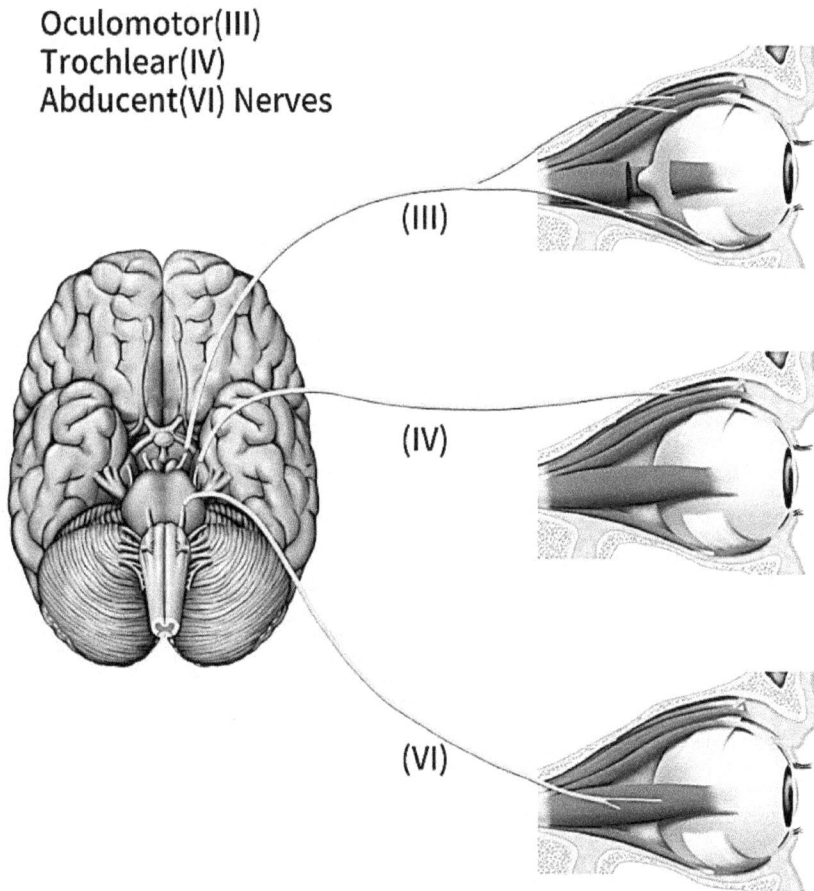

Oculomotor(III)
Trochlear(IV)
Abducent(VI) Nerves

(III)

(IV)

(VI)

Cranial nerves involved in vision

- Optic (CN II): senses light via rods and cones in retina; works with CN III to change pupil size
- Oculomotor (CN III): constricts or dilates pupil in visceromotor control mechanisms; to reflect ANS state; innervates four of six pairs of extraocular muscles
- Trochlear (CN IV): innervates superior oblique muscle which pulls the eye downward; convergence and divergence with medial and lateral movements
- Abducent (CN VI): innervate lateral rectus and partially innervates contralateral medial rectus muscle; necessary for eye tracking and gaze shifts
- Trigeminal (CN V): ophthalmic branch provides sensory to upper face, upper eyelids, and cornea
- Facial nerve (CN VII): innervates orbicularis oculi muscles which closes eye lids.

In total, *six* paired cranial nerves serve sensory, motor, and autonomic innervation of the eyes that work in sync to manifest ocular movements, reflexes, and vision.

Ocular movements – Binocular teaming

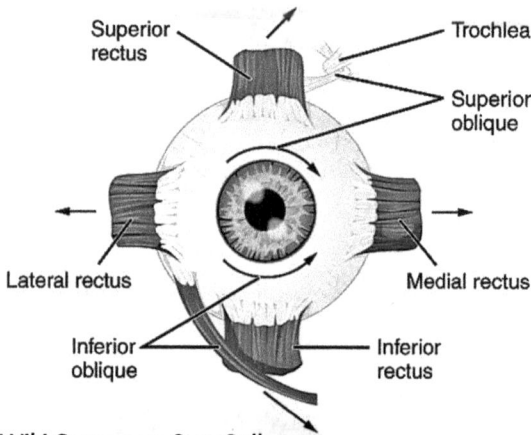

Superior rectus
Trochlea
Superior oblique
Lateral rectus
Medial rectus
Inferior oblique
Inferior rectus

Wiki Commons Stax College

A sclera (fascial layer) envelops the eyeballs from the optic nerve to the ciliary region, which serves a homeostatic function of moisture maintenance and as suspension ligament within the orbit cavity. Four rectus muscles shape, suspend, and move the eyeballs. Two additional oblique muscles move the eyeball in angular directions. One extraocular muscle moves the eyelids. Most of these muscles originate from fibrous fascia (common tendinous ring) in front of the optic canal (posterior of the orbits). From this origin, the muscles attach to the sclera of the eyeballs. Along with fat lacrimal glands, sacs, and nasolacrimal duct, there's an exorbitant amount of soft tissue and watery structures suspended in the two small orbits.

> *Bodywork for Sensory Integration* adds a novel way of revealing structural imbalances that can augment both Vision Therapy and Sensory Integration intervention

Assisting Function by Treating Structures - Vision & Oculomotor

All visual motor skills are heavily supported by complex intersystem processing from vestibular and proprioceptive/kinesthetic input. Developing fine-tuning discrete balance of movement with eye stability, matching the activity demand, contributes to varying levels of success in a developmental progression. Ocular motor muscles must be free from osseous compression and soft tissue restrictions to contribute to ocular adjustments fully. Both vision therapy and sensory integration methods address visual motor skills in their intervention models. Vision therapy has become a focused treatment for isolating and enhancing eye movement skills with some cognitively-directed balance exercises. Sensory integration therapy has vision movement skills embedded in the structured activities to entrain vestibular and proprioception. Vestibular input stimulates basic ocular motor reflexes, preparing for the eye to move. Postural intervention activates ocular motor skills to enhance whole-system workings.

Neither theory, however, addresses the possibility that structural compression can create an imbalance of extraocular muscles with connective tissue hindrance. Fascia is an extension of the tissue weaved through the muscles. The face structures and the eye orbits can experience and retain mechanical compression from the birth process or in-utero positioning. Trauma to the face, such as falling when learning to walk or getting struck by a ball, or even more severe events like car accidents or abuse during early childhood can also impose osseous compromise. The size and symmetry of the orbit, as well as equal and unrestricted eye movements, indicate these effects. *Bodywork for Sensory Integration* adds a novel way of revealing and assisting self-correction of myofascial attachments to the eyeball to augment both treatment approaches and, in a few cases, has replaced these therapies.

Besides the muscles and nerves involved in eye movement, professionals other than plastic surgeons discuss little reference to the vast amount of connective tissue around the eye and orbits. Truths are held in those tissues that can help bodyworkers understand eye movements and assist in binocularity efficiency. There's a false assumption that exercise and training stretch and mobilize tight extraocular muscles. Manual palpation of the movement process of each eye offers a subjective way to learn more.

When considering the orbits of the eyes, several bones form the opening and many soft tissues house and suspend the eye. The sphenoid is an important juxtaposition for all neurocranium and facial bones to develop balance in respective planes. At birth, the sphenoid has several synchondroses that take months and even years to mature fully. There are three sites on each side of the sphenoid where fascia compressions from birthing are concerned. Retained strains and torsions at the synchondroses are common, affecting all structures that eventually will further articulate. The intracranial membranes often play a role in orbit structural balance as well. *Bodywork for Sensory Integration* for vision and oculomotor function addresses the additional soft tissue suspension of face structures. Compressive forces from birth into the soft bones may translate structural compromise into the cranial nerves controlling both seeing and visual movements and influencing the eyeball shape.

Treating the growing cranial vault and facial bones within their suspending membranes before craniums become firmly fused and ossified assists in the self-correction of tensegrity and symmetry of orbit size, shape, and alignment.

Bodywork Treatments that help Vision and Ocular Motor

- Frontal bone decompression for superior orbit(s); treat for symmetrical positioning of frontal bone segments in relations to residual metopic fontanelle. (See p. 227)
- Decompress and balance sphenoid [unfused segments under age 8 years] to free compressions to posterior wall and optic canal of orbits (See p. 228)
- Hard palate decompression (maxilla/vomer/palatine complex) to release orbit floor; high arch palate can suggest adverse pressure into orbits (See p. 236)
- Dis-impaction of maxilla with ethmoid & lacrimal to free medial curves of orbits (See p. 229)
- Decompress zygomas to decompression lateral & inferior-lateral walls of orbit(s). (See p. 235)
- Decompress mandible anterior & inferior release CN V & VII pathways (See pp. 232-233)
- Release soft tissue restrictions surrounding sclera of eyes and extra ocular attachments (See pp. 230-231)

Vestibular – Auditory System

Senses of Head Movement, Hearing, and Body Perception

Bodywork objectives: Decompress osseous structures & balance fascia tissues surrounding the inner ear, housing the vestibular-auditory organs encased in the delicate temporal bone (petrous segment). Balance tissues forming the Eustachian tubes to optimize drainage and equalize air pressure. Release retained birthing compression or other mechanical injury to the cranial base & brainstem region, freeing structural compromise on cranial nerve nuclei and spinal pathways.

Sliding scale of sensory perception, adjustment, and adaptive response

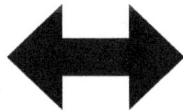

Developmental Trends and Behavioral Considerations

The human middle and inner ear house several sensory functions via two sensory pathways. The vestibular system sends sensory receptors for both angular and linear acceleration of the head. In addition, forces caused by changes in head position stimulate various pressure receptors. The auditory system is facilitated by the mechanics of vibrations and frequencies of sound waves that strike the eardrum and transduce to hearing receptors in the circular cochlea.

In communication with other systems, the vestibular system provides information about balance, gravity, and posture, activated by righting the head. With the head moving and being moved, the child is in the stages of the development process of mastering body control and coordination. To process movements, a child has to tolerate and then adapt to the sensations of their body moving or being moved. To process sounds, tolerating and locating sound sources are fundamental for integrating higher levels of auditory processing.

The bilateral Eustachian tubes are an anatomical feature of the inner ear that may occasionally get mentioned in vestibular-auditory processing literature. However, more needs to be explored about these structures' role in the vestibular-auditory complexity of processing information. Decades of sensory integration practice have observed (revealed?) a consistent correlation between a history of ear infections and subsequent vestibular processing difficulties with resulting sensorimotor challenges. It is a novel consideration of treating the bony and fascia containers that surrounds and suspends the middle and inner ears. Additionally, the structures forming and influencing the Eustachian tubes expands our understanding of when the vestibular and auditory system might benefit from some bodywork methods.

Eustachian tube function in relationship with sensory integration maturation is a concept that has yet to be explored scientifically. Maturation of the structure of the bilateral tubes draws the structures of both soft and osseous tissues into an inferior angle of the medial end that opens where the nasal cavities meet the upper throat. The change in head shape during growth influences the development of this angle. Horizontal tubes are more likely to be less efficient at fluid drainage of any backup. Another issue seldom explored in sensory integration function is the status of the sinus membranes of the nasal cavity. Chronic sinusitis can create various degrees of soft tissue inflammation and edema, contributing to narrowing and even blocking the drainage and flexibility of the epithelial membrane of nasal cavity walls. The Eustachian tube drains within the tissues of the soft palate (a part of the nasopharynx). The hard palate (maxilla/vomer/palatine complex) can have a mechanical influence on the soft palate, leading to drainage compromise of the eustachian tubes.

In addition, the structure and mobility of the tongue, especially during infancy and early childhood, play a role in massaging and facilitating good Eustachian tube motility. Any tongue-tie present may limit the tongue's elevation and the mechanical pressure on the hard palate.

Birthing compression can compress the palatine/vomer/maxilla bony complex into the soft palate and the area of the tube opening. A temporal bone compression can project structural strain medially into the whole of the middle and inner ear structures. Treating bony structures of the inner ear has been a long-standing method in chiropractic and osteopathic care and is known to reduce inner ear fluid stagnation. It has been shown to lower the risk of otitis media.

This dual system, monitored by vision and proprioception, forms perceptual adaptations to sounds and noises, effects of movement, and gravitational forces on head control. But a much more complex adaptation of these two systems involves feeling and coordinating the senses and movements of two sides, front and back, and up and down of the body. The auditory sense also increases awareness of how the body relates to the environment surrounding them. This awareness is vital for efficient coordination to develop but also for perceptual skills of knowing body sides and having an integrated sense of directions.

A sense of direction comes from the integrated and intuitive knowledge of the distribution and placement of body parts in relation to the whole. This system is constantly stimulated when a young child is picked up, laid down, and carried, and thus also processes proprioceptive input of the neck and visual information. The structural balance, the integrity of sterile air and protection of fluids within the tubes, and the surrounding osseous and fascia structures, might influence the natural inferior tilting the tube is led into through the size maturation of the head and face.

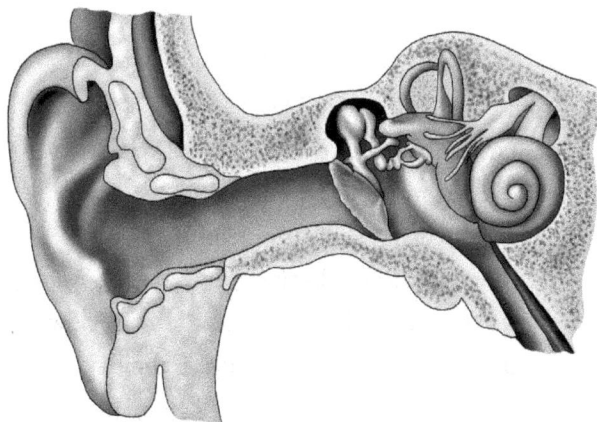

Spatial awareness of one's body is developed by blending auditory processing with complex vestibular information. Head righting, timing, and rhythm of the left and right sides match the sensations from the body so that the head and body move in harmony. This system is a team player in coordination integration of other senses that create balance, eye movements, posture, muscle tone, and attention. At rest, the paired tubes are in a collapsed state. The tensor and levator veli palatini muscles open the Eustachian tubes by contracting the soft palate's muscles. If undue tension exists in these muscles, the Eustachian tubes are not as mechanically capable of clearing themselves.

If the mucous membranes are inflamed from post-nasal drip or enlarged adenoids or other issues, fluid can build and be held to stagnate in the lower tube. In babies, tongue activity during non-nutritive and nutritive sucking repetitively activates the functional mechanism of keeping the ET mobile and optimal. By adulthood, the lengthening of the skull helps pull the tube inferiorly, having less reliance upon sucking to milk out fluids and activate tube mobility.

Adaptive Responses of Vestibular Processing

Infancy and childhood have an extensive list of adaptive responses to vestibular information involving movement skills, coordination, comfort with kinesthesia, and praxis skills. Some are:

In infancy

- Tolerance of being moved and handled under various speeds and styles, (known as gravitational security); riding in a car or stroller, clothing changes, bounced on a lap
- Coordinate head & body with sights & sounds; midline orientation
- Leveling action of semi-circular canals with eyes focused at breast coordinates eyes & ears
- Eyes focus on faces; eyes track with sounds; body follows eyes
- Head righting and body midline orientation (quality and symmetry)

In childhood

- Bilateral coordination of eyes, arms, and legs
- Body and head orientation for balance and postural symmetry
- Tolerance and control reactions to imposed movements
- Mature eye tracking and focusing skills (eye teaming)
- Focus attention and concentration as background sensations are processed for balance
- A sense of direction based on understanding right-left, up-down, front-back perception of one's own body generalized to the world outside their skin

Adaptive Responses to Auditory Processing

In infancy

- Turn towards sounds with movement responses
- Tolerates increasing variety of sounds
- Inhibit startle response (Moro) to sudden or louder sounds
- Verbal response to sounds
- Tongue moves in response to sounds

In childhood

- Discriminates between increasing sound variation
- Speech production represents auditory reception
- Engages reciprocally in social & conversation
- Understand verbal directions
- Attention to sounds: sustained attention
- Comprehension of what is spoken (matched with cognitive maturation)

Adaptive Functions of Eustachian Tube Physiology

- Protect the middle ear from nasal secretions and backwash of milk draining what might come into this space.
- Maintain stretch-activated drainage by yawning and adequate tongue pressure into hard palate during sucking and feeding
- Intermittently opens or narrows in response to pressure and maintains air-containing space, which equalizes the intra-tympanic air pressure with that outside the ear
- Supports auditory input (hearing) clarity by clearing intra-tympanic space (internal side of the eardrum)
- Quick autonomic recovery from sudden, alarming, novel movement or air pressure changes to regain homeostatic mechanics within the inner ear environment

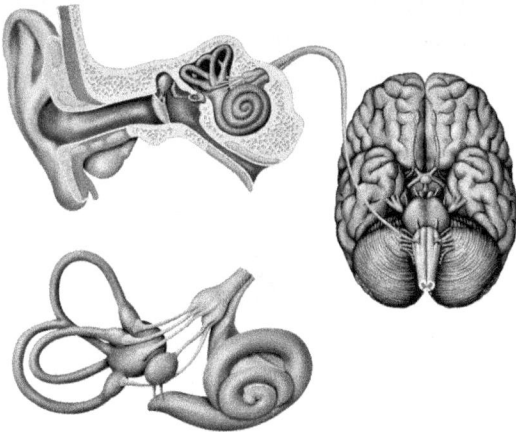

Anatomical Considerations

- Vestibulocochlear nerve (CN-VIII): serves both vestibular and auditory via branched nerve
- Semicircular canals and Cochlea
- Endolymph fluids that bath and generate sensory signals
- Lymphatic pathways
- The Eustachian tube, (pharyngotympanic tube); bone and cartilage segments
- Temporal bone's petrous portion (part of the cranium base) houses the middle and inner ear structures
- Junction of soft and hard palate (roof of mouth) as drainage site for Eustachian tubes

The mutuality of these complex systems

The vestibulocochlear nerve (CN VIII) splits into two large divisions, emerging from the pontomedullary junction (brainstem) and exiting the inner skull via the internal auditory meatus of the temporal bone. The developmental stage of the temporal bone's ossification and cranial suture immaturity can directly affect the middle and inner ear because the bone's softness is vulnerable to mechanical impressions. The middle and inner ears are housed deep within the complexity of the temporal bone. The auricle (outer ear) is a projection of vascularized skin and cartilage from the scalp over the temporal bone.

The temporal bone's mastoid cavity is a honeycombed area behind the ear canal and opens into the middle ear. The inner ear houses the vestibulocochlear organs and has two main functions: convert mechanical signals from the middle ear into electrical signals and maintain a level head with symmetrical body balance in response to position and motion.

Vestibular System

Two sensory receptor portions serve the vestibular system within a bony labyrinth—semi-circular canals and otoliths. Three semi-circular canals (bent tubes) and two otolith organs, each filled with fluid (endolymph), are lined with thin, hair-like receptor cells. These organs operate simply: the endolymph is agitated when the head is moved, and the hair cells are bent. These sensory cells differentiate the direction and speed of any movement, to detecting acceleration and deceleration, as well as rotational or linear movements. Signals are conducted to the brainstem and cerebellum, creating the perception of motion. The movement of the endolymph within both vestibular apparatus functions similarly to a carpenter's bubble level, creating a complex sensorimotor awareness of midline orientation and bilateral control of both sides of the body. The source of the endolymph fluid is unknown, or if it circulates with other branches of the lymphatic system.

The paired Eustachian tubes serves as a protectant barrier from various secretions and fluids and a pressure equalizer to the inner ear mechanisms. The two vestibular mechanisms, with feedback from both auditory system and baroreceptors in the Eustachian tube vicinity, the paired systems function similarly to a carpenter's spirit level. The displacement of fluid in the semicircular canals signal the speed and direction of head movements. Combined with information from the eyes focusing on targets and proprioception of the neck and body, the inner ear mechanism modulates and regulates tolerance of movement and balance of posture and coordination of movement.

Auditory System

The tympanic membrane (eardrum) separates the external ear from the middle ear. Its function amplifies and transmits sound vibrations from the air to the tiny ossicles (bones) inside the middle ear. The malleus bone bridges the space behind the eardrum to the other ossicles, sending signals onward toward the cochlea. Sound waves from the middle ear vibrate the cochlea, filled with watery endolymph. As the fluid moves, the cochlear segments also move, stimulating thousands of sensory hair cells (stereocilia) and transducing electrical signals to the brain via the cochlear nerve.

Sound reaches the cochlea, where the basilar membrane motion and hair cell stimulation convert specific frequencies into tonotopic information that encodes the place of cochlear stimulation and intensity. The spiral ganglion neurons receive this signal and then relay this information to the brainstem but not the cerebellum. Multiple cranial nerve branches contribute to multiple sensory processing for complex human functions involving the auditory apparatus. For example, the ear drum's inner surface is innervated by the glossopharyngeal nerve, which is involved in tongue movements and swallowing. There are both sympathetic and parasympathetic nerve branches in the auditory system.

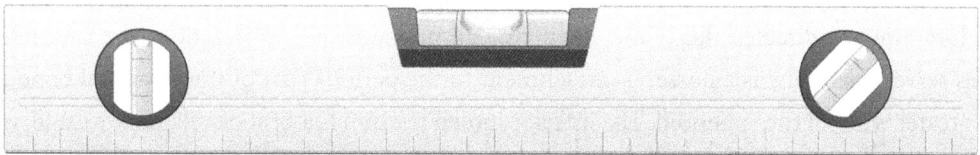

The Leveling Function of the Inner Ear Mechanism

Eustachian Tubes and Surrounding Structures

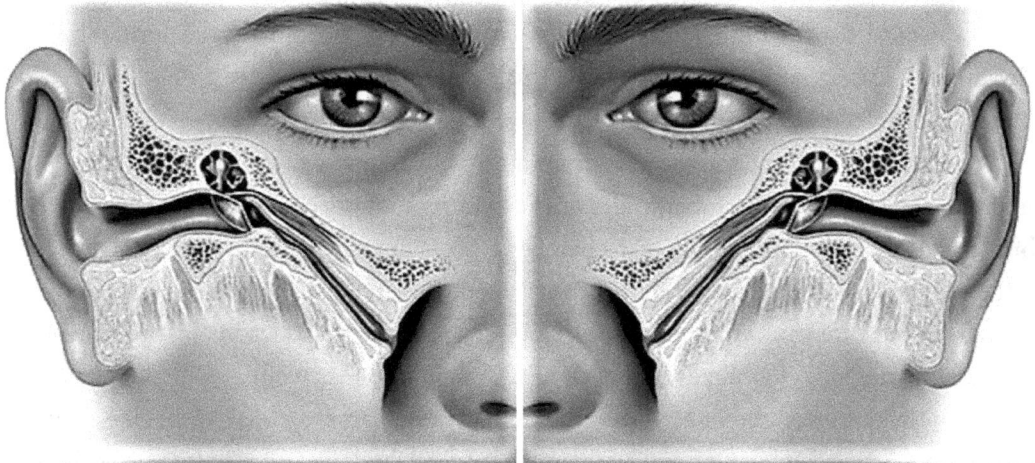

Paired Eustachian tubes are dynamic organs connecting the middle ears to the posterior throat.

Tight membranes can amplify decibels, whereas loose, flexible membranes can dampen or dissipate sounds. An elevated sympathetic state, causing sound hypersensitivity, can be created when cranial bones and membranes are compressed or torqued. Loud sounds can set off the Moro reflex, a normal startle response. Clenched jaws, retracted tongues, tight throats, and elevated shoulders can be behavioral manifestations of protective retraction to sounds.

A protective acoustic reflex exists (an involuntary contraction of stapedius and tensor tympani muscles in the inner ear set off by loud sounds) that tightens membranes. Sympathetic reactions to sounds can emerge if the reflex persists and does not relax. The reverse may also be true. Tight membranes surrounding the outer, middle, and inner ear structures may intensify sounds, leading to a lack of integration of the acoustic reflex. Behaviors of sound defensiveness can develop.

In addition, consistently tense or restricted membranes will be less than efficient in moving interstitial fluids through the region. Interstitial fluid is believed to be the source of lymphatic fluids, so it stands to reason that this same fluid source could be the source of endolymph, even if the source is localized and system-specific. Stagnant fluids can be a source of bacteria overgrowth, leading to ear or sinus infections. Compressions that do not resolve at the temporomandibular joint, the temporal bone, and the sphenoid region can adversely affect auditory physiology.

Draining misdirected fluids and equalizing air pressure are two functions the Eustachian tubes serve. Each tube is an osseous attachment to the petrous part of the temporal bone and the greater wing of the sphenoid. The inferior lumen portion is a fibrocartilaginous canal, with an opening formed at the middle ear's medial margin extending to the upper nasopharynx. The end of this cartilage portion flares within the mucous membrane of the nasal cavity walls. Tiny muscles attach to the tubes, making the lower portion a dynamic organ. These muscles are activated by tongue pressure into the palate with sucking and swallowing. The tubes are a protective barrier from reflux of nasopharyngeal secretions, bacterial formation, and labyrinthitis. Eustachian tubes are supposed to house only sterile air, but it is the potential source of ear infections if fluid gets in.

When swallowing, the tongue presses up into the hard and soft palate, which helps open the inferior segment of the tubes, milking them to promote drainage. This allows counter-force to push air into the middle ear chamber to equalize any pressure imbalance, as long as the tongue has a full range of motion with adequate strength and pressure. Tongue-ties without proper care can interfere with assisting the Eustachian tubes in this way. Sneezing and yawning naturally mobilize the length of the paired tubes as well.

Much fluid moves through, and needs to be moved, for optimal vestibular-auditory-Eustachian function. Endolymph and perilymph interstitial fluids are necessary to activate hearing and vestibular sensory nerve endings in the cochlea, semi-circular canals, and otolith organs. The sources and tributaries of interstitial fluids for inner ear function require much more research on how far it travels in circulation and how it co-mingles with neighboring fluids. Multiple lymph nodes and pathways surround the head's side, ranging from superficial to deeper nodes. The movement and flexibility of soft tissues allow for the efficient and proper exchange of fluids in a homeostatic cleansing function. Freedom of osseous restriction of the connective tissues surrounding the utricles and semi-circular canals enables the movement of endolymph to conduct the business of transmitting sensory signals and keeping the terrain clean.

The hard palate and its relationship with the tissues of the soft palate also has have a direct effect upon the distal ends of the Eustachian tubes where drainage should happen in the upper throat. Eustachian tubes can be decompressed and cleared with bodywork methods.

Assisting Function by Treating Structures – Vestibular/Auditory/Eustachian Tubes

Due to their intimate proximity, the vestibulocochlear nerve (CN VIII) and both sensory systems can be treated simultaneously by reaching the respective anatomical components of the middle and inner ear. Remember, young children's bones are still in the midst of maturation and the skeletal field contacts many sites of fascia rather than fused or ossified bone. This is especially true in the cranium. Compressions from all cranial bones can translate mechanical distortions into the vestibular-auditory apparatus housed in the petrous portion of the temporal bones. Retained restrictions in osseous membranes from birthing, plagiocephaly, or other head trauma, or a high hard palate related to ankyloglossia (tongue-tie), can tighten the Eustachian tubes and hinder optimal physiology. Lateral forces can strain into the soft segmented temporal bones. Anterior compression of the occiput can compress directly into the temporal bone's mastoid process behind the ear canal. Posterior compression through the frontal bone can adversely affect the balance of the yet unfused sphenoid segments as well as the intracranial membranes of dura tissue. The sphenoid can be pushed backward, interfering with the auditory pathway to the thalamus. Temporal bone and inner ear structural imbalances, or sphenoid compression or misalignment, can negatively affect the Eustachian tubes' structural flexibility, drainage, and equilibrium functions.

As mentioned with tongue mobility, the suspended freedom-of-movement of the hyoid bone can eventually reflect into the thyroid gland. The thyroid regulates many functions, including muscle tension. As a result, soft tissue restrictions around the hyoid bone can translate to possible Eustachian tube restrictions. In addition, when the tongue is restricted, it may be unable to develop adequate tongue pressure up into the hard and soft palate, hindering aid to establish inner ear homeostasis. Releasing pressure in the structures surrounding and suspending the Eustachian tubes allows the tubes to open and drain naturally. The number of ear infections a child has would be a prime indicator of stresses upon the Eustachian tubes.

Chronic sinusitis or other sources of inflammation can also negatively affect the function of the Eustachian tubes, creating distortions of sound vibrations due to swollen mucous membranes with secretions. Anyone recalling a significant cold or rhinitis can attest to the uncomfortable symptoms of back-up pressure into the inner ear with a sinus cold or fluid perfusion in the inner ear. This can really distort the reception of both vestibular and sound information, especially when only one side is affected. It is stressful for the brain to have to process two distinctly different signaling.

The ear is a fascia extension of the fascia sphere container and a distal extension of the tentorial membrane. Plagiocephaly and other misshapen head conditions can be reflected by asymmetry of ear placement on the skull.

Though not yet fully proven, the Eustachian tubes may be afforded a direct "stretch" with the temporal bone decompression, (relieving both fluid stagnation and pressure) as positive clinical outcomes suggest. In addition, the decompression of the maxilla stretches the soft palate. *Bodywork for Sensory Integration* has been shown to prevent and resolve inner ear infections or a profusion of fluid. Air pressure and fluid drain are key elements of healthy inner ear function.

Assisting the release of tissue restrictions theoretically prepares sensory receptor sites and cranial nerve pathways to function within normal physiological parameters. Unrestricted fluid pathways always optimize the entire vestibular-auditory-Eustachian tube systems. Lymphatic drainage massage has proven positive effects upon general inner ear workings in children with past and current history of fluid effusion in the middle and inner ear as well as otitis media (ear infections).

Bodywork Treatments that help Vestibular-Auditory-Eustachian function

- Frontal and parietal bones within the fascial suspension forming sutures and fontanelles (lambda, bregma, sphenoid, and mastoid) around the temporal margins, anterior and superior extensions of the intracranial membranes, and the dura mater. (See p. 227)
- Temporal bones are the most lateral extension of the intracranial membrane (tentorium cerebelli and dura mater); mechanical forces can impose into the vestibulocochlear nerve, eardrum, etc. Eustachian tubes. (See p. 228)
- Sphenoid; sphenobasilar synchondrosis (junction of sphenoid to occiput), and the surrounding fascia of the cranial epicenter. (See p. 228)
- Occiput condyles / Cranial base can have force from the posterior aspect of the cranium, causing strain into the mastoid process of the temporal bone and into middle / inner ear. (See pp. 211-213)
- Maxilla-vomer-palatine complex of the hard palate can hold translated compressive forces into soft palate; proximity also to the lower ends of the Eustachian tubes. (See p. 236)
- Zygoma and buccal pads of the cheeks can retain compression into cranial nerve innervation sites and other facial structures. (See p. 235)
- Mandible can hold posterior compressions towards and into the external and internal auditory meatus structures and superior compressions into the TMJ joint and cranial nerves (trigeminal and facial outlet) and vasculature (carotid arteries and jugular vein). (See pp. 232-233)
- Fascial and interstitial field surrounding all soft tissues cranial nerve pathways, innervation sites, and vast blood vessel system of the ear. Lymphatic structures and nodes that serve the petrous region of the temporal bone can be stagnant, which can translate to stagnation in endolymph and perilymph fluids. (See pp. 189-191, 192, 193)

Tactile - Sense of Touch

Somatosensory System of the Skin

Bodywork objective: Balance tensions of superficial fascia and interstitial layers woven through dermal tiers, to promote parasympathetic tone in the largest body organ (fascia and skin combined). Promote optimal lymphatic pathways for healthy fluid exchange and cutaneous autonomic nervous system detoxification.

Developmental Trends and Behavioral Considerations

There is nothing more instinctually valuable than the sensation of touch. Science has repeatedly proven healthy human touch contact is essential to well-being. The tactile system develops in the womb from the associations the brain makes through the inputs caused by the pressure-filled, inter-uterine space. Deprivation from well-intentioned touch can have a significantly negative impact on the nervous system of a growing infant. Regardless of age, the brain needs well-intentioned touch daily, not only to survive but to thrive. The right touch helps one feel safe and secure and prepares a child for better learning and thinking. One can assume that well-intentioned touch can help resolve protective retractions and resultant behaviors.

The skin houses the protective and discriminative device of the cutaneous autonomic nervous system. Special receptors determine whether or not a touch input merits a fight-flight sympathetic reaction or feeling comfortable within one's skin. Quality of life is balanced by the cutaneous autonomic system with all other systems.

Sliding scale of sensory perception, adjustment, and adaptive response

Somatosensation includes all sensations from the skin, mucous membranes, limbs, and joints. Human skin is a highly specialized organ where receiving tactile sensory information is part of the body's preservation mechanism for homeostasis, maintaining fluids and protection from microbes, and social connectedness to others. The skin is also highly involved in regulating and balancing the autonomic nervous system. The complex cellular architecture of cutaneous nerve fibers mediates these functions. Sensory nerve receptors lie within or near the interstitial spaces between the dermal layers. Some receptors function in a protective and defensive mode, while others in a discriminative learning mode. Defensive mode results in tissue retraction. An estimated 3000-9000 densely packed sensory receptor cells are within one square inch of human skin.

The vast amount of touch receptors is located at the margin where the epidermis meets the dermal skin layer. Touch sends waves of input to the cortex, activating corresponding receptive regions in the brain, which can assist in the down-regulation of an overactive limbic system as long as touch is perceived as pleasant and unstartling. The skin detects perceived dangers and will be the first organ to tighten in a protective retraction reaction to a noxious sensation, as exhibited by a response such as goosebumps. The skin is a vast communicator and knows how to protect.

Purposeful response and use of touch is demonstrated by skills of stereognosis, localization of touch to skin, and skin tolerance to worldly input. Fascia flexibility between dermal layers where sensory cells are housed are woven with cutaneous autonomic fibers. Sympathetic and parasympathetic responses, protective retraction and discriminative learning, occur at skin level. The resulting reaction, though communicates to all other nervous system.

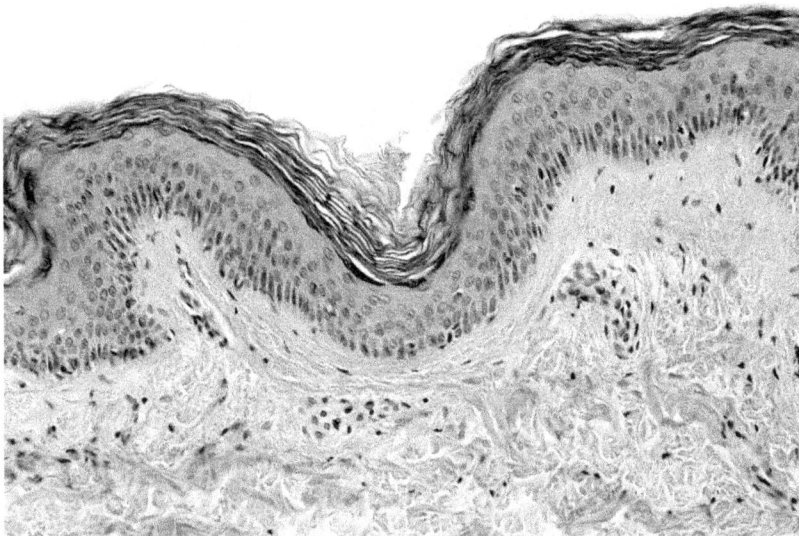

Electron microscopy of epidermis layers. Image source: Shutterstock

The skin is also the covering of the superficial lymphatic system. Vastly expansive, the lymphatic vessels begin as tiny micro vessels adjacent to the microvascular meshing of capillaries and venules. Here, tissue waste is transported away from space between cells tissues towards larger lymph vessels and eventual nodes. Eventually dumping into the vena cava and onto the liver, the lymphatic system is a primary detoxification mechanism. If the liver is not flowing well, the system can become stagnant. Organ and vessel congestion can occur, and nodes can become clogged or engorged. Underlying mechanisms of inflammation can overwhelm the system. The skin can substitute as a detoxification organ if the liver filtration is suboptimal.

What is commonly referred to as "touch" involves more than one kind of stimulus and more than one kind of receptor during one instance of touch perception. The sense of touch is a symphony of input activated by various receptors. Touch receptor cells are in different classes of mechanoreceptors. The first group is the tactile group of superficial cells in the skin. Proprioceptors and baroreceptors for interoception represent another aspect of the somatosensory system and will be discussed in the following sections. Perception of touch is the person's brain interpreting the product of that processing.

Touch reflects either a sympathetic (protective or painful) response or a parasympathetic response. A parasympathetic response ranges from neutral reactions to calming, ease, enjoyment, joy, security, and safety. The discrimination ability to discern differences between soft-hard, smooth-textured margins of shapes and other subtle sensations are the skills of somatosensory function. The integrity of autonomic nerve fibers in the skin can be assessed by quantitative analysis of cutaneous responses. Axon reflex reactions result in dilation or constriction of cutaneous vessels, sweating, or from horripilation (goosebumps) or erection of arrector pili muscles of hairs. These reactions make the skin feel uncomfortable, painful, and irritating. Cold temperatures, painful events, fearful events, and even the memory of such events can stimulate these sympathetic reactions in the skin.

The behavioral responses of skin reactions can also be behavioral responses to the state of the superficial vasculature and lymphatic systems. Rarely does the lymphatic system get mentioned in the same perspective as the sensory integration and processing of somatosensory information. Here, inflammation and other biological issues can manifest into sensory challenges of tolerance for touch.

Beyond behavioral opinions about another's reactions to touch input, sensory fibers can be assessed using quantitative testing, a concept routinely utilized in neurological sciences but rarely in the practice of sensory integration. Furthermore, skin biopsy studies have expanded our knowledge of the cutaneous nervous system fibers by examining the integrity of sensory receptors.

Adaptive Responses of Touch System

In Infancy

- A maturing absence of protective retraction to familiar input
- Lessening of primitive Moro reactions with handling, bathing, and diaper changes
- Tolerance for bathing and dressing changes
- Turns face and body towards tactile input they find perfect or preferred
- Tolerance for clothing against their skin
- Bonds and soothes with the source and intensity of tactile input
- Recovers quickly from alarming touches input and temperature changes

In Childhood

- Stays calm or calms quickly to touch input.
- Inhibits and habituates to "background" senses of air currents, movement of cloth, clothes tags and seams against skin, elastic bands from clothing.
- Engages and tolerates full scope of self-care and grooming tasks (showers or bathing, hair cutting and combing, face cleaning).
- Food texture tolerance and acceptance for a varied diet.
- Engages, seeks, or accepts hugs and other forms of human bonding expressions.
- Reaches acceptable levels of adjustments to inside and outside temperature differences.
- Lowered incidence of having goosebumps.

Anatomical Considerations

The skin is one of the largest organs of the body housing different receptors with distinct sensory functions, and is made up of three main layers:

Epidermis: The thin upper layer serves as a water barrier, protects from invasion by pathogens, and monitors sudden temperature changes. Protective retraction of skin is the sympathetic response to a "threat" perceived by the body (e.g., goosebumps, degree of tautness, flinching or withdrawal of limb from touch). The epidermis is the outermost layer of skin in mammals. It is relatively thin, is composed of keratin-filled cells, and has no blood supply.

Dermis: The thicker middle layer containing hair follicles (each with a muscle and nerve ending), various nerve projections, and many blood and lymph vessels are embedded in a collagen-rich framework. This framework is the terminal ending of the fascial network. Within the deeper dermis, thicker tissue contains blood vessels, sweat glands, hair follicles & shafts, lymph vessels, and lipid-secreting sebaceous glands.

Subcutis (or hypodermis): The deepest fat and collagen-rich connective tissue and insulating layer which contains abundant blood and lymph vessels houses more of the variety of sensory neurons. The hypodermis, holding about 50 percent of the body's fat, attaches the dermis to the bone and muscle, and supplies nerves and blood vessels to the dermis.

Collagen and elastin fibers:[4] Within the fascia connecting the epidermis layers to the hypodermis, this cellular matrix functions as an augmentative whole-body communication network at warp speed.

Sensory receptors are classified into five categories:

- mechanoreceptors (for discrimination)
- thermoreceptors (for temperature detection)
- proprioceptors (for pressure reception)
- chemoreceptors (for homeostasis of cellular environment)
- pain receptors (to alert / alarm of injury)
  - A-beta fibers
  - A-delta fibers
  - C-fibers

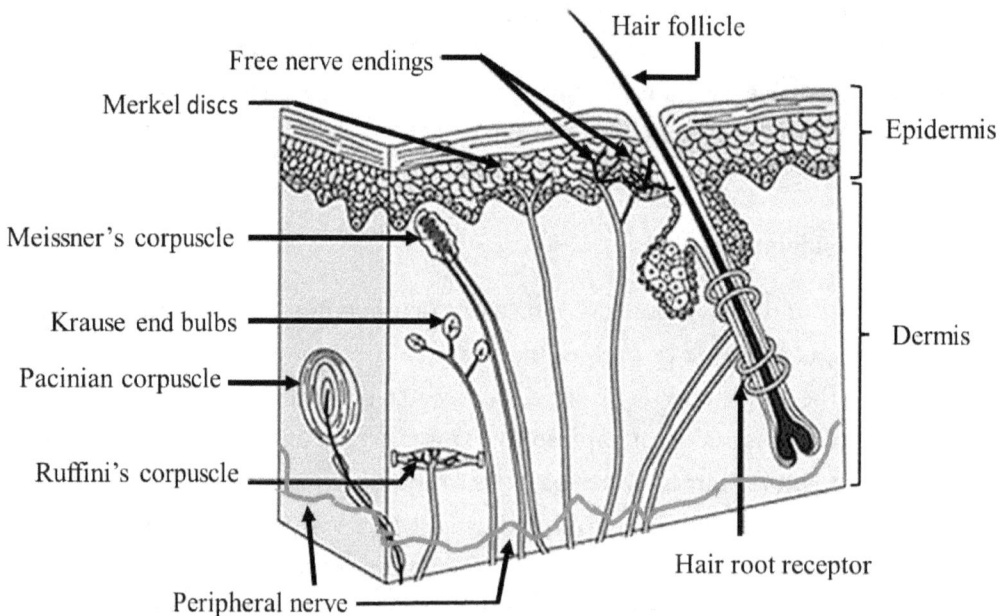

4 Sensory function of the fascia matrix network still remains to be fully understood but is now believed to be a background sensory feedback structure contributing to the maintenance of tensegrity and functional movement.

The Autonomic Nervous System Cells of Skin

The dermis has an important zone subdivision of the simple outermost layer and a deeper, more complex layer. The division is formed by the anatomical arrangement of two major arteriovenous plexuses. The thermoregulatory control of these vessels relies on input from unmyelinated C-fibers of the cutaneous autonomic nervous system. Sweat glands and smooth pili muscles of hair follicles are also effector organs of these postganglionic C-fibers, which as polymodal sensory afferents, hold rapid scanning properties for function of protection. Additional autonomic receptor fibers are thinly unmyelinated alpha A fibers (nociceptors for pain, mechanoreceptors for harmful pressure, and the thermoceptors). These sensory fibers send collateral signals to the blood vessels to generate sympathetic signals to alarm a homeostatic response. The autonomic nervous system, as alluded to previously, has its distal ends within the skin. Recall that the ANS's primary function is to react in order to restore homeostasis. For the skin's sake, that means microvasculature will be the first site of vasoconstriction and protective retraction through the superficial fascia layers. Goosebumps are the obvious sympathetic reaction, but more subtle tension can be palpated with experience. Scattered among all the other better-known sensory receptors in the skin, there are a class of cells assigned to the cutaneous autonomic nervous system. These cells are microscopic fibers, classified by size and speed of conduction, and have a direct effect upon cutaneous vasculature reactions. Most of what is known about these fibers is from adult populations with various neuropathies but little has been studied in sensory processing behaviors.

Protective (protopathic) receptors detect temperatures, protection pressure (like from a thorn), and pain (from blunt or sharp trauma). Discriminative receptors distinguish between textures, are pressure sensors for learning tool use, and connect emotions through touch for comfort and bonding. The body reveals a continuum of the sympathetic or the parasympathetic reactions to various tactile inputs. The dorsal column system and the spinothalamic tract are two major pathways that bring somatosensory information to the brain via the spinal cord. The human skin is a highly specialized organ for receiving sensory information but also to preserve the body's homeostasis. These functions are mediated by cutaneous small nerve fibers which display a complex anatomical architecture and are commonly classified into cutaneous A-beta, A-delta, and C-fibers based on their diameter, myelinization, and velocity of conduction of action potentials.

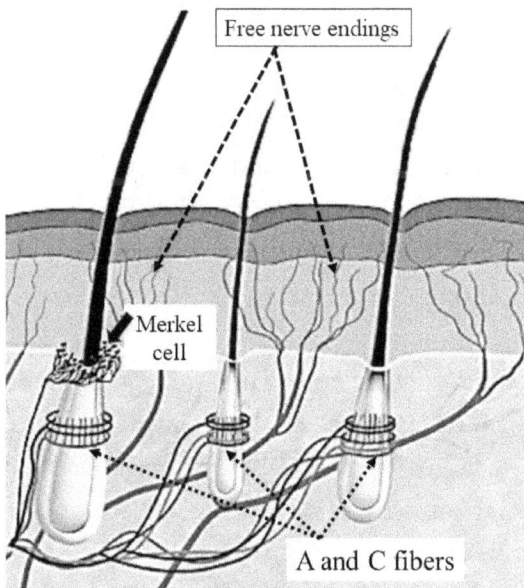

The cutaneous autonomic nervous system cells.

Types of specialized sensory receptors:

Merkel's discs: The most sensitive of all the receptors. Slow-adapting, unencapsulated nerve endings, respond to light discriminative touch. These branching receptors detect the point of touch contact with precision and differences in textures. Pressure into the edge of a Merkel disc without collapsing the fluid filled cap can still stimulate sensory perception. Found in upper skin layers near the base of the epidermis, both in hairy skin and hairless (glabrous) skin (hand palms and fingers, feet soles, and lips), they are more densely distributed in the fingertips and lips. The receptive fields of Merkel's disks are small with well-defined borders. That makes them finely sensitive to edges and they come into use in tasks such as typing on a keyboard, buttoning a button, trimming your own fingernails, **and performing non-invasive bodywork methods!**

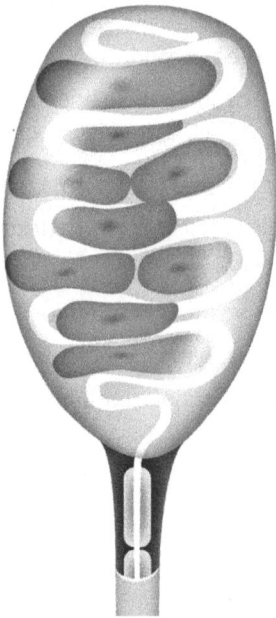

Meissner's corpuscles: Fluid-filled encapsulated neurons with small, well-defined borders that are found in the upper dermis suspended by fine collagen fibers. They also project sensory fibers into the epidermis and are stimulated by lateral motion and movement against the skin. They, too, are found primarily in the non-hairy skin on the fingertips and eyelids. Receptors of fine touch and pressure, they also detect flutter sensations and low frequency vibrations. Like Merkel's disks, they are more densely situated in fingertips and less plentiful in palms of the hands. Meissner's corpuscles contribute to exquisite sensory perception of hands for fine dexterity, with similar function as Merkel's disks. The main difference between Merkel's and Meissner's corpuscles is the speed of transmission of sensory signals.

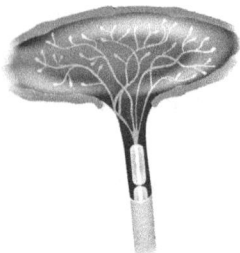

Ruffini endings: Larger sensory endings with as singular branching dendrites are highlighted by being surrounded by collagen fibers and located deeper near the base of the epidermis. Encapsulated, these receptors are slow to adapting and detect skin stretch and compression, joint activity, and warmth. They give feedback to match power of a grip based on degree of resistance or control body parts during movement, such as screwing in a lightbulb with exact pressure (to not crush the bulb). The sensory integration term (adaptive response) for this is called "grading" of movements with discrete adjustments. Found in both glabrous and hairy skin. They can be considered the gateway receptors of proprioception and kinesthesia. Ruffini endings also detect warmth. Note that these warmth detectors are situated deeper in the skin than are the cold detectors. It is not surprising, then, that humans detect cold stimuli before they detect warm stimuli.

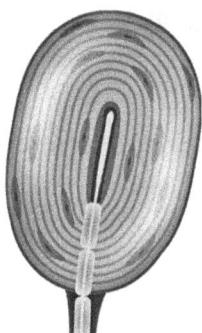

Pacinian corpuscles: Larger like Ruffini's corpuscles, sensory endings co-spiraled with fine fibers of collagen. They are fluid-filled and encapsulated receptors which also rapidly adapt. These receptors are stimulated by the mechanical compression of transient (not prolonged) pressure touch and high-frequency vibration and discern the depth and danger of pressure input. Located in both hairy and smooth skin, deep in the dermis, these sensory neurons also project fibers to bone periosteum, joint capsules, visceral organs, and genitalia. There are fewer Pacinian corpuscles and Ruffini endings in skin than there are Merkel's disks and Meissner's corpuscles.

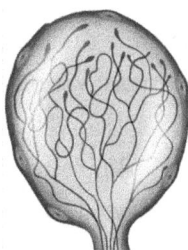

Krause end bulbs: Receptors of cold sensation, cylindrical or oval bodies at the nerve ending are formed by an expansion of the connective tissue sheath the nerve fiber is suspended by. These bulbs have been located in the conjunctiva of the eye, in mucous membranes of mouth and lips, and in the epineurium of larger nerve branches.

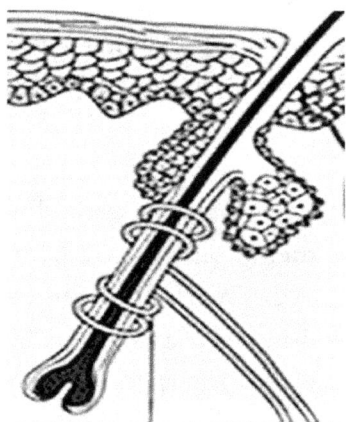

Hair receptors: Nerve fiber plexus (peritrichial) are also unmyelinated dendritic branches surrounding the roots of every hair follicle. Free nerve endings as well, they rapidly adapt as they detect hair movement and skin deflection. Some hair receptors are able to detect stimuli that have not yet touched the skin. From the pain of hair being pulled, the movement caused by wind on bare skin, or conversely, the calming effect of hair brushing or head massage, these cells are direct stimulation to the ANS. These receptors are one of the first that therapists should be mindful when making contact with a child's skin.

Free nerve endings: Belonging to the class of unencapsulated sensory receptors, a free nerve ending is as its name implies. Dendrites from primary sensory neurons in the periphery project randomly throughout the epidermis as free-floating "wires", and can be either myelinated or unmyelinated. Free nerve endings are the most numerous of all the receptors in the skin and are stimulated by stimuli that cause skin or cell injury (pain), by hot and cold, and to various degrees of light touch. These nerve endings are slow to adjust (slow to inhibit or habituate) to a stimulus and less sensitive to any change in the stimulus. Much has been learned recently about the categories of free nerve endings of pain perception.

Groups A and B (which are myelinated, larger fibers and faster conduction speed) and C fibers (which are unmyelinated, smaller, and lower conduction speed). Meaning, it takes more input to activate A and B fibers, than it does C fibers. Sharp, well-localized pain is attributed to A fibers and dull pain that may also radiate is attributed to C fibers.

Bodywork Treatments that help Somatosensory Touch

- Any bodywork method categorically falls under the heading of therapeutic touch. Regardless of the technique, therapy begins as soon as contact with the child's skin commences. The level of invasiveness and the intention of the therapist's hands will be scouted and reacted to by the somatosensory organs in the skin.
- Lymphatic drainage massage has proven time and again to assist skin detoxification and promote clearing of inflammatory properties that might be related to cutaneous autonomic nervous system cells in the skin. (See pp. 199-201, 202)
- Spinal processes and nerve roots innervating dermatomes may be compressed, eliciting facilitated segments to the somatosensory regions. All dura tube techniques directly at soft tissues surrounding and suspending spinal structures can assist self-correction. (See p. 194, 212, 213-215, 216)

Remember that the dermal layers are the distal ends of the whole organ of fascia. The tension of fascia and superficial circulation can determine the kind of sensory perception occurring in the skin even before the information reaches the brain. The experienced bodyworker can develop a keen, though subjective, sense of assessing the skin's tension by matching the skin's tone and resistance, holding space for the layers to acquiesce, and then self-correct. Behavioral patterns related to chronic protective retraction can make a person's skin less tolerable. Treatment starts as soon as hands touch the skin. Even through clothing, the therapeutic presence of a therapist's hands can be felt. Touch is everything.

The touch system is so incredibly in tune that the person receiving touch can perceive that person's intentions. Thoughts and emotions can transmit through touch, communicated as vibrations through the fingertips. The C fibers are designed to sense if contact is safe and well-intentioned. Touch not only creates vibrations but impacts vibrations. A guitar string will vibrate when strummed, but those strings can also vibrate as a reaction to loud or strong noises in the room. Fascial fibers may conduct either direct sensory input or provide sensory feedback monitoring by their woven nature. Too much pressure from the practitioner's hands can dampen the communicating vibrations of tissues, which may be perfect when pain is firing at the site. Too invasive, aggressive, or too dynamic of touch can set off protective retractive reactions.

Skin can reveal fascial restrictions at the superficial level. When massage is not helping to eliminate fussiness, colic, or sleep patterns, greater skill in discerning tissue response is warranted. Recognizing signs of overload or body retraction is vital, or the therapist risks imposing even more negative input. Sometimes the evidence of tension is clear to the naked and discerning eye through outward signs of protective retraction of the skin or asymmetries in the body. It also requires waiting to "listen" to the fascial fibers lying under the whole skin to understand how the person responds to the touch input.

Sensory integration therapists train to recognize the difference between under-responsiveness and shutdown (from overload) behaviors. Bodyworkers can gain a similar perspective from each child on the "just right" input through mindfulness of the touch contact used.

A bodyworker can develop a keen sense of assessing skin tension

The Touch System of the Therapist

To raise the level of your therapeutic competency, first consider the location and depth of the various somatosensory cells in your skin. Ponder how the area, density, and design serve you and your hands. Do you know which tactile receptors you rely upon during your routine practices? Are you cognizant that there are many levels of invasiveness of touch contact we can use when approaching a client's body? Each level of contact will yield different tissue responses. Are you depending upon Golgi tendon and joint tendon receptors to gain information about the client's tissues through more forceful touch contact? Have you an awareness of your free A & C fibers that possesses an ultra-refined ability to perceive subtle tissue changes of tension between dermal layers? A touch that is absent of pressure may not excite C fibers and a stress reaction, but might with too light of moving touch. Imagine your C fibers meeting the client's C fibers with matched tone. Analyze the invasiveness of your experienced intervention style and compare your hand pressure to a non-invasive touch

Many practitioners were unwittingly programmed to tune into and over use their own proprioceptors and kinesthetic channels. It's why we push too hard and invade tissues too deeply. It's why we consider cupping and dry needling to be suitable detoxification methods instead of realizing all layers of tissues are blocked at the lymphatic vessel level. We dam up fluid pathways with compressive touch that would send more vibratory signals if flowing freely. Tune into the receptors that can detect discrete signals, such as the movements of fluid under skin and fascia. Merkel and Meissner's discs make for better-listening hands to detect these subtle regions of tissue communication. Your "other" touch cells can become highly trained by the experiences affordable in these bodywork methods.

How patient are you waiting for a tissue response? How quickly will you jump to a solution in your treatment time based on all previous client encounters, assuming you know the problem? Waiting and allowing time for your hand receptors to receive the story from the client, you will pick up much more cellular communication. Too much pressure movement from your hands stimulates deep receptors that drown out the voice of the skin receptors. If you press too deep, you activate your proprioceptors and deep receptors, and essentially eliminate the communication of the client's cutaneous autonomic nervous system. That old concept belongs in the world of "doing things to clients," but in the world of Bodywork for Sensory Integration, the idea of self-correction of cells, tissues, and fluids requires the opposite kind of touch approach. It requires listening hands that help resolve protective retraction and not forcing a need for recovery time of microvasculature.

What are your stress levels when you approach a client? Does the intimate closeness of personal space make you nervous? Do you have any pain or discomfort in your hands or body that might interfere with the vibration transmission of sensory information? If so, your sympathetic continuum is leaning towards vasoconstriction and tension, which will transfer to the people you treat. If you're stressed and vasoconstrictive, you will likely feel less of your client's tissues and more of your tissues. Discrete sensory skills are best developed with maximal blood flow in the microvasculature of your own hands. Your therapeutic intention will reflect what's in your heart and mind, transferred via the vibration caused by your psyche and emotions. Get your own work done, and your hands and mind will be as clear as possible to prevent transference and countertransference.

The whole point of manual therapy is to assist the client closer to, or rooted in, the parasympathetic state, not to have a wrestling match with your own issues. The intention of your touch is to not invade and excite C fibers into SNS of vasoconstriction and protective retraction. Your objective should be to develop a listening touch, a waiting touch, trusting that the client's body will inform you with the knowledge for a successful session.

The person receiving touch decides if the touch is "just right."

Proprioception / Kinesthesia

Sense of body and limb position, with purposeful movements

Bodywork objectives: Balance tensions of whole-system fascia and interstitial layers woven through and around musculoskeletal system. Optimize tensegrity of this fluid-filled human body by assisting osseous-fascia to become free of retained tensions. Optimize structural symmetry and reduce mechanical strains that can contribute to retained primitive reflexes. Optimize the tone of the "background" sensory feedback mechanisms.

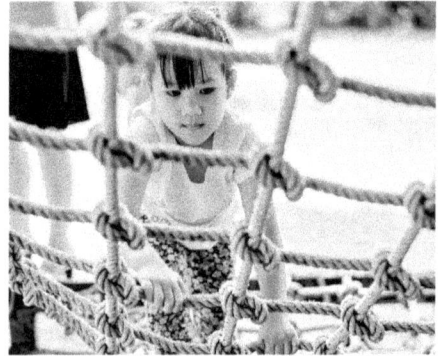

Sliding scale of sensory perception, adjustment, and adaptive response

Developmental Trends and Behavioral Considerations

Proprioception is the sense of body position and pressure. Kinesthesia is the sense of movement of body parts. The awareness (consciously and subconsciously) of how we position our body and what our limbs are doing develops by these duality-sensing systems. Proprioception and kinesthesia are fired during movement activities or by body parts being moved, but proprioception can also be activated without body part movement. Joint compression is the most common activation, so the brain feels more from the limbs. Posture and movement reactions occur in response to this system's stimulation.

In-utero experiences begin the developmental stage with pressure from amniotic fluid and then from compression of the womb walls in the later stages of the pregnancy. In the first two trimesters, the fetus has plenty of room to wiggle and react within a process of the rapid cellular division leading to growth in size and body part differentiation. The different types of sensory receptors work in concert with the result of a refined sense of touch in the tactile system.

Proprioception and kinesthesia require years of maturation and are typically stimulated and developed after birth through various stages of sensory-motor play. Both gross and fine motor tasks and sensory play, enable the brain to categorize and remember the types of sensations.

Humans also associate emotions and motivations for movement and body postures based on success and ANS responses. The natural course of primitive reflexes, stimulated by touch and proprioception input, activate the musculoskeletal system to work a baby out of a tight fetal position. Reflexive movements meet the resistance of the supporting surface as body parts push, kick, and flail. The stage of body control and movement mastery is called the sensorimotor stage of development. The increased active control over muscles inhibits each reflex, repeated use of limb and body placements and movement transitions are all driven, and movement memories are recorded through the proprioceptive and kinesthetic pathways. Maturation is further reflected by mastery of growing gross and fine motor skill sets, balance and agility, and purposeful use of complex movements known as praxis.

Adaptive Responses…

…allow for posture and balance to be effortless, autonomic, and able to make discrete adjustments. These functions both consciously & unconsciously.

- Reflex activation – response to automatic inner drive to move and become upright; gain voluntary control following reflex integration
- Raise head in gravity; body follows head
- Head alignment to shoulders (mutual work with vision-vestibular-mechanoreceptors)
- Postural alignment and symmetry
- Balance and equilibrium– with eyes open and closed; with movement imposed into body and with own generated movements
- Regulate muscle tone (muscles are ready to make an action or reaction)
- Coordination of body sides; locomotion
- Adjust mechanical forces for balance & posture
- Adjust body if mechanical compressions threaten circulation and homeostasis
- Master various skills of praxis & motor planning

These two sensory systems work simultaneously and harmoniously. Proprioception is vital for the body's stability, and kinesthesia drives and modulates muscle activation, contraction, and matching the physical demands of the task. This sense helps a body not fall when the limbs or head start engaging in an activity. Kinesthetic feedback is critical to know how hard to kick a ball, push a leg through the fabric of jeans, or exert the right amount of effort to push open a heavy (or lightweight) door. These two sensory mechanisms also work with tactile receptors to develop the necessary skills for grading motor action and pressure. These receptor teams learn and develop the body's ability to sense movement, prepare for a position change or adaptation, and determine if the surface they are on is stable or unstable. This creates "background" feedback for efficient sensory integration.

Neurodevelopmental Therapy terms of stability and mobility define the beauty of the orchestrated balance between posture and movement. Stability can be considered an adaptive response to proprioception, and joint mobility is an adaptive response to kinesthetic information.

Motor responses fed by this dual-sense system have much subcortical processing. As the body moves, walks, climbs, runs, and jumps, the child does not have much cognitive thought behind it. These skills seem to happen naturally and without a need for formal education to develop. Though movement development looks autonomically controlled, the cortex is quite active but not in the "thinking" part of the brain. The information reaches the motor cortex from the thalamus and is then scanned by the premotor cortex. These areas are behind the frontal lobe, where thoughts are generated. Much more information is processed in the cerebellum for the synthesis of coordination. Vast information is processing body data, modulating an autonomic response, and preparing a response or adjustment.

Functional use of proprioception and kinesthesia is measured by skills of postural control, balance, and coordination between the right and left sides of the body (agility). This included arms, legs, and eyes, with coordinated truck rotation to support in sync limb action. Eye tracking is a skill gained from multi-sensory processing, including proprioceptive and kinesthesia information from skeletal eye muscles activated by vestibular input and baroreceptors from gravity. Activation of proprioception and deep pressure receptors are known to inhibit sympathetic nervous system reactions and painful reception. It is why we rub the site of injury when we bump our legs or when exercise can induce a sense of well-being and positively affect our mood.

A higher-ordered "use" of these systems does develop as the child matures to engage more cognitive, executive, and even motivational and creative parts of the brain. The skills of "praxis" measure how well the body is processing the vast amount of information from these two sensory systems. Studying this complex human trait of praxis is the foundation of Ayres' Sensory Integration® theory and intervention methods.

Various skills of praxis include:

- **Motor planning** – the ability to spontaneously conceive, organize, and execute the appropriate force and angles of movement. Motor planning is based upon the current sensations and memories of previous experiences to succeed in novel activities. Example: moving body to get up on a bike and propel it.

- **Ideational praxis** - forming an idea of what one wants to do and how to move to complete a [novel] task, generalized from other experiences. Example: one thinks of a bicycle, expresses an understanding of what to do with it, and then plans the initiative to ride it.

- **Sequential praxis** – timing and rhythm for putting the motor action components together to complete the task. Example: hold the bicycle while footing the kickstand, steady the bike while raising a leg to straddle the seat, placing one foot on a pedal and balance it with the other, push off that pedal at the right moment to generate forward momentum, push in the coordination of two feet pedaling fast enough for speed, steady and guide handlebars with hands to steer the front wheel. Then reverse pedal action to activate the brake.

- **Constructional praxis** – using body parts to build, manipulate, draw, copy, and assemble items with an internal conceptualization of the finished product. The art of creating such things comes from the deep sense of the proprioceptive-kinesthesia feelings that help one know left-right, up-down, in-out, stable-unstable, symmetrical-unsymmetrical. The body developed these deeply entrenched "knowings" through the stages of praxis-building experiences. Example: building structures with blocks or sticks; successfully riding a bike. Either copied or generated from their imagination, praxis generalizes into self-driven learning and vocational aptitudes moving forward into adult life.

Praxis skill development has many pediatric therapists concerned over how much time children spend with screen time, replacing playground and imaginative play. Praxis skills allow children to efficiently plan and engage in motor activities for fun. Repetition and practice then help refine the motor skills they are engaging. Praxis abilities translate to knowing what to do and when to do it. This is an identified life role within the occupational science of human existence: fulfillment in meaningful movement tasks.

Praxis skills are the ability to generalize a movement skill from one activity to a different action (the bike riding, then relying upon similar movement abilities to ride a surfboard though the sensory challenges have varied). You'll know it when you see it when a well-coordinated person utilizes praxis skills and can learn dance steps or athletic moves quickly. Praxis skills are also a matter of degree, with different aptitude levels indicating a human trait that is very complicated to dissect, understand, and remediate when problematic. Occupational therapists trained in sensory integration principles generally are the most qualified to address praxis. The success of praxis development literally can mean the difference between becoming fully functional and independent adults with skills at organization and timing and the ability to multi-task with daily tasks one's hands and body have to complete (dressing, bathing, grooming, keeping the order of room or apartment, and even time management).

Body praxis is often overlooked, misunderstood, and misdiagnosed. Praxis is the "first step" of practice: the *first* time an action is done (kick a ball, ski down a slope, stay on a swing, or pedal a bike). Much more complex than just coordination of body sides, the brain forms a plan of action in response to what has been felt from the body and continually adjust and refine. Also, synthesizing all sensory-motor information reflects higher and more complex perceptions and intrinsic motivation to engage and perform beyond reflexes, engrams, and practice of basic skills. Oral praxis is the use of tongue movements – speech is an example of oral motor praxis, as is clicking the tongue, using the tongue to lick lips, and lateralizing the tongue in the mouth are an example.

The anatomical nature of fascia, now recognized as a sensory organ, may play a role in the background feedback mechanism that supports praxis skills. How the motor and premotor cortex coordinate information with the vastly unappreciated cerebellum remains unknown. To this day, proprioception and kinesthesia physiology are poorly represented in neuroscience because the best way to study the nervous system still requires a person to lay motionless for a brain scan procedure. When science can create a method to study the brain while a person is moving, we can all gain a much wider appreciation of the nature of sensory integration and the therapies utilized to enhance it.

New Thoughts on Muscle Tone

From a sensory processing perspective, muscle tone is a primitive response to the brain registering body position and movement sensations; followed by a signaling to ready muscles fibers to react or move. Typical babies do not have developed muscles, but they have muscle tone, indicating a state of readiness to begin moving. The pressure induced by gravity creates tension and requires a physical response to increase the arousal of the musculoskeletal system. Without that increased sensory-motor awareness, the child one is left with a state of hypotonia. Practitioners concerned with muscle tone only have subjective ways to measure it. We still don't fully understand what normal body tone is, how it is derived through specific neurological pathways, or how to quantify optimal ranges.

Following bodywork, we consistently observe a relaxation of muscle tone without a child falling into hypotonia. Bodywork observations show a balance between normal tone of readiness to move and reflexive muscle tension. Perhaps muscle tone has something to do with the tension and tensegrity of the fascia cells surrounding muscles and nerves. Our observations have been that their muscles become more vibrant and vigorous after coming around from their treatment session. When working with babies, parents routinely comment that their infant feels less tense and rigid and more like a baby should feel in their hands immediately following their bodywork sessions. Bodywork shows assistance towards balancing the body's tensegrity of soft tissues. Our experience offers a theory that the fascia matrix within tensegrity balance acts as a possible background regulator of tone.

Anatomical Considerations

- Muscles spindles – twisting receptor fibers, embedded and wrapped around striated muscle cells; activated by lengthening the muscle (and surrounding fascia) tissue
- Golgi tendon organs – embedded at the junction where tendons project off muscle mass
- Joint receptors – embedded in joint capsules
- The vestibular system is also considered a proprioceptive-kinesthetic receptor site
- Fascia matrix of connective tissue that surrounds and suspends every part of the body

The general understanding of this combined sensory system begins at the cellular receptors. The **muscle spindles** are activated by a stretch of the belly of the muscle, implying that the muscle is moved, has moved, or something moved against. The Golgi tendon organs detect the force of the muscle contraction through the tension put upon the attaching tendon. Joint receptors detect compression, twists, and torques at the union of two-bone articulating.

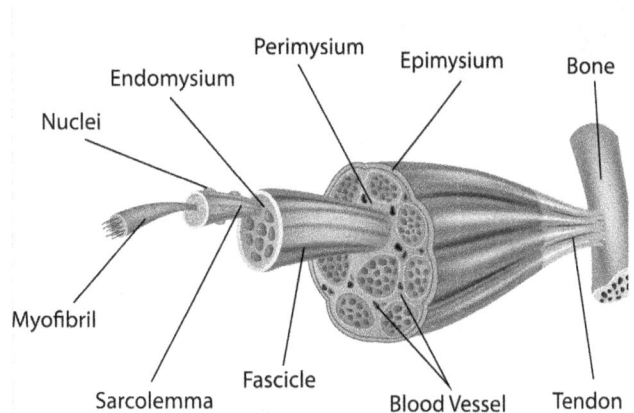

Connective tissue suspends all striated muscle mass. Endomysium, perimysium, and epimysium comprise the layers surrounding muscle fibers and blood vessels. Bone, tendon, and ligaments are extensions of the fascia matrix as well.

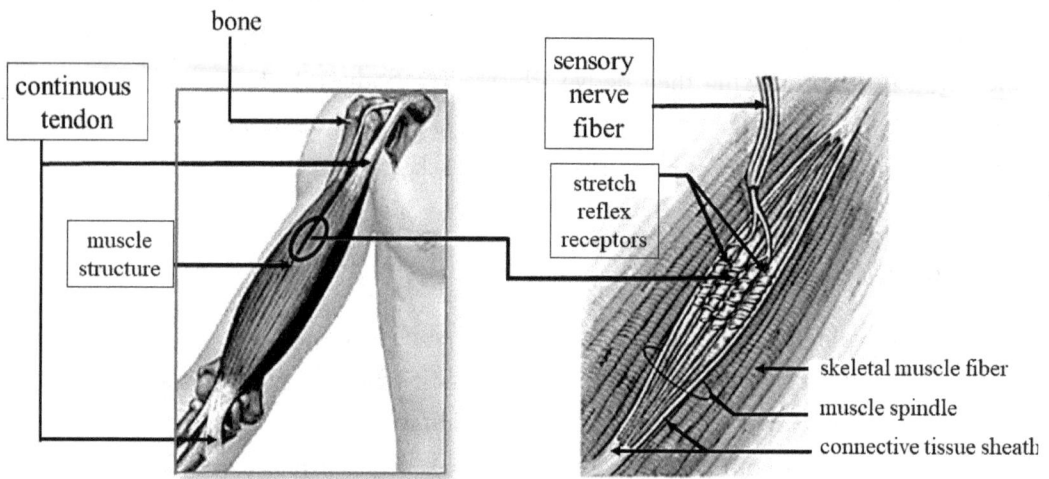

Muscle spindle – stretch receptors of muscle tissue

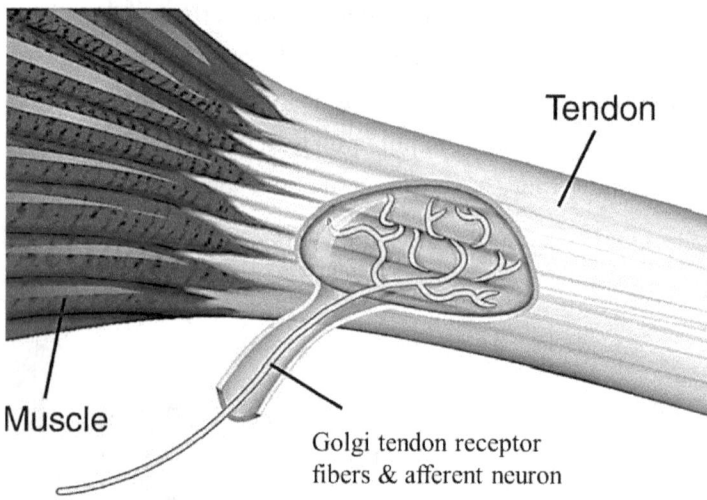

Golgi tendon receptor
fibers & afferent neuron

Golgi tendon organ woven through tendon

As fascia science accumulates, we understand an even greater appreciation of the vastness of the background structure. Tendons are connective tissue projections emerging from the weave around and through every striation of muscle tissue. Where tendon tissue emerges from concentrated muscle tissue, receptors of proprioception and kinesthesia can be found densely packed. These receptors detect the tension and tensegrity of the fascia network and the muscles it envelops. In addition, every articulating joint has pressure receptors where ligaments connect bone to bone, and these cells send sensory information about the position and pressure of the skeletal joints.

Pathways for proprioception and kinesthesia, sensing pressure and movement, travel through myelinated afferent neurons running peripherally to the spinal cord and up to the medulla. There are four separate spinal pathways: ventral, dorsal, rostral, and spinocerebellar. The first three project information up the spinal cord and brainstem to the thalamus and onto the somatosensory cortex. The latter is a pathway of the same information directly from the body's periphery to the cerebellum. Neurons are not physically connected but communicate via neurotransmitters secreting into synapses or "gaps" between communicating neurons. The cerebellum, whose full function remains unappreciated, regulates muscle tone as readiness to move and coordinates muscle contractions for coordination and equilibrium so that balance, posture, and timing of limb motions occur. The ratio of sensory receptors of these two systems significantly outnumbers all other systems.

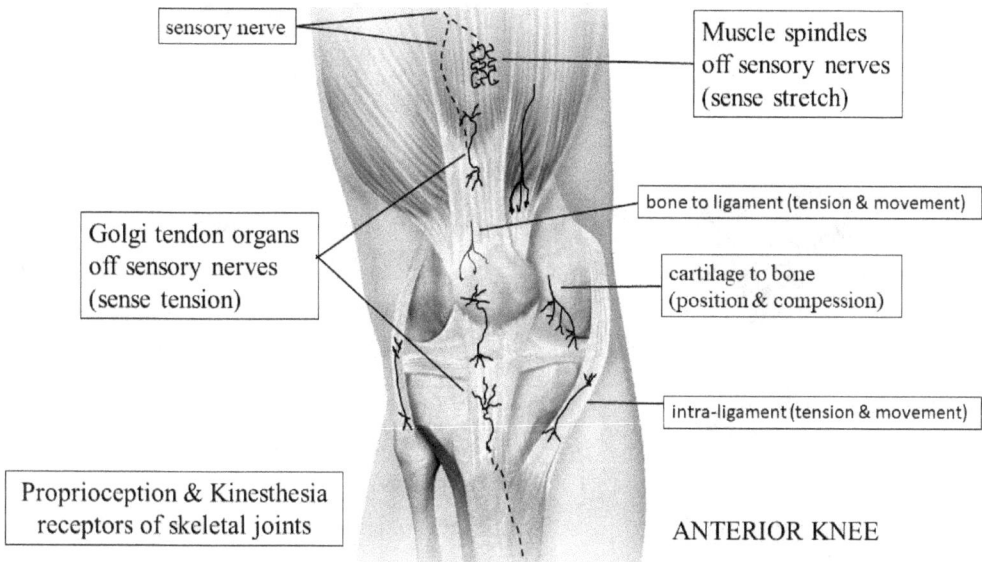

Proprioception & Kinesthesia receptors of skeletal joints

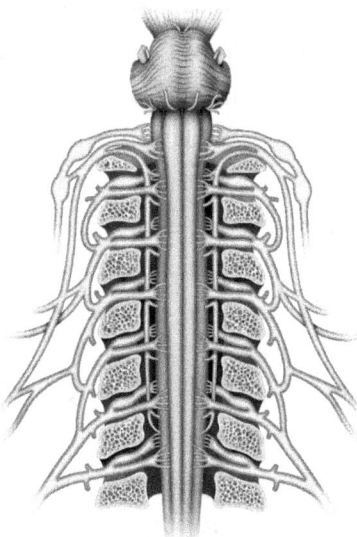

Sensory spinal nerves enter the spinal cord on the dorsal side (and motor nerves exit on the ventral side). Dorsal columns with the cervical spine. Thoracic nerve roots off the spinal cord project through the dura mater between vertebrae at dorsal rib attachment sites. Spinal alignment can create structural stress at any nerve root, as can rib alignment. Remember that bone is an extension of fascia, and the periosteum of any bone can hold adverse tension. Treating the connective tissues of the entire rib cage has a global effect on the whole thoracic cavity and all functions housed within and adjacent.

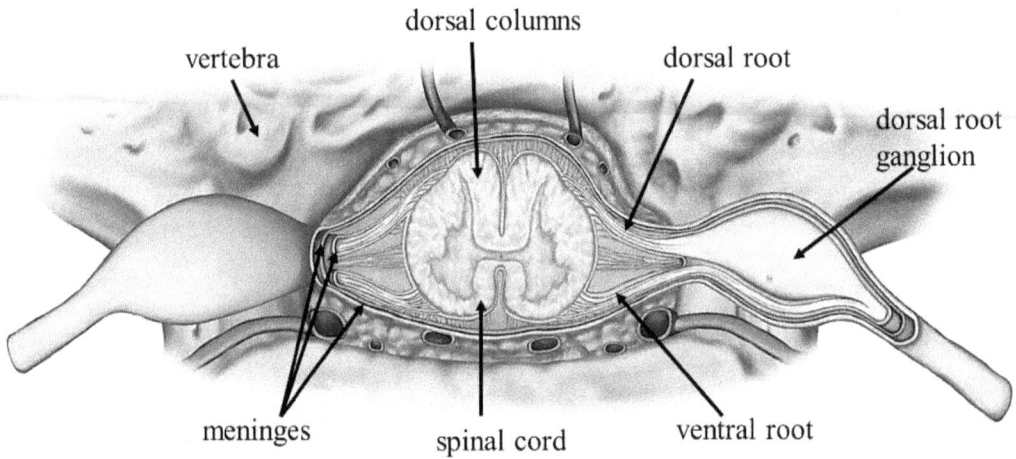

vertebra — dorsal columns — dorsal root — dorsal root ganglion — meninges — spinal cord — ventral root

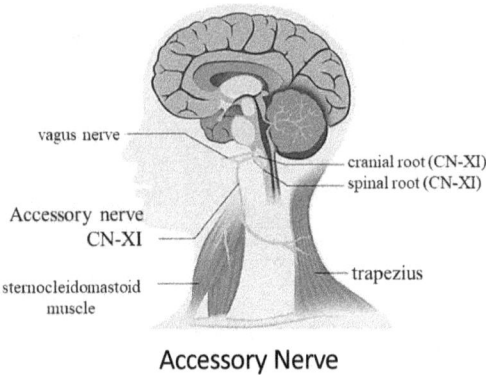

vagus nerve
cranial root (CN-XI)
spinal root (CN-XI)
Accessory nerve CN-XI
sternocleidomastoid muscle
trapezius

Accessory Nerve

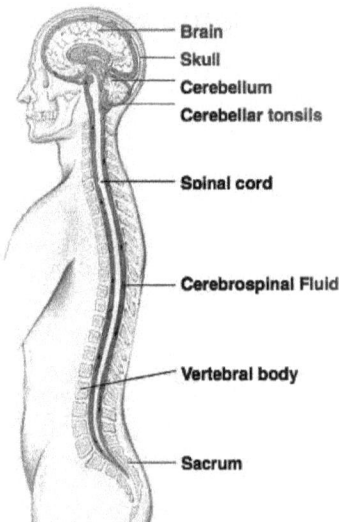

Brain
Skull
Cerebellum
Cerebellar tonsils
Spinal cord
Cerebrospinal Fluid
Vertebral body
Sacrum

Craniosacral system within the central nervous system. Wiki Creative Commons 4.0 international license. Moodle Skills BIO168035

A major component of sensory feedback and monitoring of head position on neck. A chief regulator for posture, balance, coordination, and comfort with adjusting to movements. The location and pathway of the Accessory nerve (CN-XI) is directly influenced by the tension, tone, and fascia compression at the occiput-atlas junction. The accessory nerve innervates the sternocleidomastoid and trapezius muscles, contributing to primary control over head and shoulders tension patterns. Furthermore, the CN-XI holds a similar function to the vagus nerve carrying motor innervation for the pharyngeal constrictors, larynx, and soft palate muscles. Combined with the other cranial nerves to the tongue, the accessory nerve is involved in tongue control related to head and neck control. Motor responses are the result of proprioceptive and kinesthetic information. Upper cervical compression or trauma into occiput can compromise

Neuroscience tends to appraise the number of sensory receptors in each system. For example, trillions of proprioceptive and kinesthetic receptor cells have been estimated.

Counting receptors would be enormously challenging, with a large margin of error. Nonetheless, it is accepted that tens of thousands of special receptors (eyes and ears) are compared to tens of trillions of mechanoreceptors in the somatosensory systems combined. That concept alone should convince the reader that somatosensory systems should not be taken for granted. These systems keep our body moving, upright, stationary, stable, and in control of limbs and distal body parts. The special senses work best under such stability.

Newer interpretations of how the human nervous system works, sending signals, have less to do with the number of cells and more to do with how much signaling one body has available. The number of neurons going to the brain does not indicate how many signals are in the body. Sensations can happen locally for a spinal-level response. Perhaps some sensory-motor responses don't even make it to the spinal cord. Maybe some very local adjustments occur, and that is where considering the fascia network comes into play. Also, unconscious proprioceptive signals run from the periphery via the spinal cord to the cerebellum rather than the thalamus, unlike other sensory information.

Bodywork Treatments that help Proprioception & Kinesthesia function

- Peripheral branches of fascia matrix suspending musculoskeletal stability regions: shoulder and pelvic girdle. CST and MFR to ensure that periphery is free of compressions, distortions and asymmetries of myofascial structures (See pp. 201, 208-212, 216)
- Acupressure where muscle and tendon meet for sites of held tensions (See pp. 192, 194, 234)
- Central core structures to endure freedom of spinal column and dural tube compressions, body tensegrity, cranial base integrity, dorsal columns at length of cord and at cranial base (See pp. 213-215)
- Freedom of compression of sensory and motor cortex resting under the sagittal sutures, often a site of compression from frontal bone posteriorly from birthing. (See pp. 202, 227)
- Cerebellum and thalamus communication (intracranial membranes). (See pp. 227-228)

Interoception/Enteroception/Visceroception

Bodywork objectives: Promote deepest parasympathetic state to increase personal awareness of the ANS continuum. Assist recognition of tone differences in thoracic and abdominal organs and vasculature environment and associate with sense of tension to well-being and mood. Assist self-correction of tone of enteric nervous system mesh surrounding visceral organs & vascular concentrations.

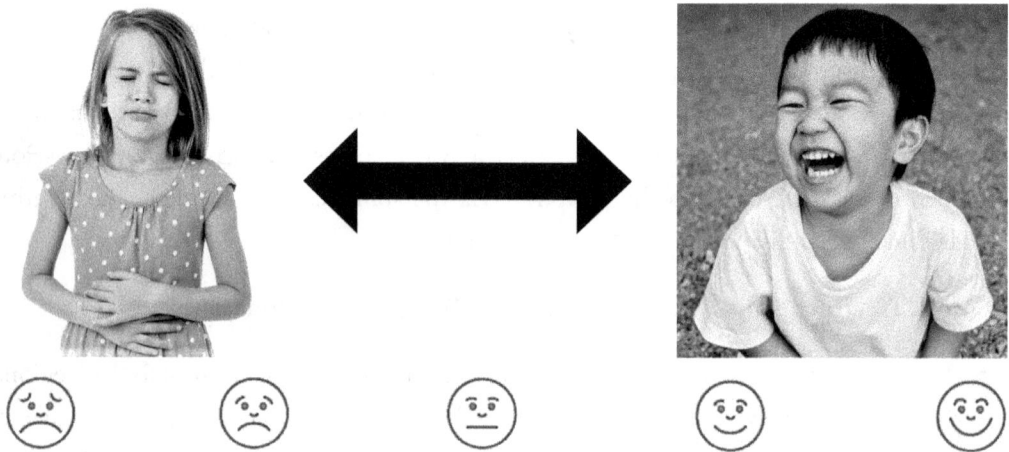

Sliding scale of sensory perception, adjustment, and adaptive response

Developmental Trends and Behavioral Considerations

Interoception is the basis for independence in self-regulation. Interoceptive behaviors create or notice a change in one's inner being; attaching meaning to physiological activity. Different categories of receptor cells of background sensation give feedback on how one is feeling (happy, anxious, or comfortable). With experience and practice, a child learns to match feelings and emotional reactions to these internal sensations, even learning a language to describe such feelings. This sensory network gives valuable information about the situation (level of safety, novelty, familiarity, emotional memories of previous experience). The physiology of interoception is automatic. The reaction, coping, and adaptive responses involve higher brain centers.

The capacity to self-regulate begins with settling into the earliest routines as a baby. The young child then gains abilities to inhibit sensory overload, manage physiological reactions to sensations, and focus attention on tasks. The physiological processes of waking or switching from sleep, adjusting to sensory alarms or reactions, and managing the biological need to access caloric intake produce alerting and sympathetic responses. The ability and speed of returning to a calm-alert-attentive state or restfulness measure the body's adjustment to such stimulating events.

As children, we learn and react to fullness and pressure in organs. Adjustments to the stretch of a full tummy, the churning of peristalsis, or the passage of food (milk) through the gastrointestinal tract, is a function of these interoceptors. We learn about hunger and how it feels to become satiated. Physiological information signals when a stomach is empty and needs food. Then interoception gives feedback on how it feels after the meal, when it's time to stop eating, if the food was enjoyed, and if emotional needs were met during the meal.

When watching a scary movie and a tightness happens in the gut, the child might react by seeking out a cuddle with a parent or pet knowing that closeness could alleviate feelings of fear. If the hug is just right to calm that fear, that ominous feeling disappears and is replaced with a better feeling. The emotions set off by a hug from a visiting relative align with how we feel about that relative. If we enjoy the hug and feel love and acceptance, touch sensations match emotional sensations.

Interoception maturation includes ease of peristalsis and movement of food/chyme/fecal matter through the digestive tract with optimal tone, flexibility, and degree of any structural resistance. We learn about our bladders first and then bowels from pressure indications; fullness means it's time to evacuate. We know the internal relief of emptying a full bladder or from a complete bowel movement. Additionally, optimal respiration and maximized breathing mechanisms of the rib cage (allowing full expansion of lungs) maintain an average body temperature, orchestrating physiology and behavior to circadian rhythms. When a cool breeze sets off goosebumps, a child might notice and grab a jacket, and then the sensations will signal if the jacket is just right for the change in the weather.

After kicking a soccer ball or bike riding, the sensations within a muscle belly register the level of lactic acid and other metabolic waste. The interoceptors give a sense of how exercise felt—enjoyable or detestable, invigorating or fatiguing. Feeling the difference between moving and standing still helps a person learn the concept of "waiting" while learning the rewards of delaying gratification.

These are all examples of how the quality of life is affected by this complex sensory system, only recently evaluated and considered in neuroscience. The connection between interoceptive awareness and emotion regulation is supported with empirical evidence. These background sensory systems are believed to be yet another complex system designed to maintain homeostasis while at the same time enabling alertness & attention to personal reactions to the internal environment. Self-awareness is developed by the degree to which the child learns to recognize the signaling of these sensations. The neuron network arises from distinctive physiological systems and not along the conventional-known nerve pathways.

These receptor cells are unmyelinated and free nerve endings, indicating rapid signaling. The fibers project to lower and deeper brain centers but not the sensory cortex (brainstem, insular and cingulate cortexes). Therefore, conscious awareness is not a feature. Instead, innate or instinctual awareness is the adaptive product.

Unlike proprioception/kinesthesia, which has traditionally been believed to be the system for understanding our body position, these interoception networks connect with motivational and emotional centers. The insular cortex creates an emotional connection with taste, reactions to visceral sensation, and mood and motivation regulation of the autonomic nervous system. The cingulate cortex has been shown involvement with higher executive functions like feelings associated with making mistakes, impulse control, decision-making, and empathy. These cellular networks exist to receive information and eventually communicate to all systems of the inner nature, both for survival and thriving.

Self-regulation is the behavioral manifestation of a system that readily adjusts on the continuum of sympathetic responses to both novel and stressful events with a return to an established level of parasympathetic recovery. Noticing how the body feels moment by moment is a motivation for purposeful, self-regulatory behaviors that is reinforced through maturation processing. Most of these receptors are baroreceptors, thermo, and chemoreceptors. Located essentially, everywhere, most treatment is focused at vascular concentrations (thoracic cavity, mesentery, and head), and nerve plexuses.

Unique types of sensory endings are baroreceptors, thermoreceptors, and chemoreceptors that scan the immediate cellular terrains. These cells act as gatekeepers and monitors, and contribute to the basic function of the autonomic nervous system. Different categories of these nerve endings rest within walls and tissues of organs, as well as in all vessels transporting fluids. The sensory function of this system is detecting fullness, hydration levels, and chemical concentration. They are located throughout the body, in all tissues, structures, and organs. Though much remains unknown, these categories of receptors include:

- **Visceroception** refers to the perception of internal bodily sensations triggered by nerves in the visceral systems: cardiovascular, respiratory, gastrointestinal, and urogenital.

- **Enteroception** is the sensation received from the hollow visceral organs; it is the sense of fullness or emptiness. Cells are both baroreceptors and stretch receptors detect pressure changes in walls of vasculature monitoring blood pressure and in the lungs to sense lung expansion. Throughout the digestive and elimination systems, organ expansion is sensed.
- **Interoception** has been defined as "homeostasis sensing" and is an orchestra of sensory reception that detects fullness, pressure, and pain. Awareness of such states can be reflected by both conscious and non-conscious behaviors and emotional associations to the resulting perceptions, especially when any need to restore homeostasis is met in a timely fashion or not. Internal organ function such as heart beat, respiration, satiety and states of autonomic nervous system are all included in this realm. Stimuli that are detected by interoceptors include blood pressure and blood oxygen level, organ distension, hunger and satiation, pain, and pressure.

For the sake of simplicity, "interoception" will be used as a collective description.

Self-Regulation within the Circulatory System – measured by Vagal Tone indicators

Self-Regulation within the Visceral Environment – measured by Enteric Tone indicators

The effect of interoception can be measured within the circulatory system by vagal tone indicators and the visceral environment by enteric tone indicators (through palpation). Vagal tone represents this balance of the sympathetic and parasympathetic continuum of the ANS and the independent regulation of the enteric nervous system. In medical situations, vagal tone most often refers to respiratory and circulatory resistance. One can objectively measure vagal tone by heart and breathing rates and heart rate variability. The heart accelerates with inhalation (to circulate the incoming oxygen) and slows with exhalation (to preserve oxygen). That basic physiology represents the dynamic balance between PNS and SNS with every breath. The more significant the difference between inhalation and exhalation heart rates, the greater the HRV ratio, which gives a higher value of vagal tone. The tone of the vagus nerve presents activation of the parasympathetic nerves. The higher the vagal tone, means the body relaxes faster after a stressful event. Conversely, the higher the vagal tone, the more relaxed, flexible, & pliable the visceral organs and fluid vessel masses become.

Clinical experiences from bodywork inform us that there is such a thing as optimal tension and tone in the visceral cavity and organs, also representing the ANS continuum. The autonomic nervous system creates a "just right zone" between sympathetic tone and a parasympathetic state with quick physiological adjustments. We have assumed that babies (and the adults they grow into) can reach minimal retraction (palpated as tension and resistance) within their tissues if allowed to relax and work at becoming calm and having self-control. However, unresolved tissue retraction in the fascial matrix can perpetuate a degree of the sympathetic state. The retracted fascia surrounding an organ feels like there is little to no freedom of movement in the surrounding space. The organs can feel pressed or drawn together with a lack of ease and glide with neighboring structures. Organ structures can feel pressed or drawn together in a chronic vasoconstrictive state, reducing the flexibility of the peritoneum surrounding the organs. All organs, especially the highly vascularized like the mesentery, are more flexible in the parasympathetic state. The rest and digest state suggests that the mobility of organs allows for better peristalsis and vasodynamics in physiological adjustments. Palpation will yield a subjective assessment of the quality of this physiological balance.

Sensory modulation and self-regulation are acquired behavioral control skills, reflecting the ANS resiliency. The palpable tone of ANS innervated organs and vessel fields can be assessed and treated directly by trained hands. Structural balance in these unappreciated areas can profoundly and often immediately change the child's basic mood and emotional adaptability. In a state of dysregulation, the organs and vessel fields can be compromised. Body signals also serve as an alert that internal body balance is off and motivates the person to do something that will restore the internal balance and help the body feel more comfortable. In other words, noticing how the body feels influences purposeful self-regulation behaviors. Visceral techniques using the approach of craniosacral therapy combined with acupressure, lymphatic drainage, and a technique called Responsive Hold (See pp. 74-75), have proven to positively affect moods and self-regulatory behaviors.

The greatest novel contribution to sensory integration and wellness treatment is that *Bodywork* directly treat the structures that regulate the ANS. *Bodywork for Sensory Integration* considers all structures housing interoceptor groups as contributors of ANS adaptability, requiring physiological balance. Self-regulation is a byproduct of structural motility and mobility of organs, blood and lymphatic vessels, and nerve plexuses innervation sites.

Organ and vessel regions have been organized here in this manner for ease in understanding the focus of the realm of direct treatment to the ANS. (For more information, see p. 283 on Self-Regulation).

Adaptive Responses (a sampling of behaviors)

- Maximize breath capacity & rib expansion, lungs, & pleura; flexibility of respiratory diaphragm
- General or specific discomfort; tummy aches, headaches, thirst
- Feeling and reacting to cold, hungry, lonely
- React to sensations that stimulate hunger pangs

- Recognize & tolerate stomach expansion and contraction; recognize empty or full feeling
- Recognize and react to signals of digestive tract motility; toileting signals
- Recognize when awake and when tired; know when it's time to sleep
- Know when the body needs movement and when it needs rest
- Feeling sensations of posture and agility success (or failure) and self-worth at achievement

Anatomical Considerations

Bodywork for Sensory Integration is focused to promote self-regulation by facilitating parasympathetic activation and relaxation in these visceral organs and vasculature concentrations, along with key lymph node pathways:

- **Thoracic, Respiratory, and Circulatory Structures**
    - Heart & lungs suspended in mediastinum (membranous partition); pulmonary artery mass
    - Rib cage & all intercostal muscle structures layered with connective tissues
    - Respiratory diaphragm with core ligament and suspending fascia
    - Thoracic dura mater with nerve root projections
    - Sympathetic ganglion chain anterior of the spine deep within thoracic cavity
- **Abdominal Organs and Visceral Structures**
    - Perimeter attachments of respiratory diaphragm
    - Liver & stomach draped by respiratory diaphragm & peritoneum; suspending ligaments
    - Biliary system & common bile duct; investing peritoneum tissues
    - Lesser curve of stomach; innervation by celiac plexus
    - Enteric nerve mesh within mesentery and peritoneum layers
    - Digestive sphincters (seven) and investing fascia
    - Kidneys and adrenal glands; with vascular and organ peritoneum tone
- **Umbilicus**
    - Remnant ligaments of umbilical cord blood vessels
    - Convergence of major fascia attachments at thoracic/abdominal junction
    - Intersections of deep front line fascia band
    - Adjacent abdominal circulatory and lymph chains (cisterna chyli, deep visceral nodes)
- **Lymphatics- Interoception of elimination & detoxification**
- **Cranial base and cervical (occipital-cervical base)**
    - Occipital condyles surrounding/forming foramen magnum
    - Associated nerves, blood & lymphatic vessels, muscles, investing fascia layers
    - Hyoid attachments of vast amount of soft tissue structures

Thoracic, Respiratory, and Circulatory Structures

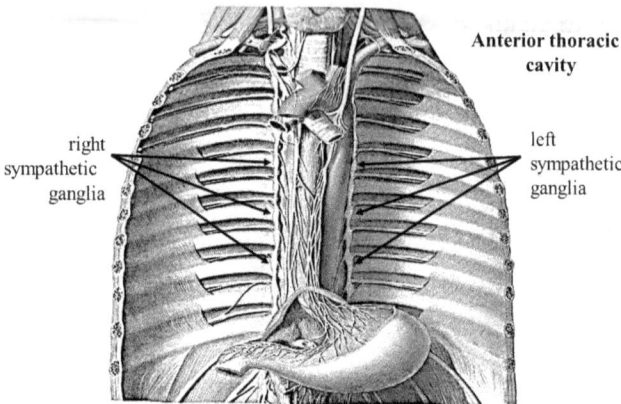

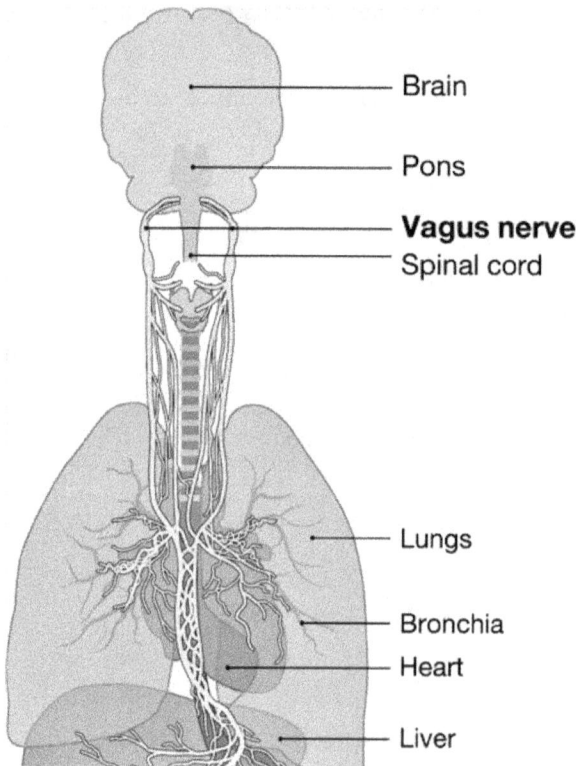

Anterior thoracic cavity

right sympathetic ganglia

left sympathetic ganglia

Brain

Pons

Vagus nerve
Spinal cord

Lungs

Bronchia

Heart

Liver

In addition, but equally important, the paired sympathetic ganglion chains located on the anterior surface of the thoracic spine play a significant role in the resiliency of the respiratory system and overall ANS regulation. Rib compression or pleura torsions can facilitate one or more ganglia into a perpetual sympathetic firing, keeping the respiratory system structurally away from a homeostatic center. The heart and lungs are suspended within the mediastinum fascia (membranous partition), anchoring to the ventral rib and sternum surfaces. Vagus and phrenic nerves meander from the brainstem, having much investing myofascial terrain to travel.

Branching of the cardiac and pulmonary plexuses is concentrated around each organ. Compression or tension anywhere in these pathways can increase stress at the organ innervations. Well-trained and experienced hands can palpate lines of pull from the nerve nuclei through these pathways. The pleura and pericardium connective tissues give the final suspension and protection to the heart, lungs, major artery & vein trunks, and bronchial tubes. Because of the tightly held relationship between the lung and ventral rib surface, these layers of fascia can be palpated to determine the quality of organ flexibility for optimal breath.

Furthermore, it has been discovered that the heart has its own intrinsic cardiac nervous system, which contains both parasympathetic and sympathetic interoceptors, forming a local circuit. Another "brain" within the body, this circuitry regulates cardiac rhythm and reactions to sensory-emotional events. It stands to reason that structural resistance within the thoracic cavity would be perceived by this network of interoceptors and convey tone of thoracic organs. The tidal volume of breath is only as good as the expansive qualities of every structural component of the thoracic cavity. Heart function is impacted by lung mobility and flexibility within the limits of the pleural tissues.

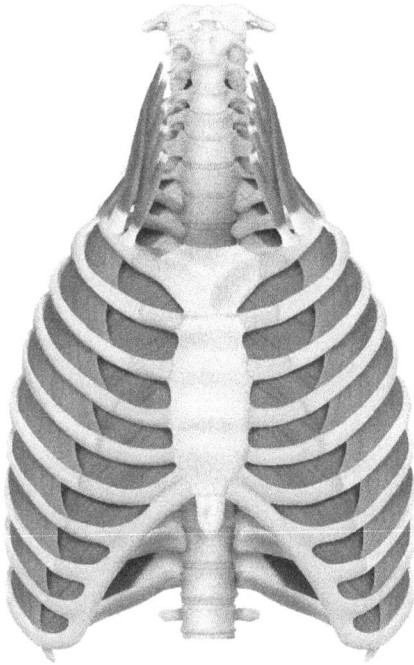

For the thoracic cavity, we begin by treating gross structures of the rib cage and the collective intercostal muscles, layered with connective tissues. The fascia influence upon the tensegrity of the rib cage is appreciated through listening palpation of the twenty-four ribs, sixty-six intercostal muscles, and all the woven myofascial and periosteum layers.

Rib flexibility motion hold functional implications for optimal posture, movement, and behavioral health. Respiratory diaphragm tensions can reflect the general tensegrity of the rib cage at the peripheral attachments. Unique kinetic patterns (of rotation) occur during inspiration and expiration. The rib attachments in the posterior have subtle movements of the adjacent vertebrae. Tidal breathing has been described in movement patterns from three distinct sections of the rib cage. Active rib motion should reflect the movement of these three sections. Rib motion motility has been shown to correlate to total lung capacity. (See more on treating rib mobility at pp. 329-330)

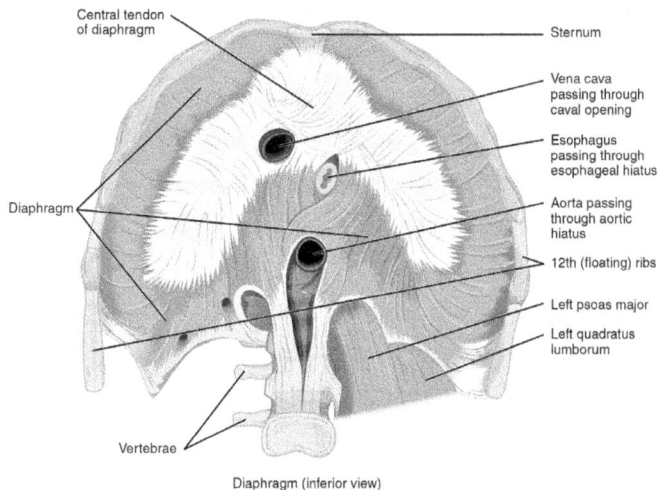

Diaphragm (inferior view)

Inferior view of respiratory diaphragm

In addition, the central tendon of the diaphragm can hold a great deal of tissue tension. At intersections of horizontal tissues penetrated by several vertically-oriented structures, the central tendon can be a crucial player in the structural readiness of ANS adaptability and resiliency. The esophagus, lower esophageal sphincter, aorta, and vena cava are some structures that had have adverse effects upon their ANS physiology by fascial restrictions within the respiratory diaphragm. The intersection of the deep front line of fascia is particularly interesting in *Bodywork for Sensory Integration* for how often tension here reflects a hindrance in self-regulation. Restrictions in the respiratory diaphragm's flexibility are rarely considered a factor in sensory modulation, self-regulation, or stress management interventions. Freedom from fascia torsion or restrictions is an assumption, but years of bodywork have revealed that retained restrictions at any diaphragm component can limit maximal breathing. Remember that the underlying organs (liver, stomach, spleen, kidneys) are suspended by vast investing fascia and visceral ligaments in the abdominal cavity.

Clinical application of the functionality of the rib cage within complex movements has historically been post-graduate learning by the interested practitioner. Though there is generalized knowledge that deep breathing can induce PNS activation, it is not standard practice to appreciate the structural readiness of maximized motility and mobility of the rib cage and thoracic organs.

Neurodevelopmental treatment training (Bobath method) included knowledge that rib cage mobility is linked to breath support in all activities: feeding, phonation, and postural control. Like the hyoid bone, the rib cage and its complexity remain undervalued for its role in stress management and self-regulation. Movement of the intercostal muscles is directly related to freedom of fascia motility surrounding and suspending thoracic cavity organs. Rib position can be affected by the pull of visceral fascia of both thoracic and abdominal cavities. Clavicles and scapulae are directly connected by suspension tissue to the upper rib cage. Their fascia and tendon or ligament suspensions can affect the ease of breath. Furthermore, movement of the tongue's floor is directly related to the hyoid muscle attachments and a great deal of investing fascia suspending dozens of structures on the anterior neck.

The stability of the rib cage during movement and speech can be reflected in the ease of breath coordination and core symmetry. Optimal breathing should occur during physical activities and speech production. If not, the sympathetic tone may exist in a constant state. The autonomic tension state affects voice production, pitch, and volume conversely with the motility and mobility of the rib cage. Though the inner layer of pleura fascia adheres tightly to each lung, the corresponding outer layer is connected with the interior surface of the rib curvature. So historically, intervention focus has been on musculoskeletal structures viewpoint for postural and movement symmetry. Muscles, tendons, and cartilage may be held in protective retraction, especially around the central hiatus of the aorta and vena cava. Little to no reference had been considered on the mass effect of vasoconstriction and vagal tone through lungs and mediastinum, especially in sensory modulation and self-regulatory behavioral intervention. Bodywork that is subtle, gentle, and non-invasive can assess and treat the structures that support breathing to improve this physiology. In turn, bodywork can positively affect the ANS behaviors of sensory processing.

> **The rib cage and its complexity remain undervalued for its role in stress management and self-regulation. It is a false assumption that thoracic structures are fully expanding within parasympathetic tone and flexibility when teaching deep breathing exercises.**

The circulatory system is composed of large, medium, and smaller vessels. The vast amount of vasculature in the human body is utterly mind blowing. The fascia's influence upon thoracic blood vessels is appreciated through thorough palpation of concentrated vasculature fields. Tuning bodywork into this level can have profound positive effects upon vasoconstriciton towards dilation, promoting ANS regulation. There are three concentrated vascular fields we focus on in *Bodywork for Sensory Integration*.

- Mesentery field: Woven through the square footage of mesentery fascia (~25-30 feet long by 6 inches wide in the adult body); approximately 20% of the circulatory system.
- Thoracic cavity: Obviously the cavity where the heart and primary branches of the aorta and vena cava are located, there is also volumous circulatory channels into and outward from the lungs. The rib cage tension has a direct effect on how well the lungs expand (raising oxygen levels) which cascades the ANS tone through these vessels.
- Brain: Between 400-700 miles (yes, miles) of blood vessels are packed into the human brain. The majority of them are capillaries so fine that the corpuscles squeeze through in single file. The smallest displacement within tissue release can have profound effect on circulation, enhancing the exchange of fluids.

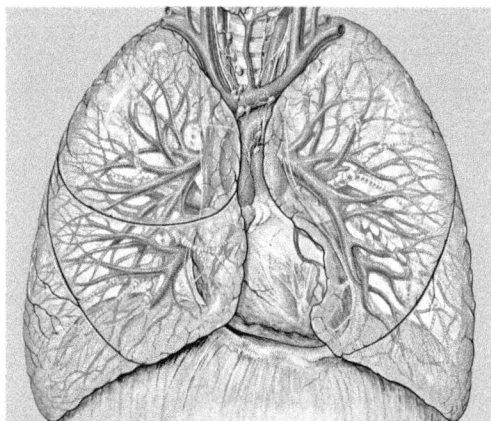

Patrick J. Lynch, Wikimedia Commons

Consider the anatomy of the circulatory system within the thoracic cavity. Arteries and smaller arterioles have proportionately thick walls of smooth muscle that adjust vessel diameter in response to blood pressure and flow demands. Arteries are lined by the smooth muscle, bonded with three layers of microscopic fascia cell layers. Endothelium, the inner layer, and an outer layer (adventitia) of connective tissue suspends arteries with neighboring structures in body tensegrity.

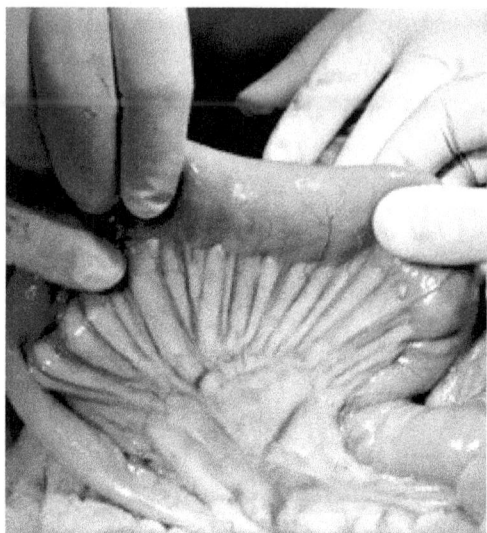

Human mesentery tissue embedded with arteries

Though obviously not held within the thoracic cavity, vasculature woven through fascia suspending the small intestines is appreciated in this photo. The larger vessels are easy to discern, but there is also vast branching of arteries and capillaries extending through the investing mesentery. Where blood vessels enmesh, vasocontraction can be palpated.

A novel concept to consider is adventitia (outer layer of vessels) is likely a part of the whole-body matrix of fascia and that the smooth muscle arterial walls are suspended in this cellular matrix.

Vasoconstriction could be a combination of fascia constriction and signals to the smooth muscles, considering the whole-body effect of protective retraction in sympathetic states. Vagal tone indicators of the thoracic organs measure self-regulation behaviors.

Abdominal Organs and Visceral Structures

The dome of the respiratory diaphragm is a robust myofascial structure separating thoracic from abdominal body parts. The diaphragm nestles with intimate contact over the top of the liver, stomach, spleen, and kidneys. Also, consider that the lower six ribs surround and protect the upper abdominal organs. (No bony structures protect the intestinal tract or mesentery, making these organs more vulnerable to external stressors). Treating the respiratory diaphragm attachments serves both thoracic and abdominal organs. Treating fascia surrounding abdominal organs goes beyond helping breathing, directly treating the ANS.

Peritoneum, Mesentery, Omentum

The abdominal fascial tissues include the peritoneum, mesentery, and omentum, along with investing epithelial layers to a large field of blood and lymph vessels. Two layers of tissue wrap each organ, suspending organs together, and remain fluid-filled and flexible so that organs glide against neighboring structures. Though named differently for corresponding organs, fascia is nonetheless a continuous sheath.

The **peritoneum** is serous (thin and watery) fascia, and has two types of tissue:
- *Parietal* peritoneum - lines abdominal & pelvic cavities (creating housing for organs)
- *Visceral* peritoneum covers external surfaces of most organs, including intestinal tract

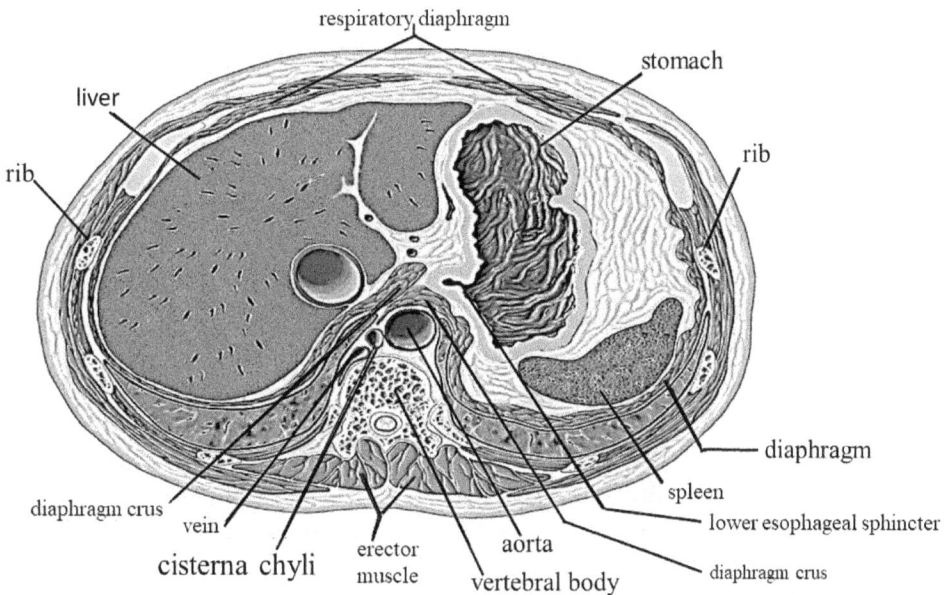

The **mesentery** extends from the abdominal walls to wrap around and suspend the great lengths of the intestines via a fused, double layer of parietal peritoneum. Mesentery is a generic term describing peritoneal extensions from abdominal and pelvic organs. However, the mesentery is now considered a separate interconnecting organ between the intestinal loops, the abdominal wall, and the visceral circulation network. An essential feature is that the mesentery serves as a suspension web for a significant field of blood vessels, nerves, and lymphatic vessels. Tension through the mesentery can reflect the ANS health of the gastrointestinal tract. Thanks to the creators of visceral manipulation, bodyworkers now have direct access to reaching these vessels within the mesentery tissues (at the fascia-vascular root) to work with the interoceptors and facilitate whole-body parasympathetic activation.

The **omentum** (greater and lesser portions) is a quirky organ made from a fold of peritoneum that suspends a large field of fat and lymph nodes and stretches over the abdominal organs. The omentum is structurally similar to mesentery tissue but has differing functions of secreting hormones, housing immune activity, and controlling infections. It is a protective layer of shock-absorbing tissue for the abdominal organs. The fascia layers of the abdominal organs are the focus of treatment to reach interoception (visceroception).

In Barral and Croibier's methods of visceral manipulation, we learned how to locate and treat the primary suspension ligaments of these organs. In *Bodywork for Sensory Integration*, these methods proved somewhat invasive for many younger children. Clinical experiences altered our approach by combining craniosacral therapy with visceral techniques to avoid alarming and setting off the sympathetic system. Our awareness was turned to the surrounding connective tissues (holding the organ's tone). Treating the vicinity of the fascia sheath that wraps the organ, or the fascia field that suspends the vasculature and lymphatic pathways, brings the body into a deep parasympathetic state without the risk of invasive touch setting off a protective retraction. In this way we globally treat the abdominal organs and vessels. In the world of sensory integration treatment, it is a long-held observation that it is counter-productive and counter-intuitive to add to the dysregulated system by adding in *more* input. Responsive Hold

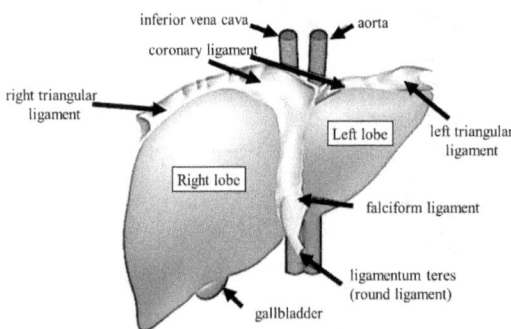

Suspending ligaments of liver has a direct effect on ANS tone

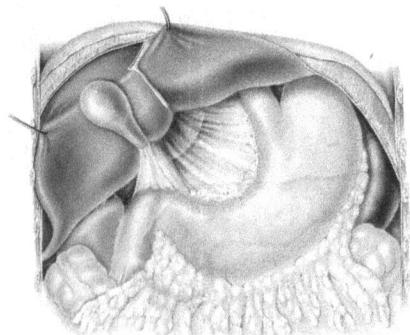

Artist's rendition of abdominal peritoneum

is the perfect match for organs held in sympathetic tone, by the use of non-invasive touch and allowing for self-correction of tissue instead of therapist-directed treatment response. The touch input of CST is the perfect level of minimal input to help the dysregulated child.

In addition, we have also discovered that the biliary system that drains the liver, gall bladder, and pancreas holds a significant role in visceroception and, thus, ANS regulation. If there is slow resistance of digestive secretions following meals, the structural stress challenges the "rest and digest" state. Often there is tension influence from the umbilical cord remnants, the hepatic plexus, or a host of other fascia tensions that adds resistance to the common bile duct. The duct is one of the crucial aspects of the detox pathways. We locate this area by following the falciform and round ligaments toward all the suspending peritoneum of the liver, starting at the navel.

The flexibility of neighboring organs sliding against neighbors has an unappreciated role in trunk control, especially rotational movements. The brilliant contribution to the science of bodywork of Barral and Croibier is the impact that reduced organ mobility has on postural alignment and symmetrical movements. The whole visceral organ contents are directly suspended off the vertebral column for structural support. The dynamic movement of organs is a core aspect of whole- body tensegrity. The movement of organs also reflects the ANS tone, with higher sympathetic and enteric tone reducing organ motility within the cavity. From the original visceral manipulation, methods which are now evidenced-based practice, to lessons learned from generalizing to various younger populations, organ suspension and freedom to move in their cavity is a chief source of sensory integration for the interoception system.

Combining the touch of CST with the specificity of VM, modified to match ANS tone, was the "just right" approach to children with sensory processing challenges

Enteric nerve mesh within mesentery and peritoneum layers

The Enteric Nervous System (ENS), a complex mesh of visceroception and visceromotor nerves woven through the peritoneum and mesentery tissues surrounding visceral organs, is considered a third branch of the ANS. Some viewpoints, though, describe it as a separate system altogether because it can run under its own power, even if the vagus nerve is severed (as proven in animal studies).

The submucosal and myenteric plexuses form this network of neurons (approximately five million). This "other" nervous system functions through a series of reflexes that sense organ fullness or emptiness and stimulates peristalsis to move food and fecal matter through the gastrointestinal tract. The neurochemistry secreted through this mesh communicates with the bacteria biome housed in the intestinal villa, constituting the "gut-brain." The ENS regulates the movement of water, electrolytes, and nutrients between the gut lumen and tissue compartments. This is accomplished by secretomotor neurons innervating the mucosa in the small and large intestines, ultimately controlling ion permeability.

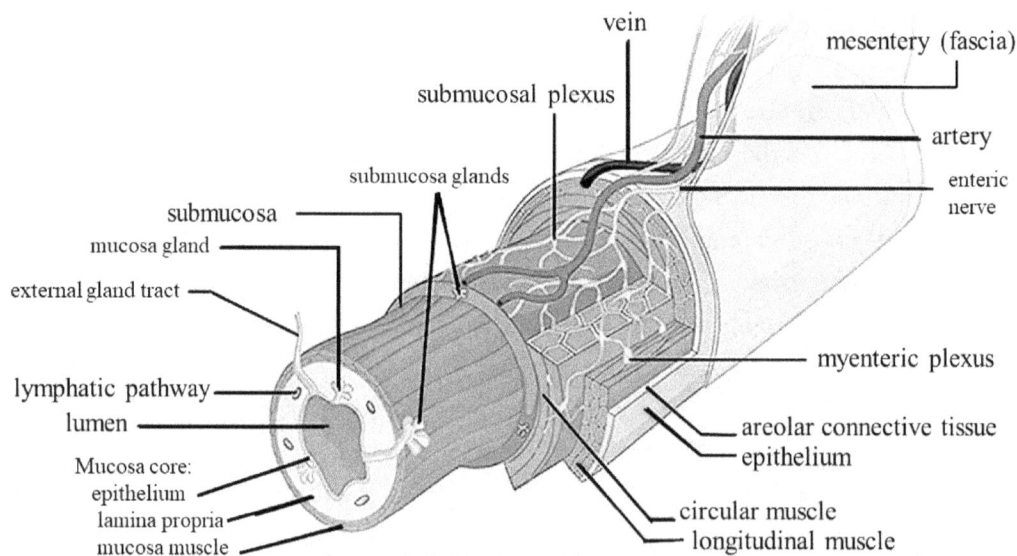

The enteric system extends into the lining of the GI system and may operate independently from the brain, spinal cord, and vagal system. However, the vagal nerve system is fed information from these receptors, though not completely understood, and probably acts as a backup modulating system to the ENS. The tension of innervation sites of vagal branches and enteric nerve mesh represents the continuum between sympathetic & parasympathetic tone, and can be palpated.

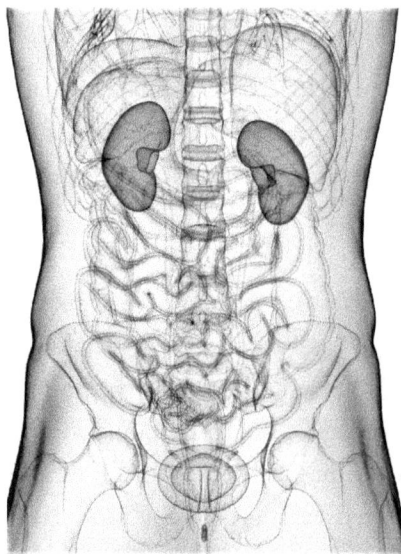

Kidneys & adrenal glands; neighboring organs

The organs in the posterior abdominal cavity are also attended to for their role in visceroception sensory processing. Like the others, the investing layers of connective tissues around the kidneys, the neighboring spleen on the lateral left, and the adrenal glands on the superior surface of each kidney are often treated in *Bodywork for Sensory Integration*. Regardless of age, positive changes in mood and temperament are observed when these organs are relieved of osseous or fascia restrictions. The baroreceptors in the walls of blood vessels and smooth muscle walls of organs regulate a change in the tension of organs and structures.

Digestive sphincters

Smooth muscle sphincters are located at the junctions of digestive tract components. Seven digestive sphincters operate as a choir in coordination with the movement of food, chyme, and fetal matter through translational peristalsis from one section to the next. The sphincters control the rate, quality, and frequency of peristalsis. As far as we know, these sphincters have both vagal and enteric innervation, and reflect the state and the behaviors of the ANS. Treating hundreds of babies over the years demonstrated an association between behaviors of dysregulated infants with sphincter mobility. In general, sympathetic activation inhibits smooth muscle motility, constricts mesentery blood vessels, and restricts sphincter-prompted peristalsis. Conversely, parasympathetic activation stimulates digestive motility and secretion activities.

Any number of structures or tissue restrictions can hinder the physiological motion of a sphincter, which in turn reduces the ease of digestive motility. Working with digestive sphincters is a direct and quick way to activate the parasympathetic state, inducing changes in coping and adaptive behaviors to all sensory input and stressors. Working with digestive sphincters is considered a direct method to treat ANS modulation, thus affecting sensory integration perceptions by change of the tone at the receptor sites.

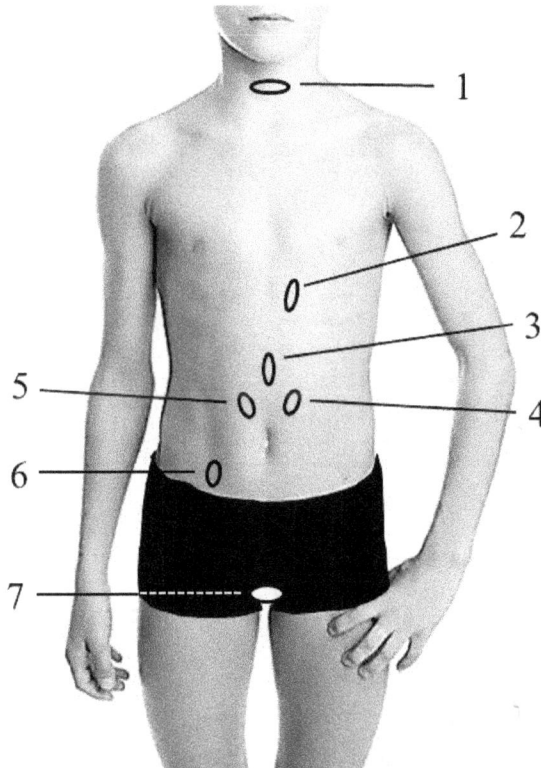

The anatomical vicinities of the seven digestive sphincters.

Autonomic Nervous System: Abdominal Nerves & Plexuses

Considering the location of the primary nerve plexuses is a great benefit in assisting the ANS reaction balance of the interoception system. Being aware of any particular plexus in the region your hand lands in, treatment can guide the treatment process with more specificity, following lines of pull to precise innervation sites of stress.

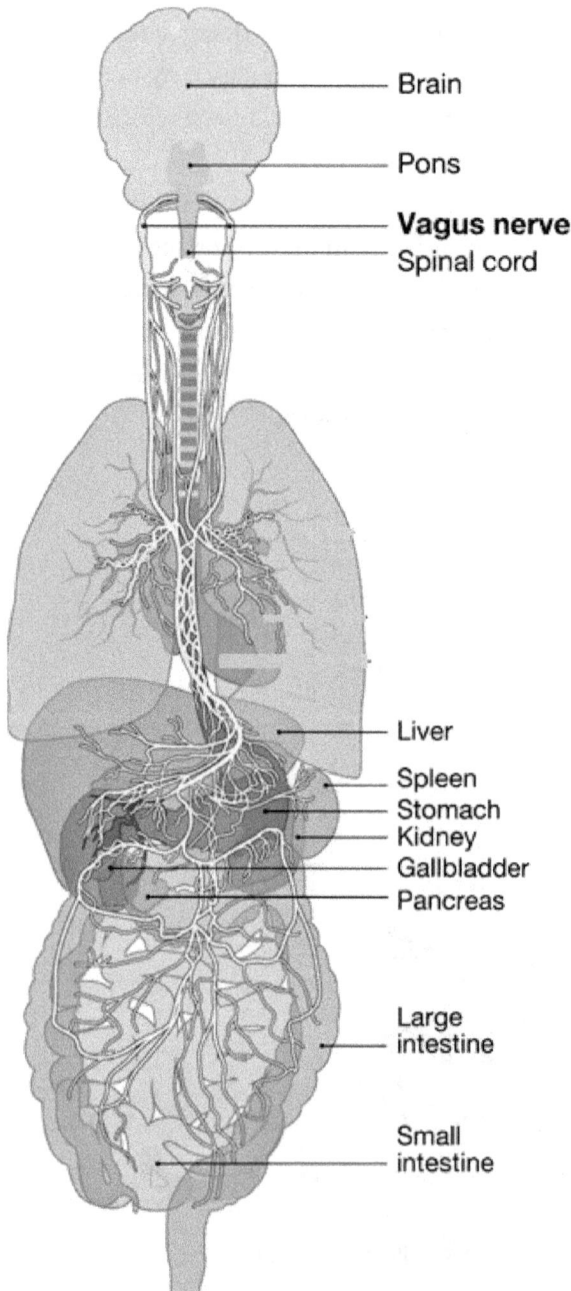

Brain
Pons
Vagus nerve
Spinal cord

Liver
Spleen
Stomach
Kidney
Gallbladder
Pancreas

Large intestine

Small intestine

- Cardiac & Pleural Plexuses
- Celiac & plexus (solar plexus) – formed from anterior and posterior vagal branches, merging with greater & lesser splanchnic nerves; buried between stomach and diaphragm at level of L1.
- [Superior] gastric plexus (coronary) – found in lesser curve of stomach; joins vagus nerve
- Hepatic plexus – over inferior surface of the liver and mingles with vagus and renal branches
- Renal plexus – a conglomerate of smaller plexuses that bundles around aorta and renal arteries/veins
- Superior mesenteric plexus – continues off celiac plexus & right vagus; surrounds mesentery arteries

The concentration of plexus branching with vagus nerve nuclei in brainstem are focal features in treatment by treating both ends to locate lines-of-pull and tension patterns. Non-invasive touch of CST combined with VM helps promote the structural indications of parasympathetic response.

The Vagus Nerve is not a simple nerve, but a system of complex branching and conjoining of nerves meandering through the thoracic and abdominal cavities, innervating organs and vessels. Branching continue to extend outwardly into the limbs and key acupressure points can influence the state of activation. Comparing the illustration on this page to the one of the previous page, an appreciation of the central plexus branching to the organ and periphery branching can be gained. The vagal system activates parasympathetic function and is a major player in the regulation and recovery of novel-alerting or scary-alarming excitation events. Both scenarios of events will likely cause vasoconstriction and possibly some emotional reaction.

The Balance of Protective Retraction

The multiple nervous systems are the most complex and confounding organ systems of the human body. In a sense, that is true, especially when it comes to nomenclature, correlative anatomy, and performing surgery. Regarding *Bodywork*, however, the nervous system is quite simple, and that simplicity is conveyed by the demonstrated continuum of protective retraction of all body parts. The autonomic nervous system reveals itself in organs and structures by retracting tissues (vasoconstriction and drawn fascia) or softly expanding (vasodilation and fascia release). Organs with sympathetic tone feel drawn inward, or "frozen" in stiffness, and parasympathetic tone feels soft, pliable, and glide easily against neighboring organs—more warmth with vasodilation with a sense of palpation and fluids' movement between layers of tissue.

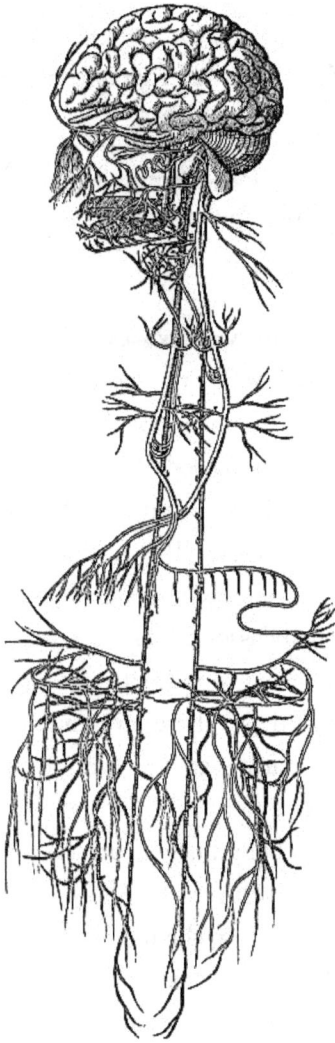

Schematic representation (by woodcut) of the cranial nerves in the Fabrica (1543) by Vesalius. The recurrent laryngeal nerves (arrows) are shown correctly as branches of the vagal nerves with the right recurrent laryngeal nerve originating more superiorly than the left one. Courtesy of the U.S. National Library of Medicine.

Through the fascial connections and following nerve pathways to innervations, therapeutic touch has a direct effect upon baroreceptors involved in such interoceptive physiology. Changes in tone can be palpated and treated with the manual therapies presented in this book.

> In *Bodywork for Sensory Integration*, our hands are placed mindfully of the concentration of nerve branching through the mesentery and key organ innervation sites.

Umbilicus and cord remnants, suspending ligaments, nerve plexuses

The navel is the body's first scar. Bodyworkers are well-aware that scars "hold" trauma stories as well as reveal fascia distortions. The shape reveals remnants of the umbilical cord strains, tensions from cord clamping, and how the umbilicus healed. Though no scientific evidence exists, theories claim a relationship between personality styles and belly button shape. The umbilicus is rarely studied in functional and behavioral therapies. The umbilicus is a region of convergence of major fascia attachments and bands at the thoracic/abdominal junction.

Retained tissue restrictions and pull lines can be visually assessed and palpated, and can project in any direction. Commonly associated restrictions are found with the ligamentous teres (round ligament) and the enveloping falciform ligament. These structures are tissue vestiges of the umbilical cord, extending three dimensionally over the liver, towards core nerve plexuses, into regional peritoneum, and towards mesentery field densely woven with blood vessels. Structures in the immediate vicinity of the belly button include the pyloric valve, sphincter of Oddi, and the celiac plexus.

Fascial line projections can extend from the urachus deep into the mesentery or as far as the central arch of the respiratory diaphragm. The deep front line of fascia passes underneath the umbilicus. The umbilical ring just below the skin is a distal representation of the fascia remnant of the umbilical cord. The residual tissue of the fetal umbilical arteries and vein turn ligamentous after birth and contribute to liver suspension.

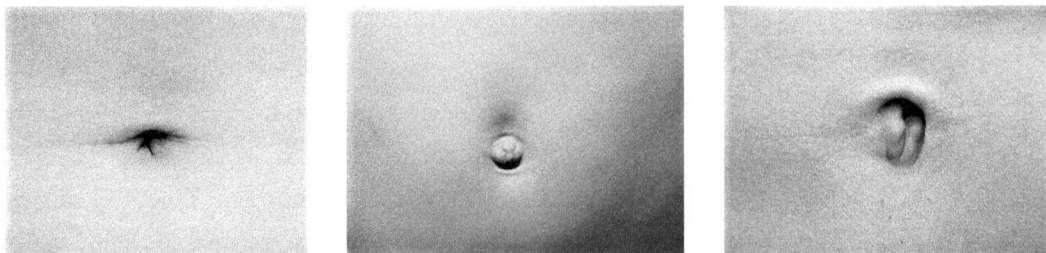

A sampling of children's navels reflect the variations of fascia distortions and strain patterns that can be found and palpated in the umbilicus region. (See illustration of ligaments of the liver on p. 191)

The mesentery vascular field accounts for approximately 20% of the body's arteries, in conjunction with major lymphatic pathways (cisterna chyli & deep visceral nodes), also contribute to the ANS status in the region. *Bodywork* provided to many, many navels has taught us that umbilical and abdominal fascia tensions can persist beyond infancy. The belly button is a site worth investigating the relationship between enteric and vagal tone related to self-regulation behaviors. *Bodywork* can assist in reducing organ stress by providing release of fascia and soft tissue restrictions, promoting parasympathetic state and behavior changes.. A great deal of structure under the belly button contributes to self-regulation and enteric tone.

Lymphatics- Interoception of elimination & detoxification

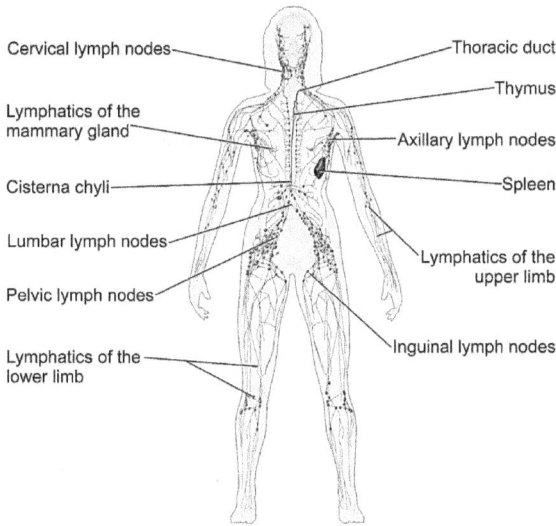

Cervical lymph nodes
Thoracic duct
Thymus
Lymphatics of the mammary gland
Axillary lymph nodes
Cisterna chyli
Spleen
Lumbar lymph nodes
Lymphatics of the upper limb
Pelvic lymph nodes
Inguinal lymph nodes
Lymphatics of the lower limb

Another novel concept for interoception is to consider the systems housing and moving interstitial and lymphatic fluids. Interoception, being baroreceptive by nature, probably has a significant role in monitoring fullness of even the smallest of fluid vessels. The lymphatic system is an extensive, whole-body circulatory system carrying interstitial fluids as an immune function and detoxification highway. Lymphatic fluid travels through vessels that parallel arteries and veins systemically through the body. The intake points are tiny branching lymph capillaries that mingle with small blood vessels. Here interstitial fluid facilitates entry into the vessel.

These microscopic vessels lead to secondary containment organs (lymph nodes made of mucosa-associated tissue). Larger lymphatic vessels are arranged in a tree-like branching with specific features. These bigger collecting lymphatic vessels contain greater volumes and have continuous movement created by staggered smooth muscle valves (lymphangions) that pump fluid more dynamically towards lymph nodes. After filtering through the nodes, lymph travels to venous blood, eventually ending in the liver.

The whole system of these valves ensures safe, unidirectional transport. This fact requires any lymphatic massage method to follow the evacuation of fluids in specific directions in the tributary flow map to not harm the lymphangions. Besides clearing waste and other metabolites from the interstitium, lymph circulation actively regulates the transportation of immune antigens and antigen-presenting cells. A proliferation of vessels occurs under inflammatory conditions in the homeostatic job of clearing pathogens and toxic agents. Lymphatics are also involved in absorbing lipids, proteins, and fluids in the intestinal villi of the small intestine. Absence or malfunction of the lymphatic system is associated with chronic inflammatory or auto-immune conditions (especially skin conditions, but is most commonly recognized as the cause of lymphedema. *Bodywork for Sensory Integration* proposes that the lymphatic system may be involved when children [people] feel uncomfortable in their skin or may be challenged in somatosensory discriminatory learning.

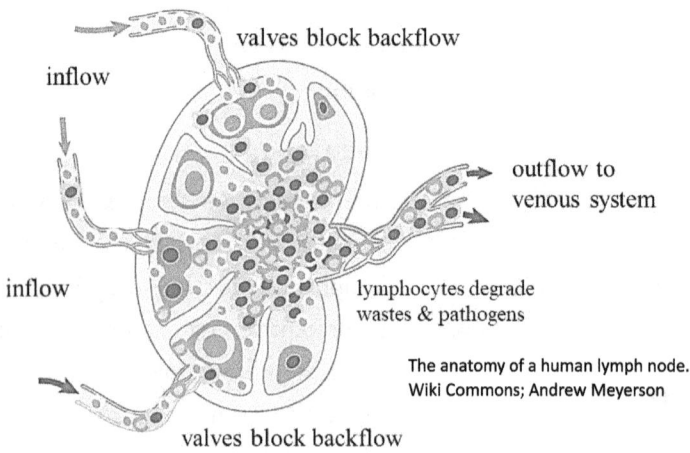

valves block backflow
inflow
inflow
outflow to venous system
lymphocytes degrade wastes & pathogens
valves block backflow

The anatomy of a human lymph node.
Wiki Commons; Andrew Meyerson

Human lymph nodes are kidney-like filtering organs. Hundreds of larger lymph nodes phagocytize pathogens and toxins, and filter sizable proteins. Lymph fluid flows in and out and can become congested or engorged from too much debris. Chronic lymph stagnation can create local or systemic inflammation as white blood cells activate histamine and cytokines to try to clear the area. Stagnant lymph field and engorged lymph nodes can be found anywhere in the body, frequently buried in muscle spasms.

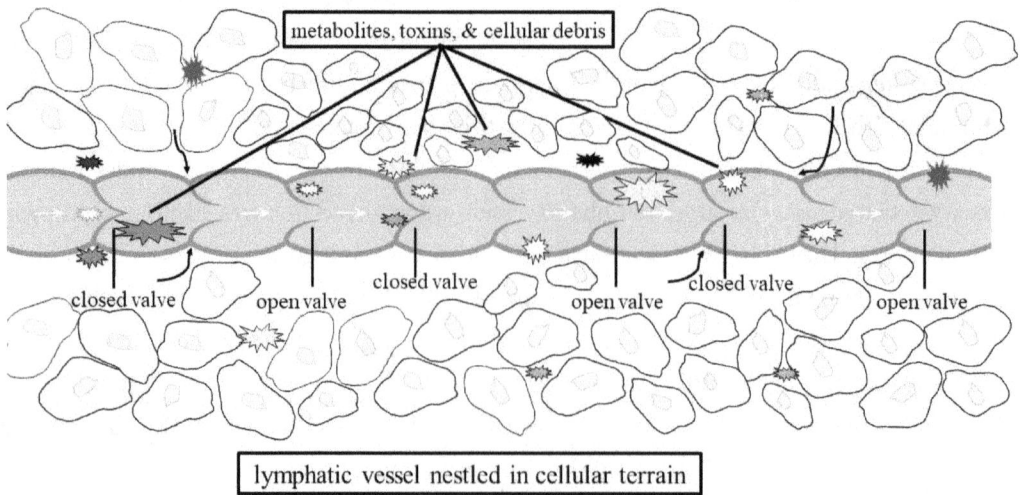

metabolites, toxins, & cellular debris
closed valve   open valve   closed valve   open valve   closed valve   open valve

lymphatic vessel nestled in cellular terrain

One-way valves (lymphangions) are tiny smooth muscles that operate under parasympathetic innervation. They reciprocally relax and contract to sweep away debris, pathogens, and metabolic wastes. These valves open with pressure from fluid volume and close to prevent backwash and regulate hydration levels for homeostasis. A lymphatic rhythm is created by the regular circulatory system (heartbeats) but differs in measurable frequency and cadence. This rhythm can be palpated and guides manual lymphatic drainage massage. Lymph flow moves in longer strides, compared to blood flow in arteries, due to the collective actions of the lymphangion valves

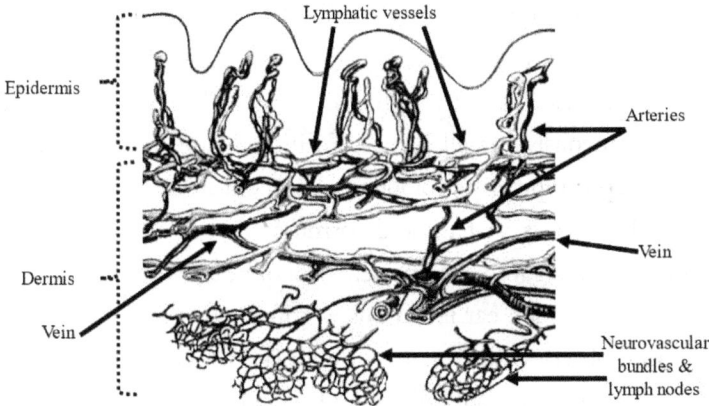

Lymphatic vessels weave and travel through the body parallel to arteries and veins. At microscopic junction, neurovascular bundles form around lymph nodes. Here is where exchange of bodily fluids occurs. Deoxygenated blood is moved to veins and oxygenated to the arteries. Lymph vessels sweep and syphon fluids from intercellular spaces (interstitial fluids), and fluids from the fascia tissue layers. Hence there are multiple circulatory systems humming aside each other.

Trauma, inflammation, fascia restrictions, toxins and pathogens can all play a role in mucking up the works. The maximal and thorough movement of the lymphatic system is believed to occur [best] when the body is in a parasympathetic state. The lymphatic system drains 15% of body and intercellular wastes, flowing to the liver. The liver with the lymphatic relationship is the body's chief detoxification pathway. When not optimally draining or moving, the body might try to detox itself through the skin. Chronic body stress and tension can interfere with the body clearing itself, giving rise to inflammation, auto immune, and perpetual alarm state in blood vessels and interstitial spaces. In *Bodywork for Sensory Integration*, we consistently observe a direct correlation to mood and temperament, along with sensory reactions, to the state of the lymphatic detox pathways.

Lymphatic drainage massage can improve sensory wellness, especially with the sense of touch for the person uncomfortable in their skin

Cranial base and cervical (occipital-cervical base)

The cranial base (junction where the occiput meets the upper cervical spine) is also where the ANS can receive assistance in regulation by reducing fascia strains and structural torsions. It takes up to eight years for the occiput segments to finalize its fusion through intramembranous and endochondral ossification. The skeletal immaturity as a stable musculoskeletal attachment, an inferior zone of the meninges, and key foramen for cranial nerve outlets makes this region vulnerable to various physical injury or trauma, regardless how benign. Head remolding from birthing can be affected by compressed cartilaginous segments. It is a false assumption that head remolding resolves itself spontaneously for all babies. Positional skull deformities are believed to be a result of back-to-sleep campaigns, prolonged placement in carrier devices, or in utero alignment. Though research suggests that these deformities do not have an adverse effect upon the brain, anecdotal evidence suggests otherwise. Torticollis, lazy eye (amblyopia), tongue ties (ankyloglossia), ear and same-sided ear infections, vestibular processing as well as other functional diagnoses have been observed to co-exist with a plagiocephaly (cranial base compressions) by developmental therapists who routinely re-evaluate the child.

The occipital condyles (lateral borders to the foramen magnum) can retain lateral compressions which can be a structural stimulus into the spinal thalamic tract, the dorsal columns or extend stress superiorly into the brainstem and midbrain where cranial nerve nuclei are housed. Ossification centers in a growing spine are separated by synchondrosis. The immature spinal column of children redisposes them to both a great tolerability to motion, but also vulnerable to intervertebral compression retention. A large head on a small neck head is supported by ligaments with greater laxity where physiological wedging of vertebral segments can put adverse tension and torsions into the spinal meninges. Meningeal tensions can extend superiorly into the brain stem, over-facilitating cranial nerve nuclei. It is common for children to have minor injury at the synchondroses between the dens and C2 body from shear forces of rapid deceleration and hyperextension events. This can occur in birthing, ill-supported car seats, or participation in physical activity with a greater risk of forces imposed on the head and neck. These ossification centers unite in midline between 2 and 4 years while neurocentral synchondroses fuse at 3 to 6 years.

Atlas (C1) vertebra develops from three ossification centers, first appearing at one year and fuse by seven years. Axis (C2) has a more complicated process, having a persistent synchondrosis below the C1 and C2 facet joint. Treating nerve nuclei has been demonstrated to be helpful as a direct treatment to down-regulated nerves and pathways out of hyperactivation. As a nerve is structurally stressed, increased tension and vibration can be palpated through body tissues.

Situated on the anterior neck, the hyoid bone is a site of significant investing fascia and myofascial attachments. Sources differ in opinions on how many have a direct or indirect attachment to any surface of the hyoid bone. The tongue, the throat, the esophagus, the lungs, the aortic arch, and the upper rib cage sympathetic ganglion are major ANS sites responding to the baroreception and visceroception of tension and tone. One of the reasons our throat tightens when we are stressed is the number of structures attached to the hyoid that can tense under sympathetic activation. Upledger referred to the anterior throat and hyoid region as the "avenue of expression" because of the complex and interrelated responses to the interoceptors and ANS responses. Working in this area using craniosacral therapy not only helps relieve structural strains and stresses but also has proven to aid in the person's ability to verbalize internal perceptions of pain, trauma, or even joy and emotions.

Responsive Hold is the optimal method to treat these structures for interoceptors to have structural balance for signaling. Most structural strains can be found in the fascia tissue surrounding and suspending these osseous structures. Manual work is fascia and soft tissue work. We do not adjust or manipulate spinal columns in *Bodywork for Sensory Integration*. Instead, we assist soft tissues into a self-correcting release process that promotes vasodilation, nerve innervation relaxation, and resets tone of all the nervous systems.

Concepts to Understand About Autonomic System - the interoceptors' role

1. Sympathetic nervous system activation causes vasoconstriction of most blood vessels (within skin, digestive tract mesentery, and kidneys). Chronic stress creates the potential for persistent vasoconstriction. Where there are artery concentrations there is potential for fascia restrictions and lack of motility of neighboring structures.
2. Keep the entire vascular system in mind when palpating for protective retraction.
3. Too invasive of touch can also cause vasoconstriction as a form of protective retraction, negating the effects of the therapeutic intervention.
4. The enteric nervous system operates separately from the vagus nerve, though it is still not fully understood how the two function together.
5. The enteric nervous system is a major division for self-regulation within ANS continuum. Vagal tone is measured by structures innervated by central and peripheral nerves, not the enteric nerves. Functionally and behaviorally, the enteric continuum for the" rest and digest" system is monitoring gastrointestinal tone, regularity, and urinary retention. A direct method to measure enteric "tone" of the visceral organs is through manual palpation; and vagal tone is through palpation of vascular masses. Both yield a subjective assessment.
6. A history of colic, reflux, breastfeeding difficulties, sleep, self-soothing and behavioral challenges are consequence of a lack of balanced reactivity and resiliency of the ANS. Infancy is likely the starting point for ANS imprinting. Adverse childhood experiences can also influence baseline ANS responses.

7. Be aware of body symmetry, or lack of, and organ placement within thoracic cavity: respiratory diaphragm balance, rib cage, lungs and heart within their fascia confines

8. Ease of structural motility of the digestive tract has an immediate and direct effect on self-regulation. Movement of food and chyme through the gastrointestinal tract segments, and the absence of chronic resistance of the normal physiological motion of organs, is the desired outcome.

9. The structural and neurochemical function of the enteric system has expanded our knowledge of a "gut-brain" and its relationship to behaviors, cognition, and emotional regulation. The neuronal mesh of the enteric system functions through a series of three reflexes that sense organ fullness or emptiness, stimulates peristalsis, and moves food and fecal matter through the system. This all can be palpated and assessed through trained, non-invasive palpation of visceral organs.

Bodywork for Sensory Integration **has become a valuable tool in balancing the behaviors of the ANS**

Bodywork Treatments that help balance Interoception / Enteroception / Visceroception function

- Treating Umbilicus and cord remnants, suspending ligaments, nerve plexuses (See pp. 176, 178-180, 181)
- Treat Thoracic & Respiratory Structures: plexuses & vagus nuclei in brainstem, pleura, lungs within pleura, intercostal layers, movement of all ribs with breaths, sympathetic nerve ganglion, esophagus and sphincters resp diaphragm, middle of diaphragm, mass of blood vessels in lungs (See pp. 175-177)
- Treating Abdominal & Visceral Structures: plexuses & vagus nerve nuclei in brainstem, sphincters, diaphragm drapes over liver & stomach, peritoneum around neighbors, biliary system for organ drainage and detox pathways, mesentery & blood vessel concentration, psoas suspensions to intestines, kidneys (adrenals) & renal plexus (See pp. 178-180, 182-183, 184, 185-187)
- Treating the Cranial base, dural tube of spinal cord, and vagal nerve projections (See pp. 171-174, 211-213)

Part III

Manual Treatments to Facilitate
Sensory Wellness

Part III

Manual Treatments to Facilitate Sensory Wellness

Clinical application of craniosacral therapy, visceral peritoneum/mesentery/vascular techniques, lymphatic drainage massage, interstitial mobilization, & other manual therapy generalized to young populations

Four categories for clinical applications:

1. Autonomic Nervous System innervation – vagal, organs, & vasculature fields pp. 170-194
2. Soft Skeleton Matrix – superficial & peripheral connective tissues pp. 195-202
3. Core Structures of Soft Skeleton – longitudinal & horizontal fascia bands pp. 203-218
4. Cranium – bones, meninges, cranial nerves pp. 219-236

Key Points of Treatment

- The methods presented in this book summarize the generalization of techniques to the population of sensory processing challenges.
- **This book augments professional training avenues. This book is not be considered the training source to learn these methods, nor should be relied upon as a singular point of professional training. Refer to professional training resources. (See pp. 348-349)**
- Look for the commonalities between methods and keep an open mind to gain eclectic skills. Blend CST/MFR/VM/LD/acupressure and other methods that teach non-invasive touch. The journey that laid the foundation for this book began with Upledger CST training, by which all other methods were compared to.
- Compare and contrast: The degree of invasion and aggressiveness, or the lack of, with touch contact defines a technique. Understand the intention of a technique. Does it promote self-correction verse require a therapist -controlled intervention. CST is not osteopathic medicine just as myofascial release is not CST or lymphatic drainage. The differences between methods have been firmly established. These methods can be practiced by any professional with a license to touch and within their scope of practice.
- Assessment is simultaneous with treatment, for as soon as hands connect with a person's body and the practitioner's intention is honorable and just, the therapeutic effect begins.

- The first consideration is to evaluate for, and then prevent setting off protective retraction. This means to "listen" for the tension continuum of the ANS with your hands, no matter where your hands land. Taking time to listen to tissue allows the child's body to scan your intentions and come to trust your level of intrusion. This is accomplished through the skilled use of Responsive Hold.

- Match your touch pressure to tissue tension and not more (when starting). Like the repetition of words spoken by someone whom you're conversing shows the act of active listening, matching your touch to tissue tension conveys your active presence with the body. The level of your touch pressure should not be greater than the pressure state of any tissue or organ you are treating. Never try to push, pull, or move a spot (until the tissues indicate it's time for that to be done). Instead, show the client's body that you are good at listening through quiet, noninvasive hands.

- Resist, or grow away from, the habit of approaching the body in the act of touching to apply a protocol method. You wrestle with your own ego (trying to do thing right) when only applying techniques, but not listening to the client's body. The client is in charge of the tissue releases. This is a huge paradigm shift and frankly, the difference between levels of clinical outcomes. Instead, touch with the intention to "match your hands to any resistance" within tissues. Matching the resistance reduces the risk of increasing protective retraction. Meeting the resistance and waiting for the tissues to guide your hands is how the tissues of your client trusts you. The faster you learn this, the faster your clients reach the level of trust required for bodywork to maximize self-correction. Don't waste time battling with tissues. Wait with tissues to self-correct.

- Moving hands can set off alarms and sympathetic protective retractions. We may misinterpret tensions in body as stress, but actually it may be protective reaction from too invasive or too much movement, or too much prodding pressure.

Quiet Listening Hands Find More than Moving Hands

- Consider all the structures under your hands at all times: bone & periosteum, muscles & fascia layers that become tendons, the fascia walls of blood vessels, the movement of neighboring lymph nodes & pathways. Most importantly, understand that the whole of fascia is the communication connection.

- Trauma effects can be held within body tissues. All future experiences can be filtered through an imprinted memory but one's reactions can also be decreased in the charge held by the electromagnetic field of the body.

- The variations of method an individual manual therapist uses are embraced both by interest as well as what properties of methods speak and resonate with the professional. However, in the situation of children (and adults) with sensory processing challenges,

"LESS IS MORE" is a very valuable principle. Sensory processing challenges denote ANS tolerance and interpretation differences that interfere with comfort as well as sensory registration (which often is from overload or shutdown). When registration is indeed at the root (lack of response to a sensory channel) experience shows us that peripheral or central pathways may be compressed or blocked.

- Recommended that the reader has some experience, or interest, in craniosacral methods. This method by far, when applied properly, is the least of the less imposing manual methods. Self-correction is a consistent theme in *Bodywork for Sensory Integration*. Be mindful of what methods dictate therapists do things to organs and structures, and compare those to therapies that instruct therapists to blend, meld, and wait for self-correction the organs and vasculature of the ANS. The therapist is actually taken out of the equation of making the changes happen when employing and fusing CST to other methods. CST-inspired visceral methods, one of the lowest risks that exists.

This book serves to share the clinical application of CST, VM, LD, acupressure with pediatric populations who have the developmental task of achieving SENSORY WELLNESS

Autonomic & Enteric Nervous Systems

Key Objective: Activate optimal vagal tone to promote parasympathetic activity of multiple innervation organs. Down regulate sympathetic branching and nerve plexuses. Match tone and tension of vagal, enteric, and vascular system.

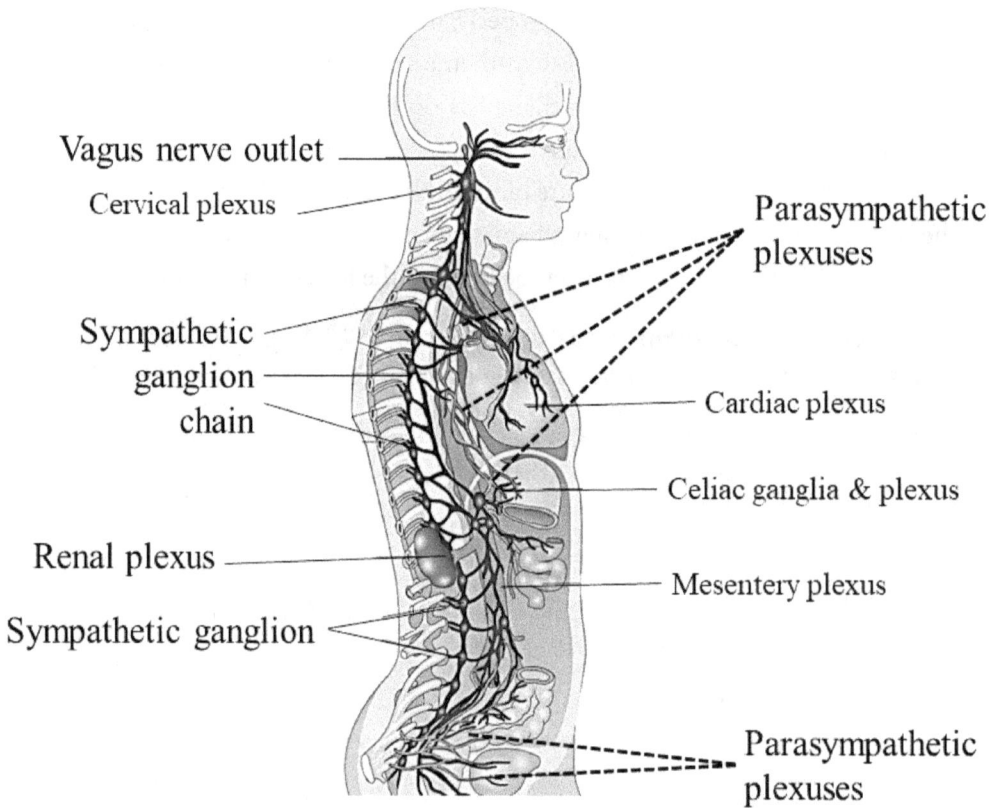

Vagus nerve outlet

Cervical plexus

Parasympathetic plexuses

Sympathetic ganglion chain

Cardiac plexus

Celiac ganglia & plexus

Renal plexus

Mesentery plexus

Sympathetic ganglion

Parasympathetic plexuses

Organized for clinical application

- Global Vagal & Enteric Systems pp. 171-173
- Sympathetic branches of ANS pp. 172-175
- Respiratory & Circulation Center p. 176
- Abdominal Center and Enteric mesh pp. 178-180
- Remnants of umbilical cord p. 181
- Elimination Centers – kidneys and biliary system pp. 182-184
- Digestive Sphincters pp. 185-187
- Supplemental techniques pp. 188-194

Global Vagal & Enteric Systems

Vagal Nerve Pathways

Several stages of treating the vagal pathway & other cranial nerves. Treat globally and then focally into separate structures dictated by tissue response. Tune into one end of the nerve's nuclei (brain stem-medulla oblongata) and one of any of the major vagus branching, as well as main vagal and accessory nerve pathways; investing myofascial triangle on lateral neck; cervical nerve plexus; and cervical peripheral nerve outlets.

Cervical Plexus and associated motor and PNS nerves

Hand One on back of head: Fingertips just off midline (at external occipital protuberance) on same side as Hand Two. Be aware of structures underneath: scalp and superficial fascia, occiput, meninges, cerebellum with fourth ventricle underneath...then the brain stem; (Medulla oblongata). Target is cranial nerve nuclei within brain stem. Palpate CN-X control central and ANS tone can be felt through lines-of-pull to innervation sites.

Hand Two on side of neck: Gently place hand the length of lateral neck. Tune into path of vagal nerve with index finger along the ramus of mandible, & cervical plexus by spreading other fingers. Fingertips can connect with innervation sites, or whole hand covers site of tensions within myofascial structures invested through lateral neck triangle; one fingertip finds hyoid edge along its greater horn.

Both Hands: Apply Responsive Hold. Tune into awareness to detect any lines-of-pull that connect both hands, (where both hands feel the two ends of a tension pattern). Hold both ends of the line and wait for self-correction in tensions. Follow paths of nerve branches inferiorly into neck. Treat vicinity of investing fascia of neck triangle.

Treat both sides. Hand One fingers move to corresponding side of occiput; **Hand Two** to side of neck, touching jaw and hyoid margins for a full appreciation of tensions in tissues.

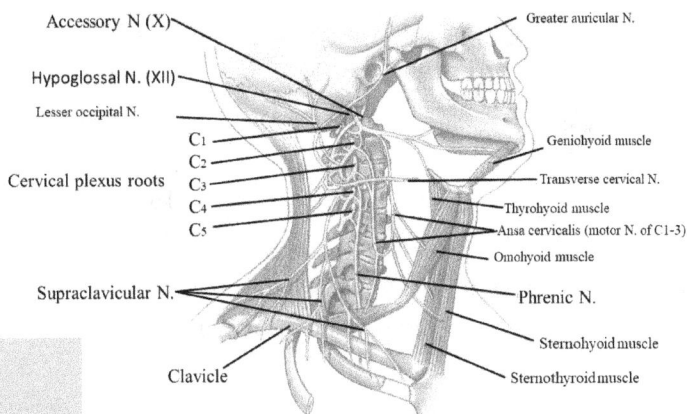

Accessory N (X)
Hypoglossal N. (XII)
Lesser occipital N.
Cervical plexus roots C1 C2 C3 C4 C5
Supraclavicular N.
Clavicle
Greater auricular N.
Geniohyoid muscle
Transverse cervical N.
Thyrohyoid muscle
Ansa cervicalis (motor N. of C1-3)
Omohyoid muscle
Phrenic N.
Sternohyoid muscle
Sternothyroid muscle

Can replicate this method for any cranial or peripheral nerve and pathway towards innervation site.

Sympathetic branches of ANS

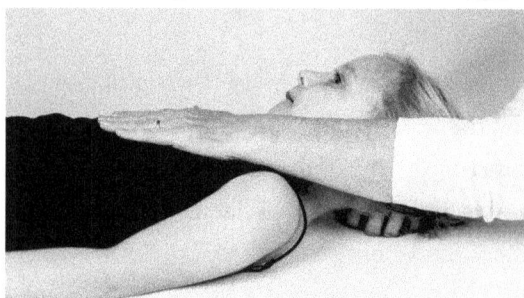

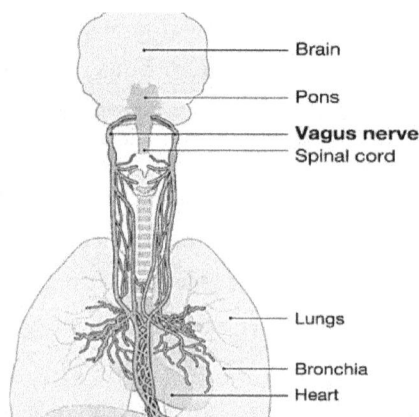

- Brain
- Pons
- **Vagus nerve**
- Spinal cord
- Lungs
- Bronchia
- Heart

Thoracic organs, cardiac & bronchial plexuses Abdominal plexuses, peritoneum, and nuclei

Hand One: remain on same site off-midline of occipital protuberance; remain focused on the communication link toward brainstem and pons housing cranial nerve nuclei.

Hand Two: now thoracic (cardiac and bronchial) nerve plexus, pleura fascia, tone of bronchi, lungs, mediastinum. moves down with index finger near midline and fingers lateral, covering the whole mediastinum. Vagus and phrenic nerves continue from the neck on the surface of the scalene muscle and along subclavian vasculature. Pericardium and pleura fascia house both vagal and phrenic branching. Tune into the tone and tensions of these connective tissue layers, apply Responsive Hold awaiting tissue response, and connect awareness of both treating hands to detect any lines-of-pull. Hold both ends of these lines and wait for self-correction to tensions.

Treat both sides. Hand One fingers move to corresponding side of posterior skull

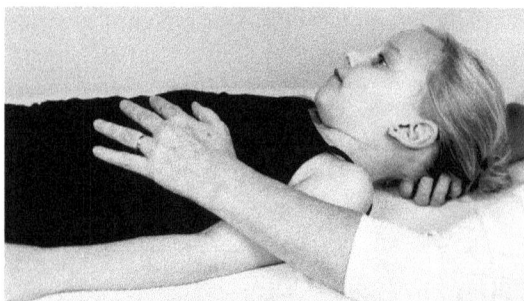

Hand One: remain off midline of occiput protuberance; remain focused on communication link toward brainstem/pons housing cranial nerve nuclei

Hand Two: moves down each central plexus: celiac, splenic, hepatic, and renal. Tune into tone and tension of connective tissue layers. Apply Responsive Hold, awaiting tissue response & connect awareness of hands to detect any lines-of-pull. Hold both ends of these strain lines and patiently wait for self-correction of tensions.

Mesentery and Enteric Nervous System with brain stem nuclei

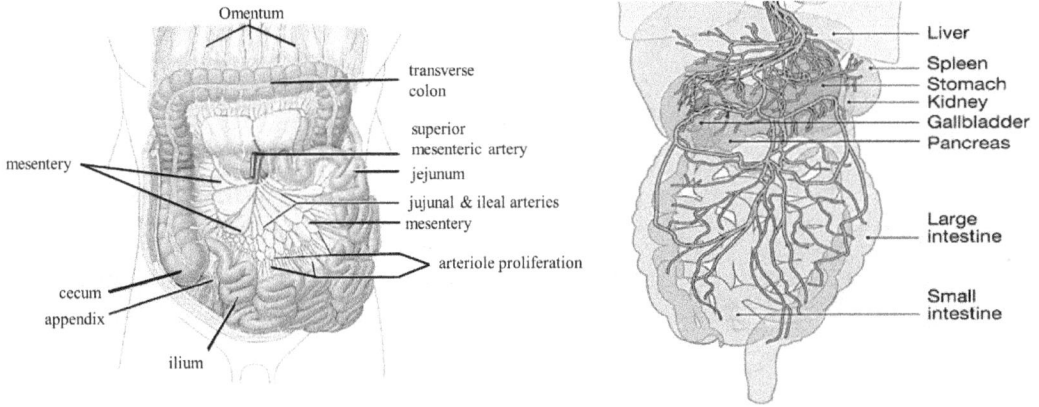

Mesentery attaches intestines to abdominal wall, stores fat, supplies blood, lymph, nerve innervation to intestines

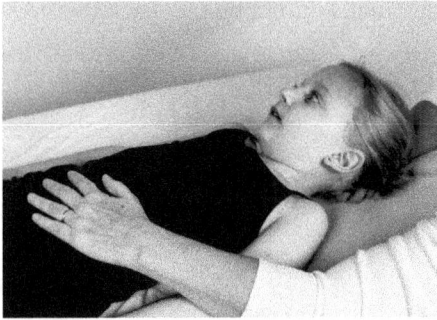

Hand One: remains off midline of occiput protuberance; remains focused on the communication link in toward brainstem/ pons housing cranial nerve nuclei.

Hand Two: moves down to the global distribution of the mesentery branches; can treat in middle or each lateral side of the mesentery region. Tune into the tone and tensions of these connective tissue layers, apply Responsive Hold awaiting tissue response, and connect awareness of both treating hands to detect any lines-of-pull. Hold both ends of these lines and wait for self-correction to tension.

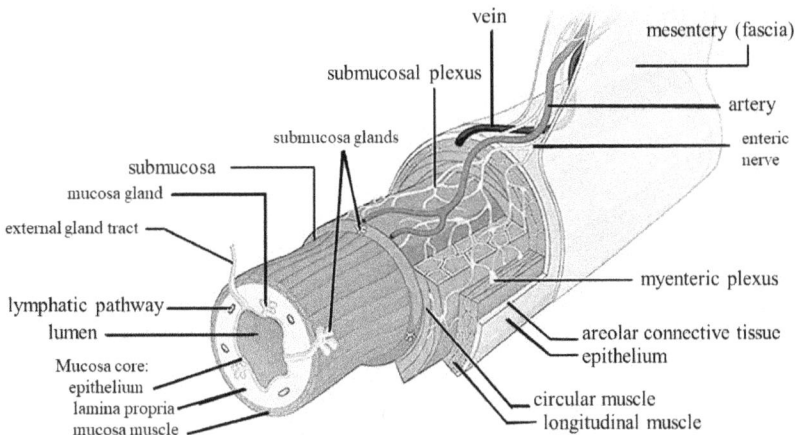

Self-Help

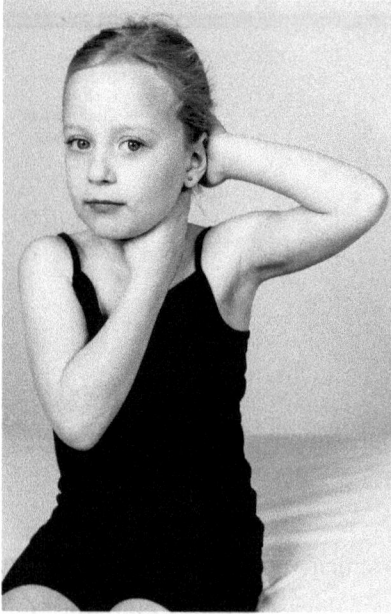

Hand One: Back of head

Hand Two: Placed on each region of vagus nerve pathway. Hold space at any region child feels most comfortable. Teach child how to rest hands for as long as possible (without straining arms) until things feel relaxed inside. Can be done in sitting or reclining. Teach child to do this at times when they get a neck ache, a tummy ache, or fear an overwhelming emotion that tightens their gut.

Bodywork for Sensory Integration is a child-directed approach to raising awareness of the autonomic nervous system for the individual. Akin to biofeedback methods of old but with greater depth of body awareness, we are routinely teaching self-care compliments of maintaining calm-alert and ready states for sensory health and wellness.

Not all, but many of the techniques in this book can be taught to children, or their caregivers, for carry over in home and school.

Thoracic Sympathetic Ganglion Chain for down-regulation of organs

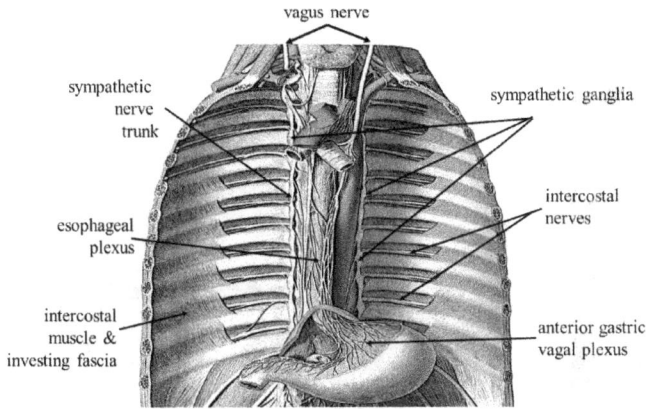

A chain of sympathetic ANS nerve ganglia is positioned on the ventral region where each rib meets its corresponding vertebra. Each projects fibers (pre- and post-ganglionic) to thoracic and abdominal organs. This technique is one of several that can be generalized to any sympathetic ganglion to down-regulate organs, identify lines-of-pull, and promote sympathetic state in localized areas (of pain or discomfort). Each ganglion is the control center of each vertebral level.

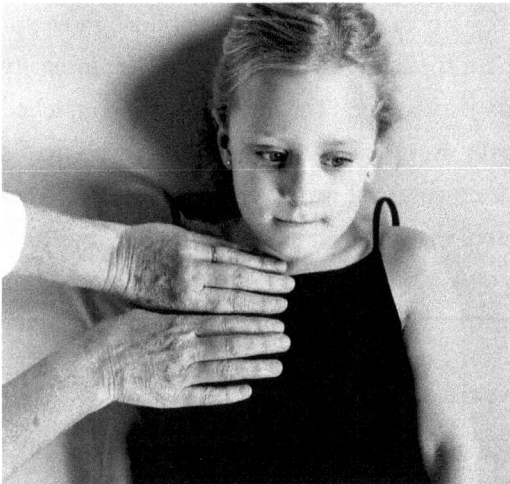

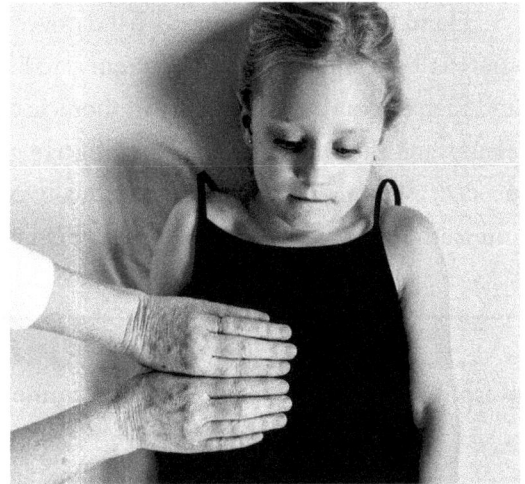

Hand One & Two: Finger tips approximate the span of chain of sympathetic ganglion within the thoracic cavity (T1-T12) on right side; setting the intention to palpate anterior surface or upper margin of the costotransverse joints of vertebral segments. Of course, each ganglion cannot be touched directly, but the effect of its tone radiating away from the spine can be felt in generally in the mediastinum, as well as projected lines of pull along each individual rib. Pericardium and pleura fascia house both vagal and phrenic branching, whose tone can also be appreciated. Tune into the tone and tensions of connective tissue layers and connect awareness to any lines-of-pull. Combine Responsive Hold with line-of-pull tension release and wait for self-correction to tensions. Breath will likely slow and deepen, warming or release of heat from tissues, heart rate drops, and increased tissue pliability through respiratory apparatus are common results. Very common to hear clients state "I've never felt this relaxed" or expressed this state through changes in behaviors or temperament.

Respiratory & Circulation Center

Global rib cage, intercostal mass, invested connective tissues, pleura & pericardium

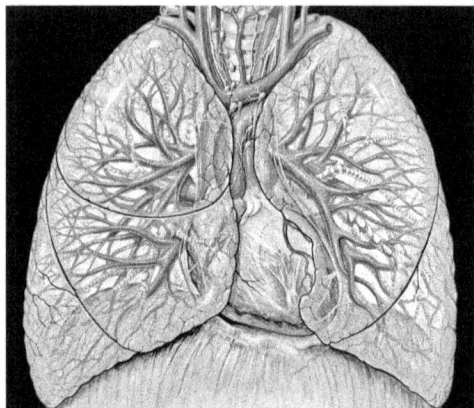

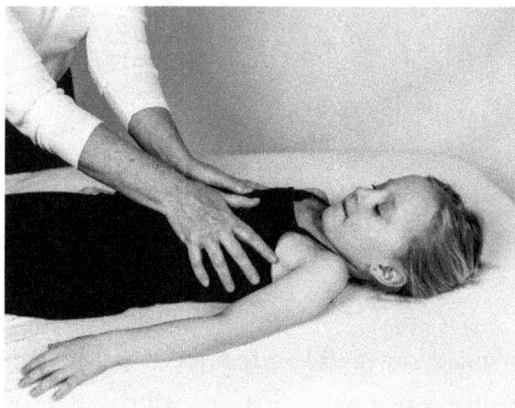

Hand One & Two: Spread hands over entire rib cage. DO NOT mobilize, move, or stretch ribs from each other. The intent is to listen for the shape and flexibility of the tensegrity and then respond to the entirety of thoracic cavity tissues. Treat the intercostals and attached pleura, and the whole fascia of the rib cage globally. Responsive Hold is primarily all that is needed, especially for a child. Occasionally precise myofascial release might be needed at an intercostal space, visceral peritoneum, or an external myofascial line of pull.

Respiratory diaphragm and invested connective tissues

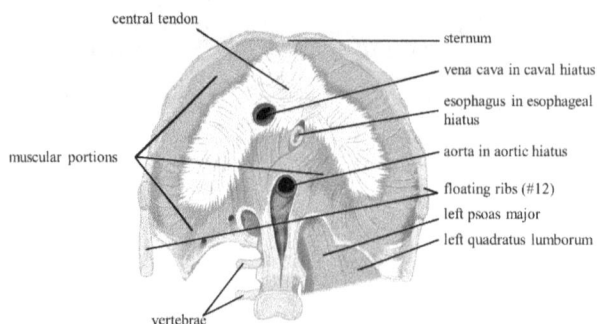

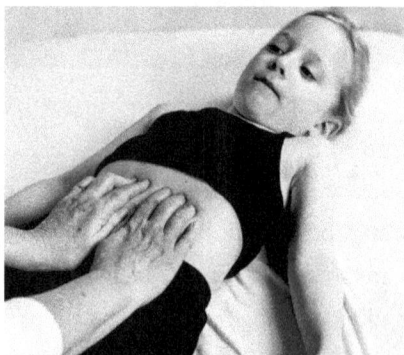

Hand One & Two: Fingertips span the inferior dome of the diaphragm, in meeting space with liver & stomach. The diaphragm is a musculotendinous sheet (sternal, costal, and lumbar parts), all inserting into the central tendon of diaphragm. The diaphragm is shaped as two domes, with the right dome positioned slightly higher over the liver. The diaphragm has two surfaces: thoracic side (contact with heart pericardium and lung pleura tissues) and abdominal side (contact with liver, stomach, spleen & upper portion of kidneys). It fills space below lungs and acts as a passageway from major vessels & esophagus. Perimeter attachments are the xiphoid process and costal margin, the 11th and 12th rib shafts, and to the vertebrae (L1-L3) by a tendinous extension called the crura. The ventral curves of these ribs can pull the diaphragm. Any tissue restrictions of the respiratory diaphragm, rib cage, or umbilical cord remnants can alter the resting placement of the LES.

Abdominal Plexuses and Enteric mesh

Hepatic Plexus, peritoneum & blood supply

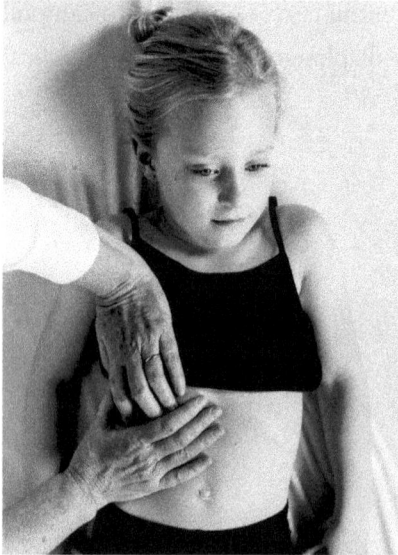

Hand One: Spread widely to cover the mass of liver, middle finger at xyphoid process and heal of hand at the curve of 10th rib meeting costal cartilage. Middle of hand tunes into the costal cartilage curve and space where liver and respiratory diaphragm meet. Visualize surrounding peritoneum and wide spread vasculature of the organ.

Hand Two: Fingertips panned in distribution of liver mass to assess hepatic tone and plexus branching (lines of pull).

Surround and support the liver. Hands can adjust position to consider the liver tone of fascia suspension by covering the 4 right lower ribs. Apply only Responsive Hold, locate global or focal spots of tension. Tune into peritoneum movement relationship with stomach, respiratory diaphragm (draped superiorly), & ANS of circulatory branching. Hold space long enough to allow self-correction of tone & movement against neighboring organs.

Don't spend more than a minute here or you run the risk of overloading and alarming.

Celiac plexus, peritoneum, & blood supply

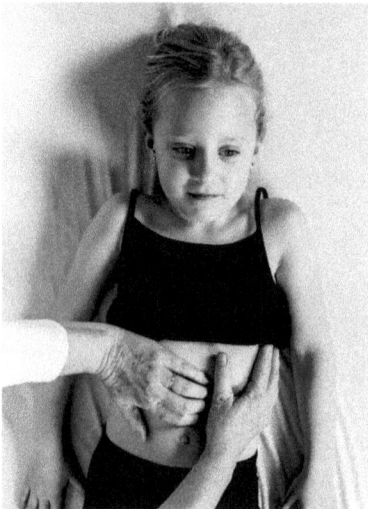

Hand One: On lesser curve of stomach, at medial curve connecting LES & pyloric valve. Spread fingertips over vicinity of celiac plexus.

Hand Two: Covers mass of greater stomach wall & four lower ribs. Assess tone of organ.

Surround & support stomach; listening to peritoneum, tone related to celiac innervation, and movement relationship with respiratory diaphragm draped over superior surface. Apply Responsive Hold, giving adequate time for self-corrective relaxation, change in organ tone, and movement with neighboring organs.

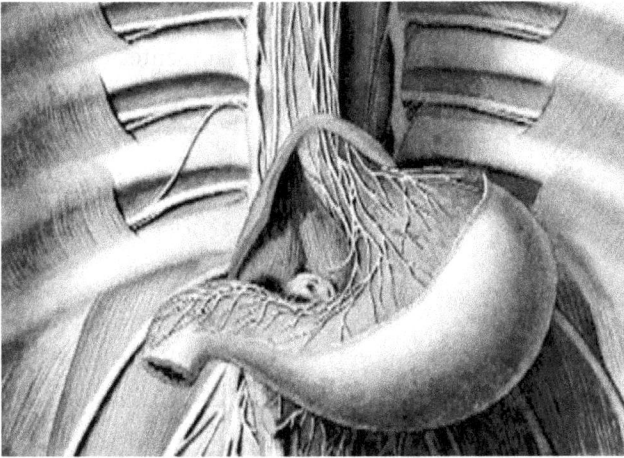

The portal triad (hepatic artery, portal vein, and bile duct) is surrounded by ANS neurons and ganglia of the hepatic plexus. It is considered an extension of the celiac plexus.

Mesentery: Enteric tone of intestines, mesentery vasculature, circulatory branch mass –2 parts

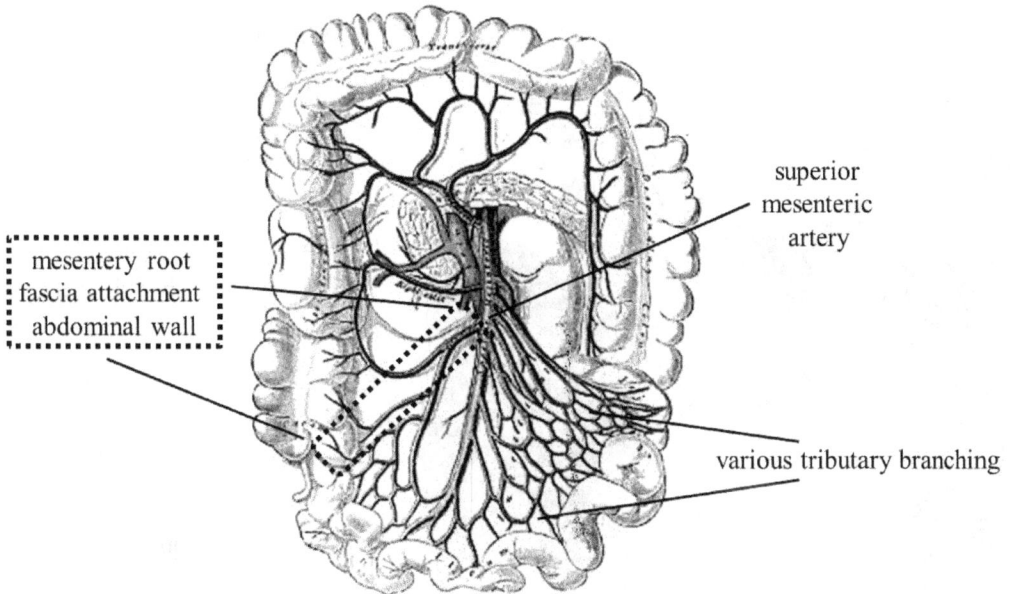

mesentery root
fascia attachment
abdominal wall

superior mesenteric artery

various tributary branching

Estimates of the percentage of vasculature woven through mesentery connective tissues range between 20-25% of total body concentration of arteries. That makes the mesentery a perfect place to treat the ANS by promoting vasodilation and reducing whole-body protective retraction. It is not critical to have hands on precisely any particular artery. Arterial walls are connective tissue extensions of fascia, both types of tissue respond collectively. There'll be a collective tone for the region and that is the focus; a mass vasodilation and fascia softening effect. Balanced Enteric tone is the goal.

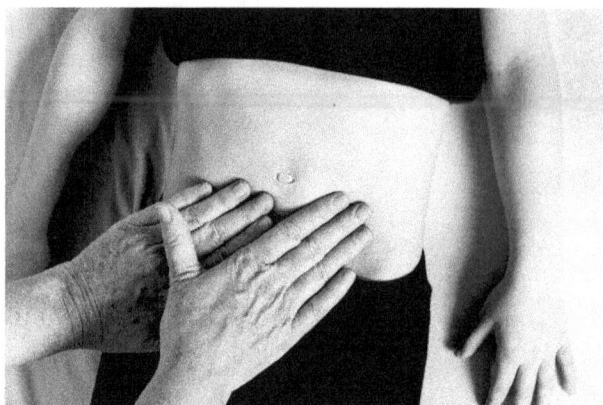

Both Hands: Gently rest both hands over the mesentery vicinity; the space inferior to navel, above pubic bone and between iliac crests. Wait to be informed by the Enteric nervous system of tone and tension of intestinal walls, and appreciate the ANS vasculature continuum. Apply Responsive Hold before any other visceral techniques to allow self-correction of protective traction, either caused by a chronic condition or from the tactile intrusion into their system. Then tune into the tone of smaller regions with a bit more precision. Detect and treat any specific lines-of-line, particularly radiating from umbilicus, but only after a global parasympathetic response of relaxation occurs.

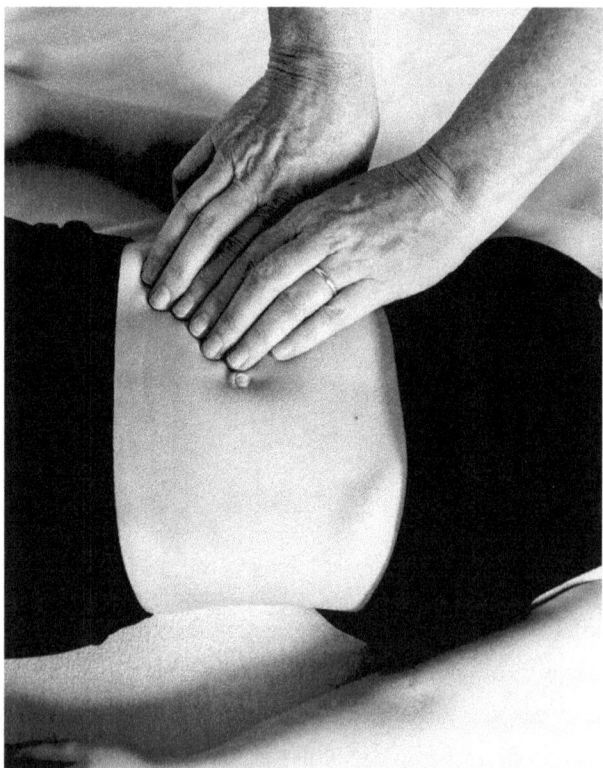

Both Hands: Place fingertips on the line between navel and iliac crest of right hip. Resting over vicinity of fascia suspension field of main mesentery artery, you are also on the central mass of the mesentery connective tissue.

Apply Responsive Hold over this line and wait long enough, refraining from any invasive touch, for as long as needed, to ensure a full relaxation, softening response. Hands can then follow any tension patterns off of the root line. Exquisite softness is required or a protective retraction will most likely occur due to the volume of investing fascia for blood vessels. In children there is seldom a need to mobilize structures or lines-of-pull unless a surgical scar or adhesion is present. For ANS behavioral changes, waiting for self-correction of protective retraction is usually all that is needed. **If tension increases, you've been on too long or your pressure is too deep.**

Remnants of umbilical cord
(vestiges of retracted veins, arteries, suspension ligaments & investing fascia)

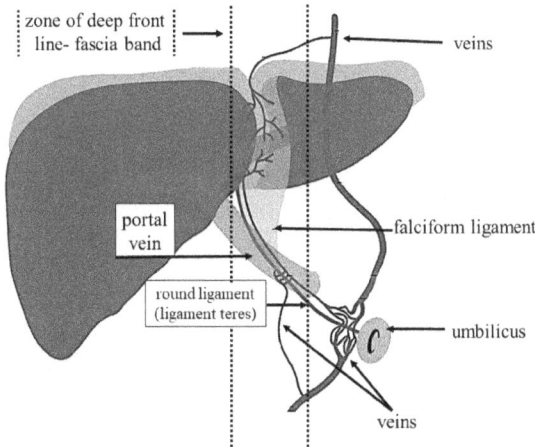

Drawing by David Gendy: Creative Commons Attribution-73812

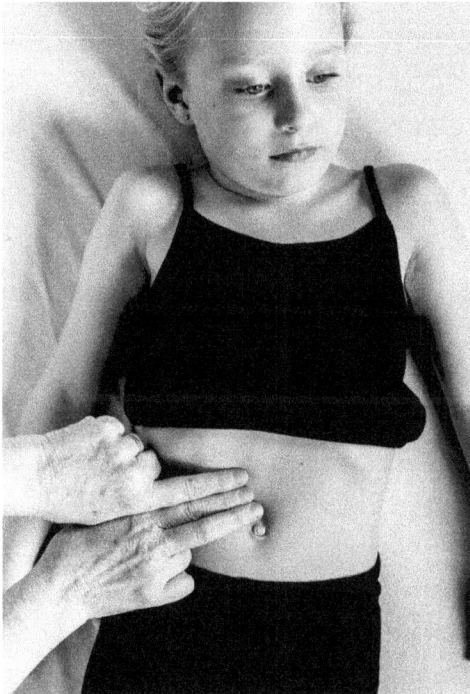

A correlation to restrictive tissues at umbilicus to behaviors of self-regulation was a significant discovery when working with babies. Tensions can have a restrictive effect on optimal respiratory diaphragm movement and tidal breath, and ease of stomach & liver physiology. Falciform & round ligaments and projecting fascia extending to superior border of liver, attached to respiratory diaphragm can be Such tensions can be retained umbilical cord pulls, in utero positioning, or deep front line of fascia. A history of colic, digestion, torticollis, and tongue ties are often related to restricted ligaments, deep front line of fascia, and investing visceral fascia & peritoneum. Furthermore, leaky gut, latent infections, less than stellar foods, and pathogen load in the gut is often correlated with organs tension & vasculature constriction. Nutrition and functional medicine considerations should be considered when structural tensions persist in this region.

Both Hands: Fingertips extending superiorly and laterally right off navel forming a continuous line over vicinity of investing fascia around falciform and round ligaments.

Apply Responsive Hold and treat lines-of-pull with non-invasiveness of touch input to avoid alarming organs and vasculature into protective retraction. No need to directly stretch these ligaments in children. Self-correcting and parasympathetic response has proven to relax and soften the region, reducing organ and vasculature stress. Be aware of the exquisite sensitivity due to concentration of nerve plexuses and condensed vasculature under your fingertips. In addition, the location of the cisterna chyli (visceral lymph node and chain for organ drainage) can also be structurally compressed or torqued by fascia restrictions from this point towards the midline diaphragm.

Elimination Centers– kidneys and biliary system

Kidneys, adrenal glands, & renal plexus

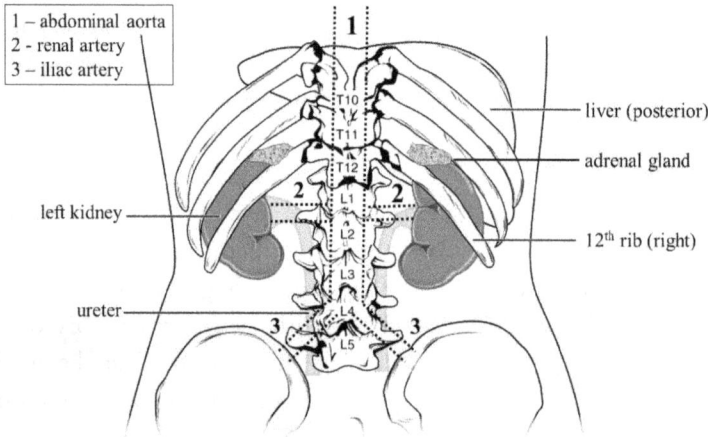

The kidneys, nestled posteriorly within the curve of the 11th and 12th ribs, are located over the medial aspect of the psoas muscle & lateral aspect of quadratus lumborum. Adrenal glands sit atop each kidney. Both structures surrounded by fibrous perirenal fascia and surrounding cushion of perinephric fat. Each kidney are layered structures woven with massive microvasculature. Several spinal nerves innervate kidneys and adrenals that form a complicated renal plexus (T12-L2), scattered between the two organs. Prominent renal arteries and veins form a structural "bridge" between both kidneys via investing fascia.

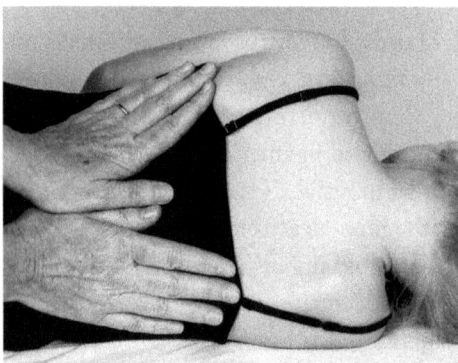

Hand One remains on central occiput: Remain focused on communication link in toward brainstem/ pons housing cranial nerve nuclei. (not pictured).

Hand Two: moves to same- side posterior of body and covers the last four ribs, just lateral to the midline, embracing the vicinity of one kidney

Alternative: Both hands cover vicinity of kidneys and adrenal glands. (Shown in photo).

Apply Responsive Hold and tune into the periphery of each kidney and adrenal gland. Appreciate baseline tension of investing tissues, holding space until routine types of tissue release occur (softening, warmth, tension lessens). If any tension is revealed medially at the fascia bridge of the renal arteries and veins and renal plexus, focus fingertips with specificity there to promote tension release. Continue with Responsive Hold and wait for indications of complete release.

Biliary system: promote organ drainage & open detoxification pathways

A quick way to facilitate parasympathetic tone (and promote lymphatic flow) is to facilitate liver drainage. Stagnation in the liver can be a result of systemic inflammation, pathogens, food byproducts (such as high fructose corn syrup). Inflamed walls of small intestines (leaky gut) can also restrict optimal drainage through the sphincter of Oddi (hepatopancreatic sphincter through ampulla of Vater). Chronic vasoconstriction can result as a protective measure to prevent inflammatory foods, proteins, and pathogens from entering blood stream from intestinal tract. Higher tone of the liver and adjacent digestive organs can also result from structural impediments or postural stress.

Treatment is focused at the vicinity of investing fascia for the common bile duct and the biliary branches. Assisting drainage of organs (liver, pancreas, gall bladder) through the biliary system not only assists self-correction in a detoxification process, but also routinely demonstrates changes in calmer behaviors and happier moods. With *Bodywork for Sensory Integration*, especially for children, we are not mobilizing suspending ligaments, merely assisting a self-correction by promoting vasodilation and expansion of interstitial tissue, which in turn facilitates movement of fluids.

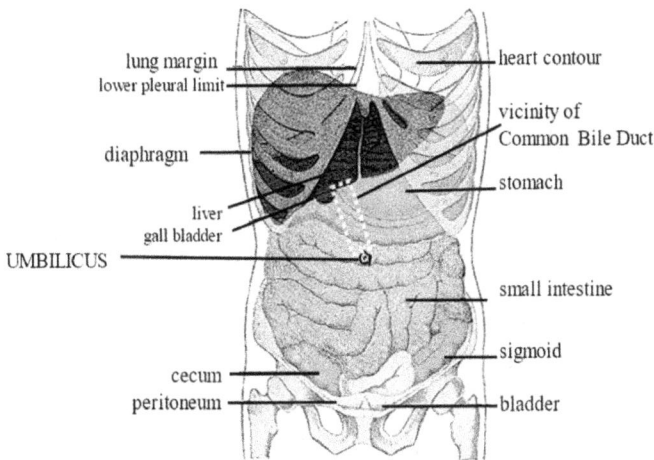

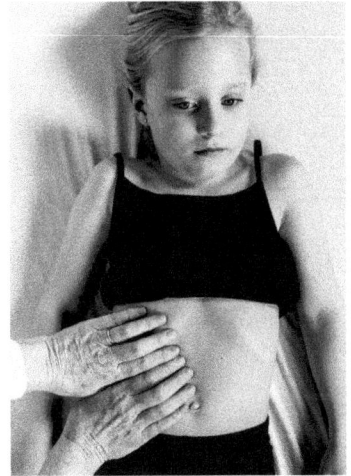

Both Hands: Line 6 to 8 fingertips over line between navel and midpoint of right costochondral arch (central point of liver mass). **Hand One** over the path of the upper bile duct and **Hand Two** fingertips spread over fascia vicinity suspending common bile and associated ducts. Sphincter of Oddi will likely be palpated by the 3rd or 4th finger of Hand Two. (The falciform ligament projections can also come into play with tension found here). Also, keep in mind that the biliary system is nestled behind the liver mass; your hands are reaching the ducts energetically and relying upon liver to assist the tissue communication. Apply Responsive Hold, refraining from any invasive prodding. Focus is on treating surrounding fascia, not a specific structure. Wait long enough, hold space, and patiently wait self-correction. It will feel like waiting for a stick of cold butter to melt from the warmth of your fingertips. Common response is audible fluid exchanges and drainage sounds once tissues self-correct, expand, and allow better drainage.

Given the organ crowding in the area, it is not easy to palpate the common bile duct directly without invasive touch. But as other techniques have proven, treating the vicinity of structures and the investing fascia around a structure consistently yields positive results (especially in young children). What appears to happen during treatment is touch contact within vicinity of connective tissues can have a global softening and relaxation effect. Routine observations during and following treatment include: audible sounds of drainage, reduced liver fullness & bloating, less stomach aches, improved regularity of bowl movements, and an expressed sense of lightness and feeling unburdened. Some adults have even reported immediate weight loss following treatment. We've had several clients have changes in lab tests such as bilirubin and lipid levels change following such treatment. We observed through our trials the consistent change in demeanor and mood. Used often in *Bodywork for Sensory Integration* to assist the body's clearing of inflammation and agents that may be contributing.

Digestive Sphincters

Fascia bands and peripheral connective tissue restrictions can hinder optimal organ motility and mobility by adding structural resistance to the action of peristalsis. The "rest and digest" state of the parasympathetic system means that peristalsis action happens readily with an absence of organ resistance. The ANS can be treated directly at the vicinity of each digestive sphincters to promote self-regulation and modulation.

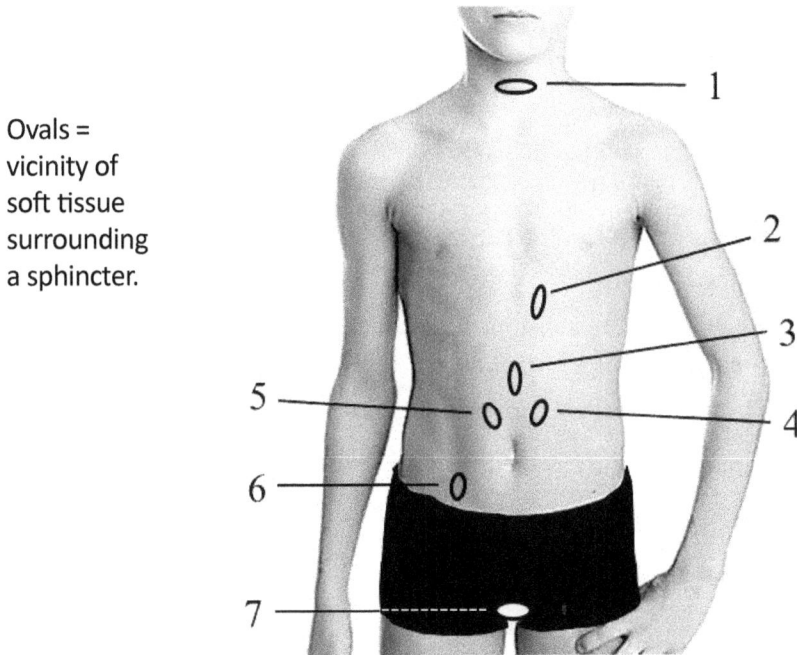

Ovals = vicinity of soft tissue surrounding a sphincter.

Upper Esophageal Sphincter [UES] (1) – vicinity is midline of throat union with neck, just under mandible; fingers are very lightly over superior portion of esophagus; UES is posterior to the larynx and hyoid cartilage position within the esophageal tube.

Lower Esophageal Sphincter [LES] (2) – vicinity is at the junction of stomach meeting esophagus, lightly lateral to sternoxiphoid joint and left 7th sternochondral joint; usually a bit left of midline; verify placement by sliding a finger along length of esophagus (starting at UES and moving caudally through thoracic cavity to vicinity of stomach entry; or travel lesser stomach curve from pyloric. A fascia connection of the length of the esophagus can be appreciated by gently holding both UES and LES.[5]

---

5 Both UES and LES can be assessed and treated along with a general sense of fascia suspension of the esophagus in relation to the deep front line fascia band using all fingers following the length of the esophagus embedded behind the breastbone and within the investing fascia of the rib cage. Palpating UES & LES, can assess and treat the esophagus length. Detect a swallow and follow the vibration of movement down its length through thoracic cavity.

Pylorus valve (3) – vicinity can be found beginning at xiphoid process and moving a finger inferiorly along midline towards the navel; easily palpated as a horizontal bump right at midline, anywhere from ½ or 2/3 distance between xiphoid and navel; verify location by feeling stomach's greater curve around the inferior margin around to the pyloric valve.

Duodenojejunal Flexure (4) – vicinity generally mirrors sphincter of Oddi's location on the left side of the midline; vicinity is along the apex of the left tenth rib curve, then approximately ½- 2/3 distances from rib toward navel; or a finger's width above the navel along the left midclavicular line; influences chyme movement.

Sphincter of Oddi (5) –vicinity is along a line between apex of 10th rib curve and the navel, then approximately half-way to 2/3 distance from mid-rib toward navel; or a finger's width above the navel along the right midclavicular line; end of ampulla of Vater off common bile duct, attached to small intestine wall.

Ileocecal Valve (6): vicinity of mid-point between navel & right iliac crest; enhances fecal movement into ascending colon.

Anal sphincter (7) – vicinity around anus; treat superficially over diaper or clothing for propriety; not a common area in need of treatment

Facilitating parasympathetic state treating the vicinities of digestive sphincters

Hand One or Both Hands: Locate each sphincter with minimal poking and prodding to avoid setting off sympathetic alarm. Rest 1-3 finger pads over sphincter and assess connective tissue tension in the vicinity. Employ Responsive Hold as if sitting atop a cold stick of butter; wait for tissues to warm and soften under your touch. Self-corrective response will occur if enough time is allowed. Anus is seldom treated in *Bodywork for Sensory Integration* unless a specific medical condition exists. You can treat one or two sphincters at a time. Due to the investing fascia connections, deeper sphincters usually respond without invasive touch or direct palpation into the actual structure. Tune into region of nearest nerve plexus for further down-regulation of tone if need be. You can treat sphincters in any order. Best completed on bare skin.

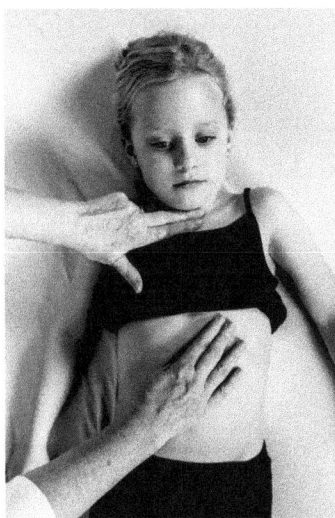

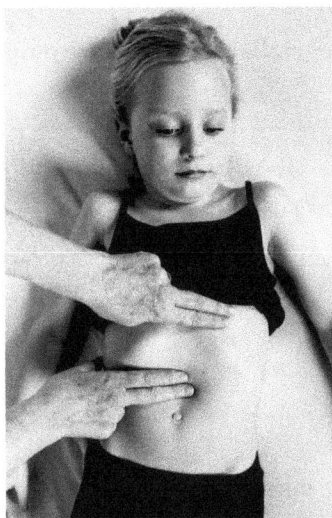

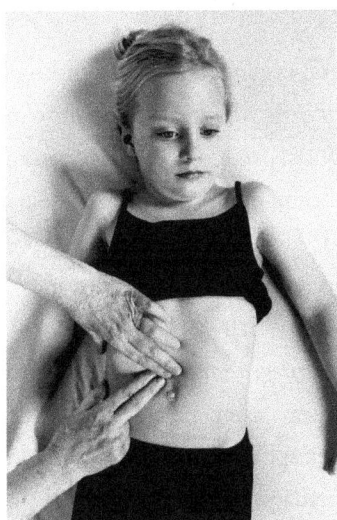

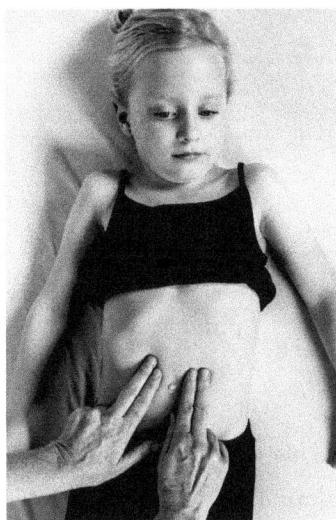

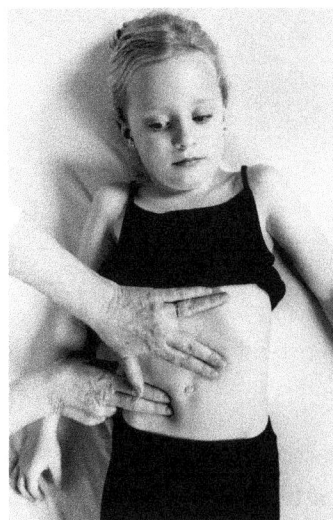

Supplemental Techniques

Acupressure + Responsive Hold = Responsive Acupressure

Combining acupressure with Responsive Hold makes these techniques "Responsive Acupressure". There's not time or space to review all the possible acupressure sites or reflexology charts for nerve meridians. These acupressure points, however, are used routinely in *Bodywork for Sensory Integration* because they can be taught as self-care and be done within the child's school environment.

- Ear auricles
- Thumb web space
- Soles of feet
- Investing fascia in iliopsoas muscle space

Ear auricles – generalized acupressure points

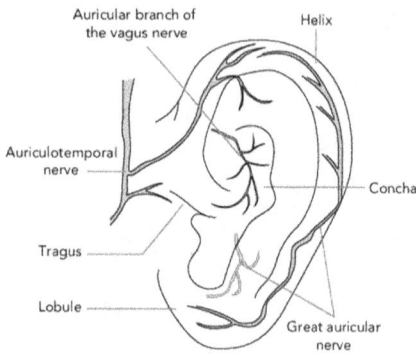

Auricle Innervation

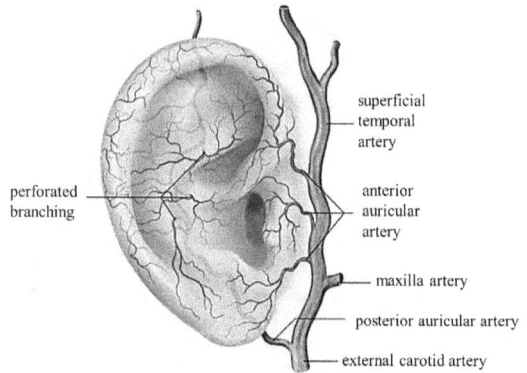

Auricle Vasculature

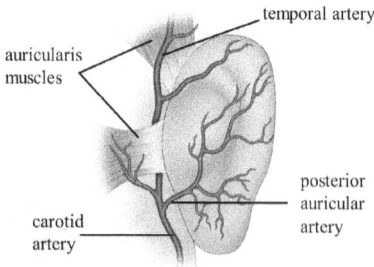

Posterior muscle attachments to skull

Many if not all visceral organs have nerve endings in the highly vascularized ears. Acupressure points can induce a relaxation response, reducing body tension and regulate ANS tone. Some points can relax sinuses, inner ear pressure, and facilitate drainage of the Eustachian tubes. Observed also to reduce dizziness, tinnitus, and irritability (especially with movement). In *Bodywork for Sensory Integration*, we have found it less important to know what points assign to a corresponding structure. Instead, we treat the auricle globally at spots of tension and rigid cartilage. End goal is to increase flexibility of the entire span of cartilage.

Ear Auricle #1

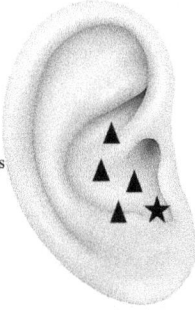

▲ = Digestive
tract organs

★ = Heart and Lungs

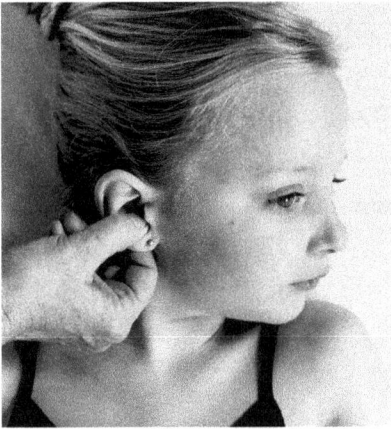

Both Hands: Place pads of thumbs deep into the valley of the intertragal notch, behind both tragus and antitragus ridges. Thumbs should contact the concha. Apply acupressure with Responsive Hold. Match the resistance of the tissue and wait for a response of tissue becoming soft and pliable. The pressure of acupressure should reach the boundaries of resistance but not too invasive to cause pain. Don't just apply pressure. Instead, make contact superficially and await cartilage & flesh release, allowing time for any protective retractions to ease. (See also pp. 64-65)

Ear Auricle #2

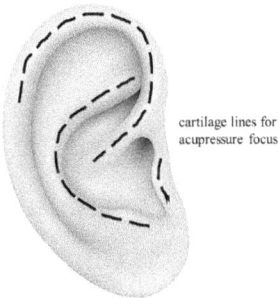

cartilage lines for
acupressure focus

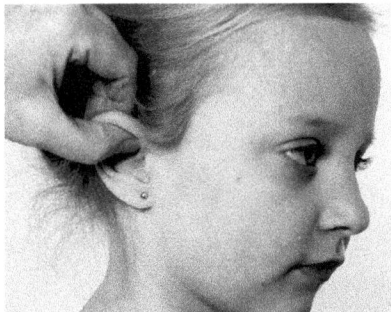

Both Hands: The auricular branch of the vagus nerve innervates ear canal, tragus, and partial ear auricle. Two other nerves innervate skin, blood vessels, and other connective tissues of the outer ear. Contact with these nerve endings is direct treatment to the vagal meridian pathways. Sites of firm cartilage may be long-held protective retraction. Practice-based evidence observes that reducing tone and tension through the cartilage of each ear reduces sound hypersensitivity in some people and promotes generalized parasympathetics. Apply acupressure in the realm of Responsive Hold at firm points of cartilage, hold until a softening and auricle flexibility is achieved. (See also pp. 64-65)

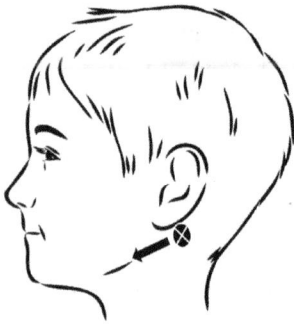

Triple Energizer - 17

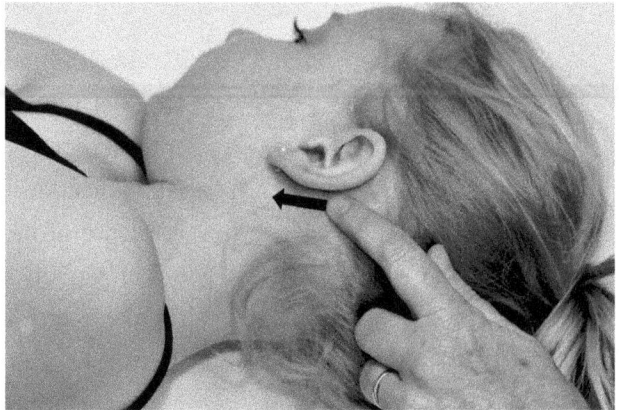

Located behind ear (TE-17): Palpate depression halfway between mandibular angle and mastoid process. Place index finger on site, apply no pressure at first, matching resistance given by the tissue. Wait for tissue response of softening. Treat both sides. Do not apply too deeply as the facial nerve exits nearby. Too much pressure can elicit a vasospasm response, eliciting the opposite of target objective. This point can help soothe face pain as well as ear pressure, tinnitus, & promote drainage.

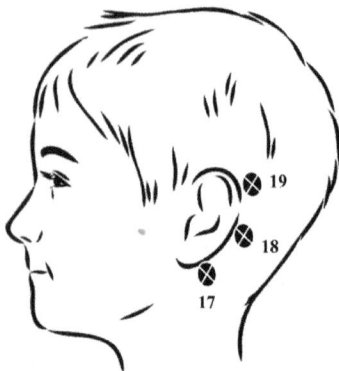

Triple Energy points

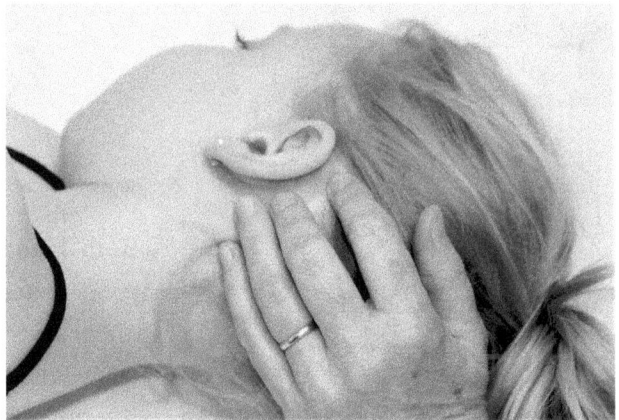

Two other points related to TE-17, can be treated individually or simultaneously. Spasm Vessel, (TE-18) right behind the ear, in the center of the mastoid process and just posterior to TE-17. TE-19, just above TE-18 1-3 fingers to locate the sites, and rest finger(s) on site(s) awaiting tissue directives. Apply acupressure that matches tissue resistance. Hold for 4-5 seconds, or long enough to wait for self-correction. Headache, tinnitus, and ear drainage are commonly helped by these acupressure points.

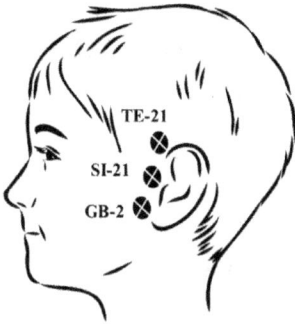

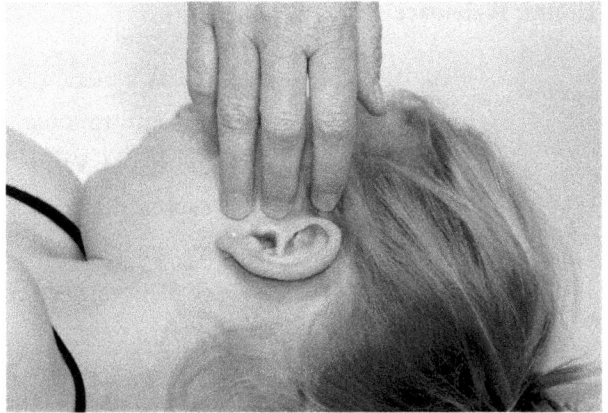

The Ear Gate acupoint (TE-21) is the top point in this three-some, located behind the mandible ramus margin. It is easier to palpate this acupoint with the mouth opening to feel the jaw movement and location. Use index finger in the depression in this area. pointer finger, palpate the depression in this area. Just inferior is the SI-19 point (small intestine meridian), and the GB-2 (gall bladder). Applying mild pressure to 1-3 of these sites for 5-10 seconds on each side is believed to soothe pain linked to TMJ syndrome, tinnitus, and otorrhea. There points can also relieve pressure inside the ears and stimulate sinus drainage.

Thumb Webspace

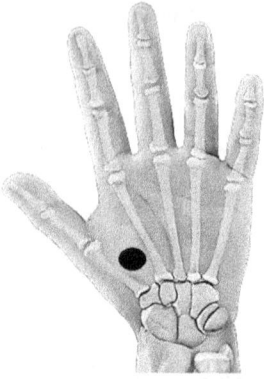

A general down-regulation point for the ANS. Often a point to treat headaches, most commonly known as the Hand Valley Point. This point has proven helpful for nausea, dizziness, general pain & inflammation. It is helpful for people with motion sickness, vestibular processing challenges, head rushes following bodywork treatment.

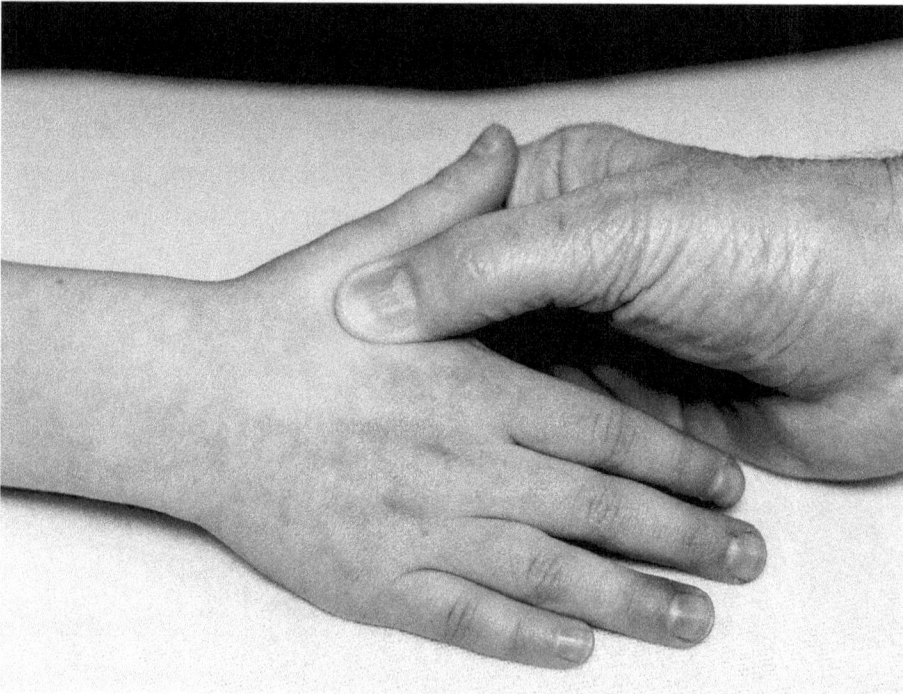

One Hand: Make contact with thumb and index finger into the most proximal tissue of the thumb web space. Thumb on dorsum of hand between 1st and 2nd metacarpal bones, into the most proximal tissue of the thumb webspace. Just distal to the ends of the metacarpal heads. Evaluate and match degree of tension and tone, and wait long enough to allow protective retraction to ease. Match tissue resistance, applying acupressure with Responsive Hold until muscle mass softens, expands, and eases its tension. Too deep and aggressive of pressure will likely alarm the sympathetic and cause pain. This is not an exercise of dominance over tissue, rather an exercise in patiently waiting for self-correction of the ANS. Teach this as self-care.

Soles of Feet

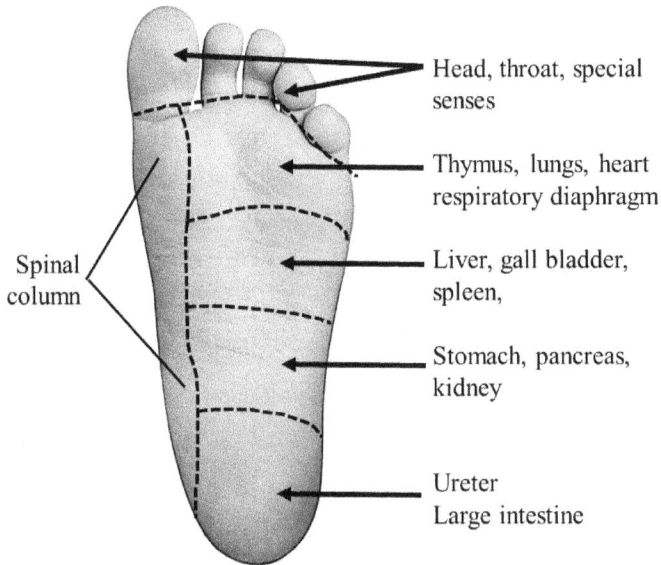

Head, throat, special senses

Thymus, lungs, heart respiratory diaphragm

Liver, gall bladder, spleen,

Stomach, pancreas, kidney

Ureter
Large intestine

Spinal column

Reflexology (acupressure based upon acupuncture principles) involves applying pressure to specific points corresponding to various organs and body structures, according to ancient Chinese and Eastern health practices. Believed to have multiple health benefits including general stress reduction, assisting digestion, and promoting sound sleep. All factors that contribute to efficient sensory integration.

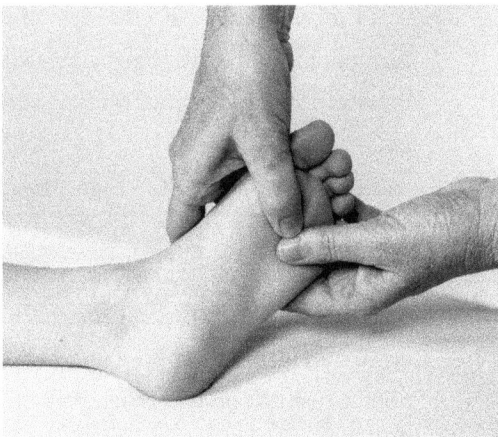

Both Hands on One Foot: No particular point. Gently push into skin surface to detect localized tension, drawing in, or a spot of higher tone. Reframe from massaging. Apply Responsive Acupressure and wait for tissue to self-correct and release. It is very common to observe release effecting the entire body and the client may report feeling other areas reacting.

Iliopsoas muscles with extensive investing fascia

The fascia in this vicinity projects to the lateral aspects of small and large intestines and the mesentery. The lower branches of the lumbar plexus and the substantial femoral nerve innervates and travels through the region. Practice-based evidence commonly reveals a tensegrity relationship between the iliopsoas complex and mesentery flexibility, motility of the lower digestive sphincters, and often with projections up into the respiratory diaphragm and celiac (solar) nerve plexus. Peritoneum of neighboring organs can also have a structural pull this muscle mass (kidneys and adrenals are common restrictions with the psoas minor).

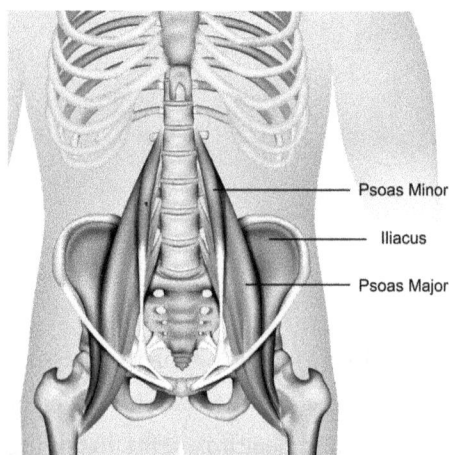

Psoas Minor
Iliacus
Psoas Major

A slew of behavioral manifestations related to iliopsoas restrictions can be decreased comfort with eating & eliminating, enuresis, and overall self-regulation and modulation. In some Eastern traditions, the psoas group is said to be the seat of the soul, storing trauma and emotionally charged tension. Can affect posture & pelvic mobility with quick postural adjustments, balance reactions, and coordination. Child will stand in lordotic posture if tight & restrictive. Investing fascia borders peritoneum, mesentery, and blood vessels. False assumption that all children work out tight hip flexors with time spent on tummies and other movement experiences. Babies taught us this muscle group can hold key remnants of fetal flexion with fascia restrictions, leading to structural stress on visceral organs, posture, and balance.

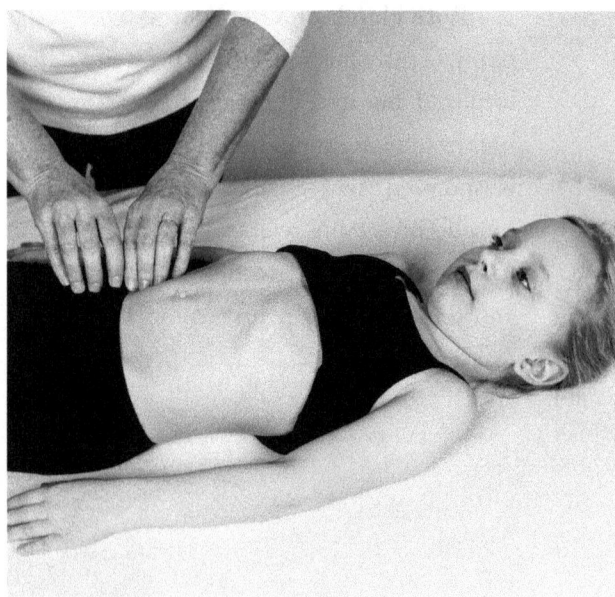

Both Hands: Focus on the surrounding and suspending fascia that invests within the layers of the iliopsoas structure. Combine Responsive Hold and acupressure with CST and treat any line-of-pull that is detected. Wait as long as it takes on skin's surface, matching tissue resistance. Be exquisitely gentle with palpation and do not rush to add additional pressure for risk of setting off sympathetic alarm. The psoas group plays a key role in fight or flight response, especially if chronically tense.

Soft Skeleton Matrix: superficial & peripheral connective tissues

Organized for clinical application

- Skin and scalp pp. 196 & 202
- Vascular walls of epithelium p. 196
- Lymphatic system pp. 198-201

Key Objective: Facilitate a parasympathetic state in superficial fascia, the lymphatic and vascular mesh, and the massive number of touch receptors. In skin, reduce activation of C fiber receptors through releasing microscopic fascial tension and promote vasodilation. Balance the tone of arrector pili muscles around hair follicles. Enhance fluid exchange for detoxification and cleansing cellular terrain.

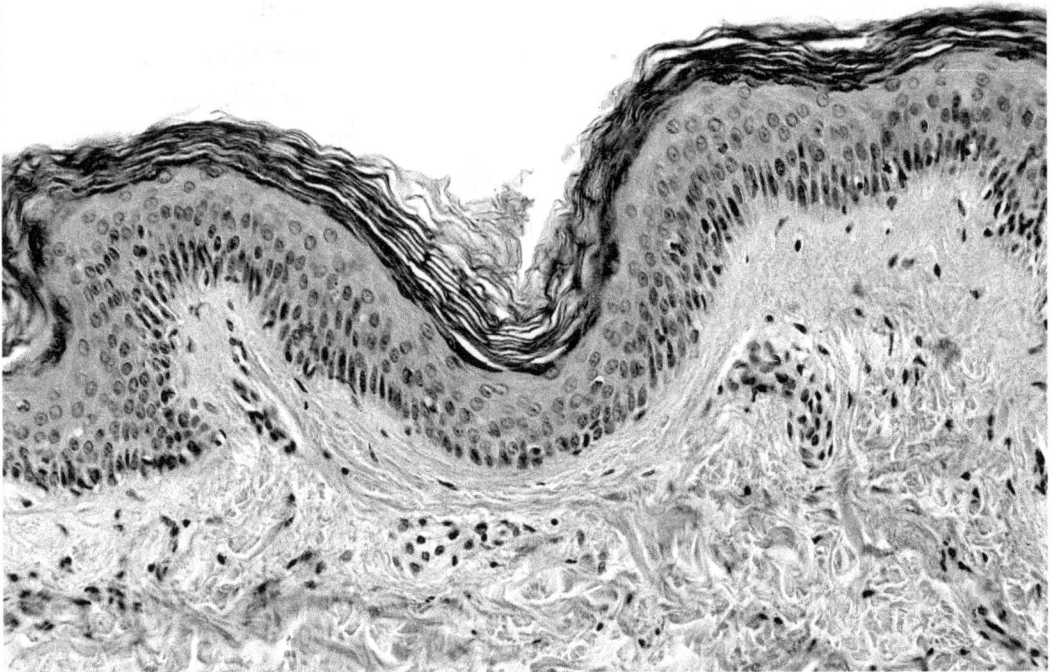

Electron microscopy of the layers of human skin.

Skin and scalp

The skin is a significant gateway to the ANS because of its size. Along with a massive number of touch receptors, sympathetic receptors (C fibers) also await contact with the outside world. C fibers are fine, free nerve endings that wrap around the root of hair follicles or can be suspended freely in the dermis. These ANS cells are so sensitive that they are activated by vibrations in the air or from the slightest contact with clothing. Skin contact can be relaxing and calming through the physiology of touch, or contact can elicit the opposite effect.

The level of touch intrusion is an essential factor to consider with the child, their level of protective retraction, and their system's preferences. Tension from protective retraction has not been routinely considered a factor in sensory behaviors, but it can be revealed by skin tautness. When we only think of muscle tension as an indicator of stress, we miss subtle tension in the skin. ANS states and behaviors can be felt, respected, and assisted toward a global parasympathetic response, starting at the skin level. The subtleties of epidermal layers over superficial fat, fascia, and muscle can be appreciated by the therapist. It is best accomplished by hands that don't move and don't invade. Activation of C fibers can set off global and microscopic vasoconstriction at the skin level. C fibers are wrapped around or near hair follicles and stimulated by mechanical or temperature changes. Painful input and temperature extremes are the natural enemy, but when C fibers are damaged, obliterated, or inflamed (as shown in children with autism and touch defensiveness), the nervous system can be held in a perpetual sympathetic state of hyper-reactivity. The child's sensory tolerance dictates their comfort with your approach toward their skin. (See more about Touch Defensiveness at p. 288)

Vascular walls of epithelium

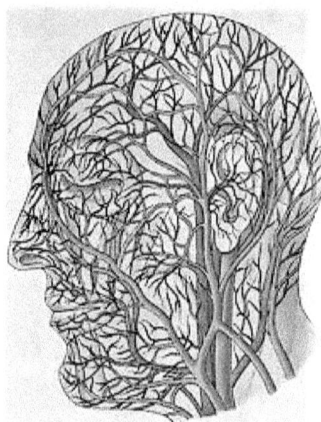

Arteries and veins of the cranium and face.

The walls of arteries, veins, and lymph vessels are similar in construction of epithelium or endothelium. Arteries are lined with an additional layer of smooth muscle to aid blood mobilization, whereas veins and lymph vessels have valves to protect against backflow and stagnation. The commonality of these fluid byways is the surrounding and investing fascia. Within existing knowledge of ANS regulation, measured by vasoconstriction and vasodilation, questions abound over the role the systemic field of fascia plays in vascular homeostasis. Protective retraction may be a whole-fascia reaction. Working with all the systems simultaneously, *Bodywork for Sensory Integration* often brings us to locations where vasculature is concentrated so that we may directly assist ANA. This aspect of anatomy gets little attention in managing disruptive behaviors related to sensory issues.

The venous system is a key detoxification pathway of the skull, with close proximity to the encircling lymphatic vessels. Between the layers of skin and connective tissues of the skull, there is a vast tributary of veins that channel between the spongey bones (diploic venous system). Neighboring arteries and lymphatic networks share the suspending fascia network. This complex vascularity is what makes the head prone to bleeding with any head injury. Lined with a single layer of endothelium supported by elastic fascia, these veins take two full years to develop in infancy. The layers of the scalp dermis and epidermis sustain a tremendous amount of pressure during birthing. It is assumed that tissues naturally decompress to help the drainage mechanism recover. Bodywork for babies proved that resolution of scalp decompression is not always complete. Scalp work is routine in *Bodywork for Sensory Integration* for that reason.

This new perspective of considering both therapeutic touch and challenges of tolerating touch input was gained by working lymphatic system with different clinical populations. Lymphatic mobilization has consistently contributed to changes in behaviors associated with irritability, clothing and grooming intolerances, food intolerances, personal space comfort. Stagnant lymph can indicate areas of past and current infections, die-off pasted to vessels, dehydrated vessels, and tightness that may be related to neuritis and neuralgia. Lymph nodes are in the vicinity of every neurovascular bundle in the body. Engorged lymph nodes have been found to be directly related to headaches, digestive dis-ease, muscle cramps and spasms, generalized malaise. This has also been observed related to both hyperactivity and hypoactivity in children. Lymphatic drainage of head can help with inner ear problems. (See more at pp. 188-191, 202)

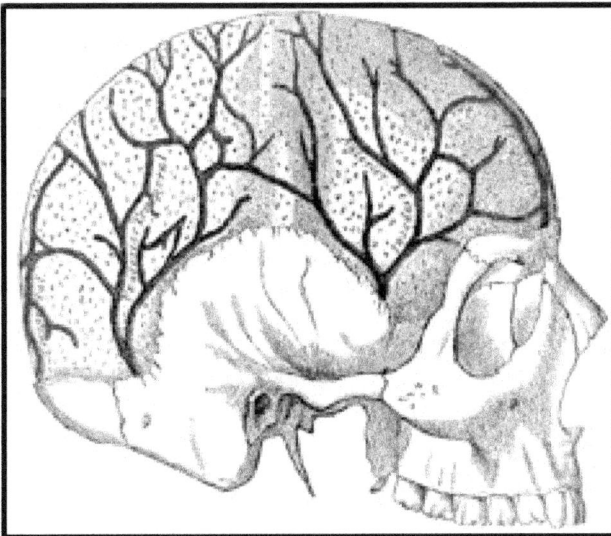

Diploic venous system meanders through porous cranial bones; drains into the dural venous sinuses.

The human lymphatic system

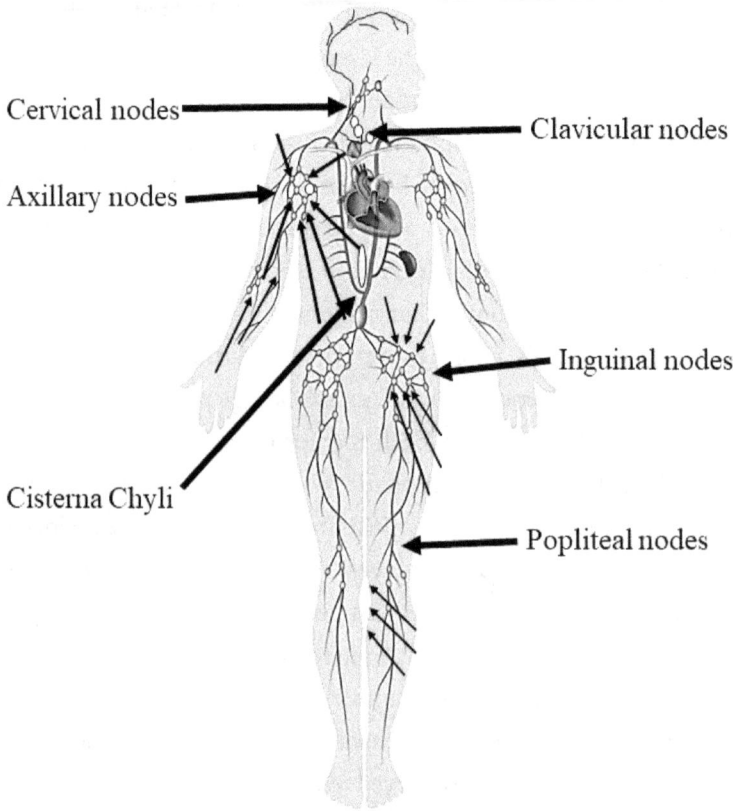

Cervical nodes

Clavicular nodes

Axillary nodes

Inguinal nodes

Cisterna Chyli

Popliteal nodes

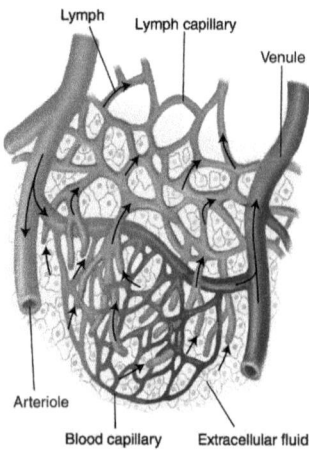

Lymph
Lymph capillary

Venule

Arteriole

Blood capillary   Extracellular fluid

Woven relation of lymphatic network with capillaries of arteries and veins

This book offers ONLY cursory knowledge of management of lymphatic system. It is imperative that the clinician be formally trained to feel the proper depth and rhythm of evacuation strokes. Of all the methods included in this book, lymphatic drainage has the most contraindications to be aware of, and the Chikly Method is preferred by the author.

In addition, clinical experiences have discovered that lymphatic drainage massage not only reduces touch defensiveness, but also can reduce agitation by finding source of pain from stagnant fluids and pressure into organs. Locating and evacuating engorged lymph nodes in muscles have also proven to reduce idiopathic toe walking.

Lymph massage should evacuate fluid in one specific direction toward lymph nodes; but not be moved in circles. Rhythmic strokes mimic the lymphatic system's movement, touch should match tension of the system without pressing hard on the vessel field. This method is replicated throughout the whole-body lymphatic system. Little literature exists that considers inflammation as an undercurrent of sensory disturbance of skin. However, lymph massage has shown consistent help in sensory disturbances of the skin (in children and adults). (See more at p. 298)

Lymphatic drainage: feel for skin motility and match the resistance

Axillary nodes: Static touch (begins); Moving touch (just enough to move water)

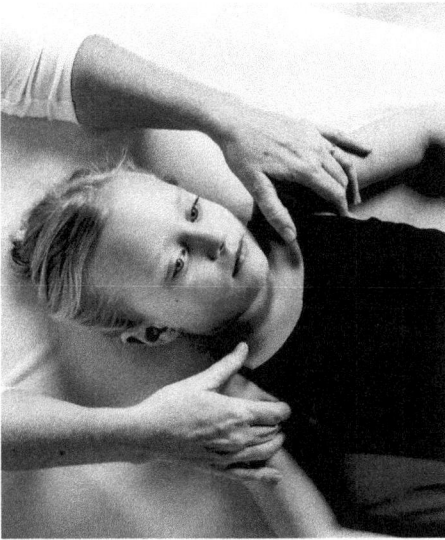

Both Hands: With soft pressure, rest 2-3 fingers in child's axillae. Before beginning, rest and listen with fingertips to discern degree of fullness, node engorgement, and stagnation. This first stage also allows for protective retraction to ease. Gently, with rhythmically sweeping stroke, evacuate any congested fluids in the axillary nodes towards the axillary cavity. This moves fluids towards the subclavian node chain and the subclavian vein. Clearing axillary nodes help create a siphoning effect for each upper body quadrant. Its most ideal to perform the common bile technique to drain the liver first (see p. 183) to facilitate a whole-body detox movement.

Lymphatic Drainage Inguinal modes

One Hand: Rest flat hand over the surface of the inguinal lymph node cluster. Do one side at a time to manage the degree of fluid volume of the lower extremities. Hand stays lightly on surface with non-compressive touch. Stay above the muscle tissue. If hand is making impressions into flesh, your pressure is too hard and invasive, risking compressing lymph instead of evacuating and creating the optimal syphoning effect. With rhythmic strokes to match normal lymph movement, direct the evacuated fluids towards the xyphoid process / vena cava / heart. **Treat the other side**

Abdominal: Cisterna Chyli for global effect

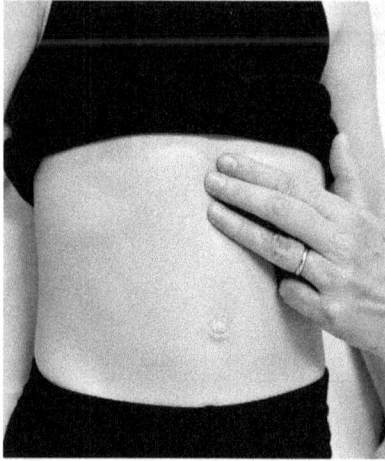

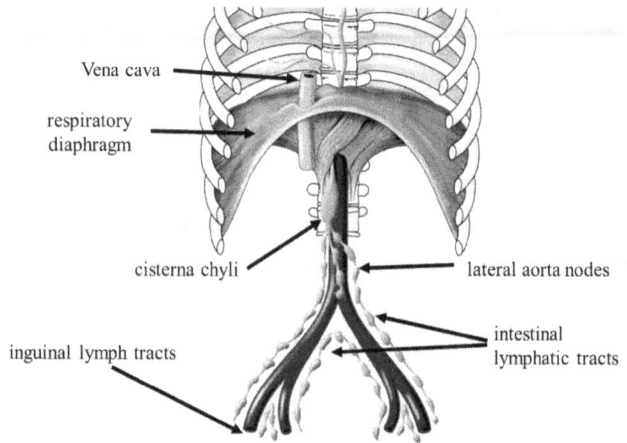

Vena cava
respiratory
diaphragm
cisterna chyli
lateral aorta nodes
intestinal
lymphatic tracts
inguinal lymph tracts

Hand One: Locate cisterna chyli 2-3 fingers rest just below xyphoid process, at or a bit right of midline; contacting inferior curve of liver. With exquisite gentleness, assess fullness. Then evacuate node fullness superiorly towards heart (via thoracic lymph duct). Proceed inferiorly towards the navel to then clear abdominal chains. Cisterna chyli is at the caudal end of thoracic lymphatic duct; it receives drainage from abdominal & pelvic viscera & lower limbs. Nestled deep in the middle of respiratory diaphragm dome, on thoracic vertebra L1-L2; posterior to liver mass. This deep node servicing organ regions can be compressed or hindered by adjacent structural distortions and fascia pulls, or stressed from gut dysbiosis, leading to stagnant terrain.

Supraclavicular chain

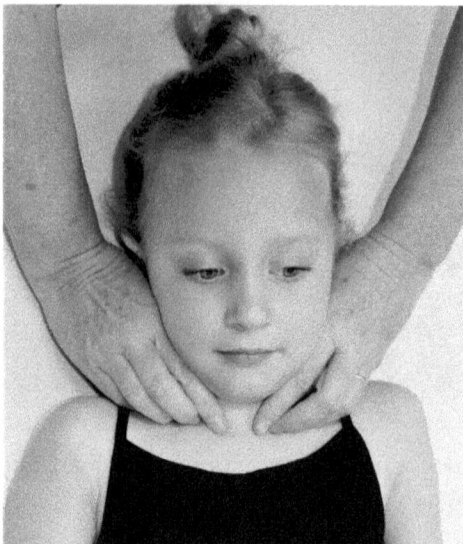

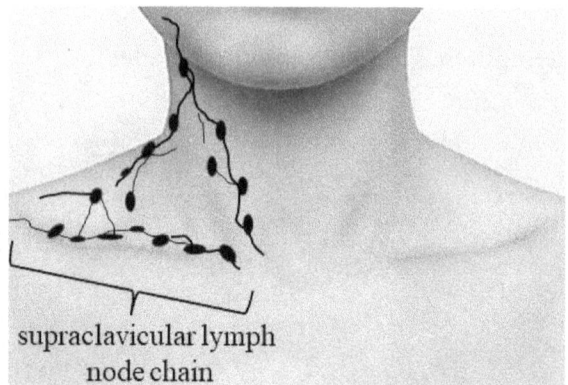

supraclavicular lymph
node chain

Both Hands: Place index fingers gently at sternal end of each clavicle, just lateral to edge of SCM muscle. Ease finger pads into soft tissue of supraclavicular space. Locate most medial lymph node of supraclavicular chain. Align other fingers along supraclavicular fossa slightly posterior to clavicle length. Fingertips are in soft tissue, not bone. If lymph nodes are congested, they are easy to palpate. If nodes are empty, the space will feel empty. Using gentle and rhythmic sweeping strokes, evacuate medial node towards subclavian veins (under sternum). Move to the adjacent node, evacuate in same manner, moving laterally the length of node chain to acromial end of clavicle.

Superficial pathway

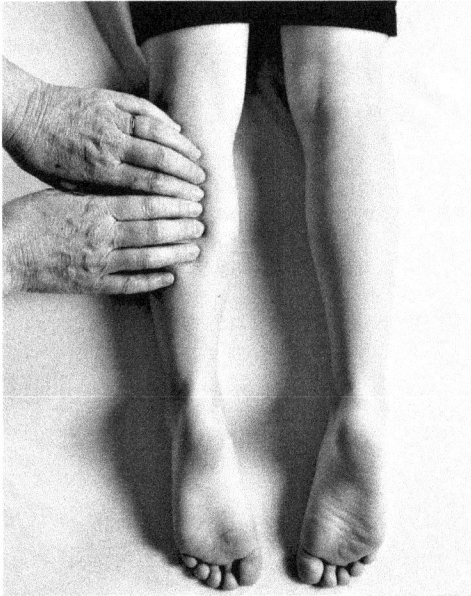

Both Hands: Though traditional lymphatic drainage was created to treat and manage lymphedema of extremities, clinical application to other population has proven beneficial for improving health and wellness. Many illnesses or toxicities can leave debris stuck in lymph tributaries and nodes, creating a scenario of stagnation and immune-responses to clear sites. Less than stellar food, food allergies, latent infections, or die off from infections can create lymphangitis. More disease and dysfunction of the lymphatic system are being recognized. We find lymphatic stagnation in muscles with growing pains and toe walkers.

Contraindications for lymphatic drainage massage:

- untreated acute infections
- malignant tumors
- patients undergoing chemotherapy
- thrombosis or phlebitis; even a risk existing
- open wounds
- conditions where increase fluid volume would stress an ailing heart
  - valve dysfunction, congestive heart failure, cardiomyopathy

One should seek formal training to learn the anatomical mapping and the directional flow, with the parallel relationship with the venous system. Lymphatic drainage has the most contraindications of all bodywork methods.

Scalp

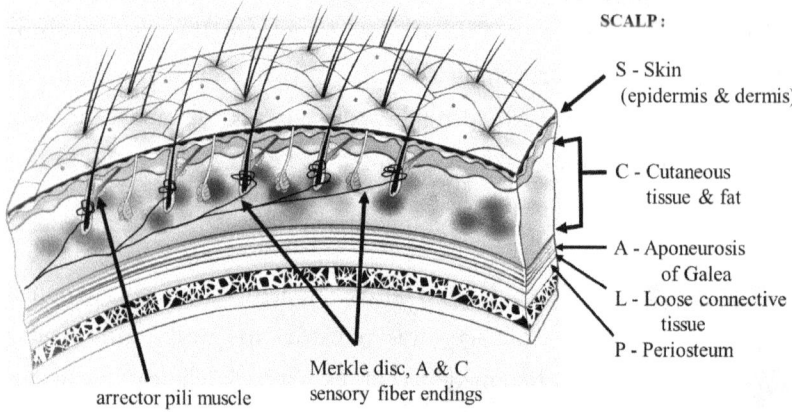

SCALP :

S - Skin (epidermis & dermis)

C - Cutaneous tissue & fat

A - Aponeurosis of Galea

L - Loose connective tissue

P - Periosteum

arrector pili muscle

Merkle disc, A & C sensory fiber endings

The scalp is a common place for concentrated inflammation and stagnant lymph, leading to hyper-sensitivities. Hair grooming and cut-ting can be prob-lematic in such a scenario. Thinnest of the layers with little else to differentiate between structures makes the scalp an easy place to reduce strain on C fibers. Furthermore, reducing structural stress on arrector pili muscles at site of hair shaft density can induce significant ANS change.

Globally mobilize dermal layers from bone, vascular network, nerve fibers

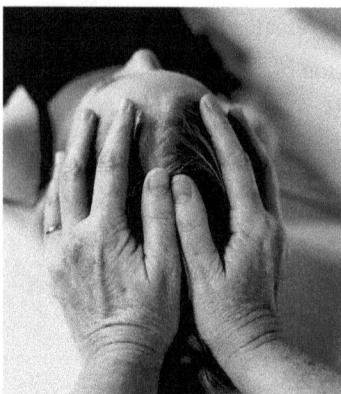

Both Hands: Cover the expanse of the scalp. Assessing tensions in the scalp and any lines-of-pull translated anteriorly and inferiorly. Tissue tension (protective retraction) reflects suspended blood vessels, cranial nerve pathways and nerve fiber endings is a contributing factor in autonomic nervous system state.

Both Hands: Hold scalp & ask child to gently turn face/nose to one side, moving into any resistance barrier. Assess degree of any adhesions to bones and hold the barrier gently in the opposite direction until tissue releases. It's surprising how many fascia adhesions are discovered in our clinic. Any history of traumatic or prolonged births, self-abuse (head), or head injury would benefit.

This technique can be taught as Self-Care

Core Structures of Soft Skeleton

Longitudinal and Horizontal Fascia Bands

Key Objective: Employ various methods to fascia lines and connective tissue networks that constitute the soft skeleton, maximizing pliability and body flexibility. Improve posture & balance and maximize sensory conduction by releasing retained compressions or structural distortions. Bodywork for Sensory Integration can enhance proprioception, kinesthetic, and tactile processing by ensuring all receptors sites are free of adverse mechanical factors.

Organized for clinical application

- Horizontal bands and lines of fascia pp. 206-212
- Longitudinal bands and lines of fascia pp. 213-215
- Lines-of-pull through investing myofascial fields pp. 216-218

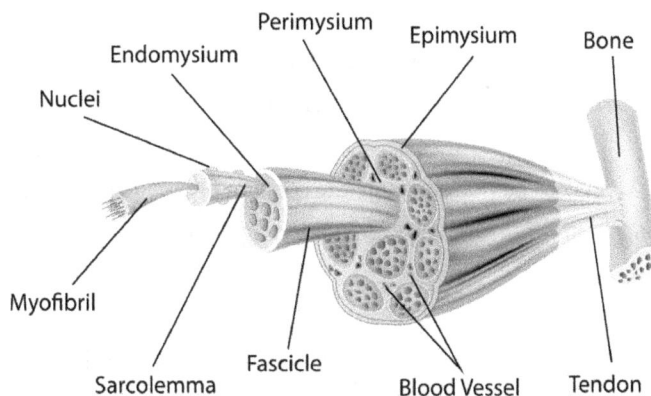

Lines of fascia enclose and weave muscle tissue. Each muscle is comprised of multiple types of tissue in a complex arrangement of blood vessels, lymphatic and nerve pathways, contractile fibers (myocytes), and multiple sheaths of connective tissue (epimysium, perimysium, endomysium). Tendon to tendon, a muscle really is a band of fascia woven with contractile myocyte fibers.

For centuries there was a perpetuated belief that fascia lacked value; it was assumed only to be the glue that held creatures together. Often discarded on the floor of anatomy labs in dissections to "reveal the real organs," the recent discoveries of fascia's purpose and function have erased such short-sightedness. Every anatomy book that does not include the organ of fascia should be considered outdated for what the world now knows about connective tissues.

All structures and organs are suspended in fascia, but there are bands (lines) that are more distinguished from other connective tissues. Connected to the finer fibers of muscles, vessels, and organs, the bands communicate local sensory signals to coordinate tone, movement, and physical effort. Each band has a regional role in orchestrating tensegrity and monitoring background sensory information. The fine fibers of structures suspended near these bands have sensory information processed through the network of fascia bands. Fascia, forming the soft skeleton, gives shape and form to the body. It is not the osseous skeleton that allows humans to stand and hold its shape. Instead, the weave of one continuous fascia field encompasses muscle cells, creating the support frame for a muscle contraction. Fascia dictates, monitors, limits, and allows for the entire excursion of muscles. There is no muscle contraction without the length of the fascia mainframe providing structural support between the bones the tendons attach to. Much sensory information is received and transmitted for the function of posture, balance, adaptive responses to forces and movement.

Meridians of fascia tissue consists of a varying ratio of collagen, elastin fibers, retinacula fibers (attachment ligaments), and extra-cellular substance (basal lamina). The basal lamina may be another circulatory system of interstitial fluids, moving nutrient-rich solution through fascia layers. Compressions, dehydration, inflammation, or injuries to the basal lamina is theorized to be the etiology for tissue adhesions. Healthy connective tissue moves easily, allowing organs to slide on each other. Adhesions are the enemy of symmetry, ease of movement, and maximal organ relaxation.

Anatomist Tom Myers has helped elucidate the concept of fascia lines and the theory of fascia's role in mobility, strength, and posture. Myers describes specific myofascial lines "hidden" within the historical bias of muscle tissue. Myers' hypothesis is that fascia forms kinetic lines for muscles to be suspended by and fascia-based interconnections affect muscle activity collectively. Upledger described horizontal bands of fascia as "whiskey barrel rings holding a barrel together. Horizontal and longitudinal with restrictions often palpated at intersections.

Myers has identified specific lines serving particular functions. For example, the superficial dorsal line extends the back into legs, extending to the plantar ligament of each foot.

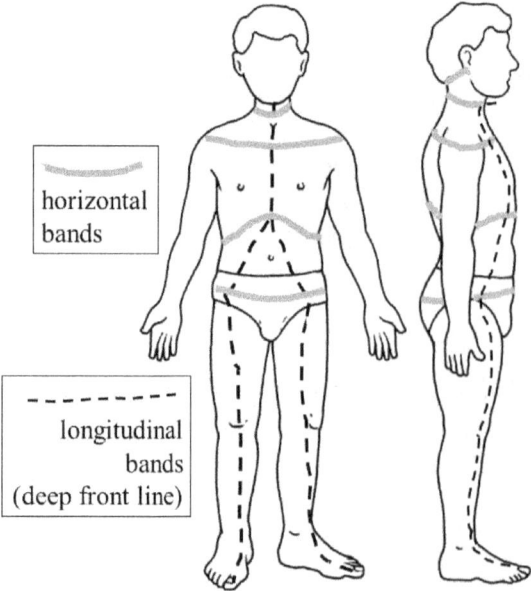

horizontal bands

longitudinal bands (deep front line)

Problems here are often propagated further up the line, possibly contributing to retained primitive reflexes, toe walking, or other background for quality of posture and coordination. Another line buried between outer fascia bands the deep front band includes the body's myofascial core. Spanning from the big toe and instep of each foot, this band passes up the legs, through the hip and pelvis, and continues superiorly along the vertebrae. Continuing upward through the respiratory diaphragm, the deep front fascia band branches to thoracic organs and up to the cranium. The midline fascia band of the tongue is a distal extension of the front band.

There is also a posterior band surrounded by peripheral spirals and functional lines from extremities and limb attachments. Housed in this posterior region is the craniosacral system within the spinal column. There are lateral band lines, as described by Myers, that frame the sides of the body, extending the length of each leg. The lateral bands crisscross the torso and rib cage, eventually attaching to the cranium after traveling around the shoulder and sides of the neck. These lines act to stabilize the torso and legs in movement relationship with head action, prompting balance and coordination. These matching side bands act in a side-bending manner, lifting hips and stabilizing the opposite side. (This structural explanation aligns with the NDT philosophy of one side moves when the other body side stabilizes). Rotation actions for posture propel a child into skill mastery of complex movements with this kind of stability - movement co-existence. The deep bands provide counterbalance for the pulling of limb movements.

The deep front band of fascia is one of the most prominent. Its function is in stabilizing the core, but clinical experience recognizes that the deep front line can be tight, shortened, or restrictive. In cases of torticollis, colic, and tongue tie, we routinely are treating for fascia restrictions in this band.

Horizontal Fascia Bands: divides and defines body cavities

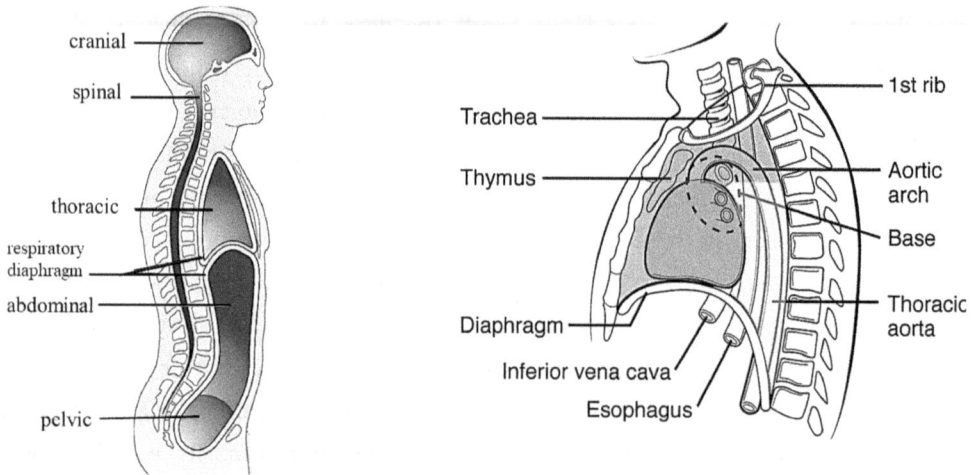

Body cavities. Source: Medical Dictionary

Cavities house organs suspended by connective tissue that forms the walls of the space. Connexions (11496 Creative Commons license 3.0: Wikimedia Commons

There are five main horizontal swaths of tough connective tissue. Upledger referred to them diaphragms. Myers calls them line of fascia bands or trains. Others have referred to these regions as transverse diaphragms. These bands create separate body cavities in which specific organs are housed and suspended. The lower one, the pelvic band, separates the urogenital and large pelvic girdle. The respiratory band constitutes the woven fascia through the respiratory diaphragm structure. The abdominal and thoracic cavities are divided by this and the organs within are influenced by the dynamic movements of this diaphragm. The hyoid bone creates an attachment for dozens of muscles, tendons, and ligaments forming a horizontal band through the neck and throat. Below the hyoid is the upper thoracic cavity and above is the oral cavity. Hyoid movements, or lack thereof, influence the movement and actions of organs in these two cavities. The last main horizontal band is considered to be the cranial base, or the fascia field suspending the occiput segments, the vast amount of suboccipital connective and myofascial tissue, and a newly discovered structure called the myodural bridge.

There are spiral fascia bands looping around the body in opposite circles, connecting the skull to thoracic spine and opposite shoulder, then down through the torso to the hip and legs. The head bone is most certainly connected to the hip bone. Rotational movements for balance and posture, and complex actions for motor planning praxis can be influenced by these spiral bands. There are also isolated fascia bands in all the extremities. These fascia bands and connective tissues may play a role in foundational muscle tone. The just right tension of fascia bands might be what creates the just right corresponding foundation for muscle tone, which will result in adaptive readiness for movement and gravitational forces.

Anatomy (fascia trains) legitimizes the science of human body tensegrity, giving a deeper appreciation of how the body postures, balances, moves, and even holds fluids and membranes up against gravity. The musculoskeletal structures theory is obsolete. Other sciences need to quickly adapt and reconsider the theoretical constructs such as reflexes. Sensory integration reflects how well a person moves in response to sensory input and meeting the demands of our physical world. Just how much fascia plays are a new paradigm to be considered (last sentence needs more work or can be deleted).

In CST we treat these bands (called diaphragms) as parts of the whole, not just treating singular bands. If the matrix is responsive and dynamic, then the tension of the bands can also be influenced by treating the whole. We are treating the whole cellular network of connective tissue fibers that lead to and away from these bands. The bands are anchoring points for fascia fibers to weave towards creating the tensegrity frame. Invasive touch here can also create protective retraction, leading to a misinterpretation of body tension, like tradition forms of working with fascia with too invasive of touch.

Considerable assumptions were made that it was possible to "stretch" fascia, but is has been solidly proven no therapeutic method lengthens fascia. How can improvements then be explained with the worldwide support of manual and myofascial methods? The discovery of the structure and function of interstitial and the cellular construction offers a new hypothesis. Gentle sustained hold (or stretch) reverses the compression of interstitial cells allowing fluids to seep into compressed interstitial cells. Noninvasive methods, avoiding compression of the interstitium, may be assisting in self-correcting rehydration of the connective tissue substate. Rehydration of connective tissues then might be what elicits the softening, pliability, and relaxation effect. Pain may be at sites of desiccated tissue, or stagnant fluids which create localized inflammation. Hydration and fluid movement may be an interoceptive sensory background signal for sensory wellness for the myofascial system.

The state of myofascial environment has rarely been considered in the realm of sensory integration practice, despite decades of treatment helping children feel their bodies improve and facilitate coordination. Sensory integration therapists have assumed the structure is always ready to conduct the business of sensory processing. Our observations indicate that structure equates to sensory function and that functional features are revealed in fascia tension and tone. *Bodywork for Sensory Integration* reveals the collective observations that structural challenges affect the entire nervous system including neurological management of sensory input. *Bodywork for Sensory Integration* has demonstrated to be helpful in "waking up" background sensations as evidenced by improved adaptive responses (often in less time than traditional therapy).

Horizontal bands

Horizontal band through pelvis

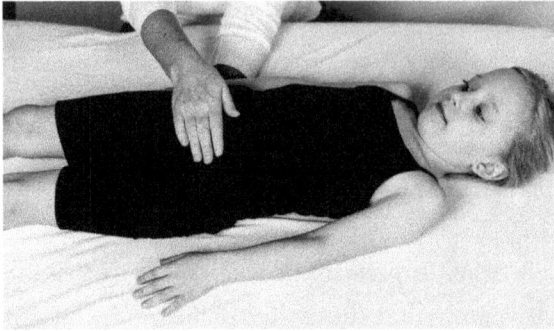

Hand One: Anterior surface of span of pelvis, between iliac crests

Hand Two: Supporting the pelvis with palm over sacrum

Begin with adequate time for any protective retraction to subside (first stage of relaxation). Apply Responsive Hold, tune into intersection with deep front fascia and pelvic spiral lines. Wait thoroughly for all the structures in the region to be palpated & acknowledged for need. Lumbar fascia lines, fascia intersections, and visceral structures such as sacral plexus, bowel and bladder suspension. Both hands are on lower margin of mesentery with fascia connections projecting into abdominal cavity. The bilateral iliopsoas complexes are lateral margins of mesentery and intestinal walls. Sacral plexus projections reveal lines-of-pull anywhere, such as lower lumbar spine, periosteum around femurs, or visceral vasculature field. Paying attention to the global effect often takes you to other organ compression, twists, adhesions. (Don't leave the area too soon before giving every structure a moment of your time or adequate time to fully benefit from your therapeutic presence).

Horizontal fascia abdominal-thoracic cavity division

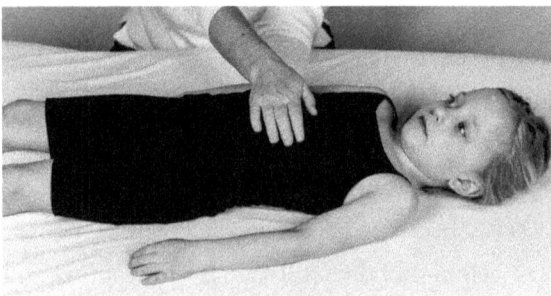

Both Hands: Repeat the same procedure as on the pelvic girdle. Bands placed on the circumference of lower ribs.

The respiratory diaphragm (between T12 and L1, L2 and L3)

Anterior neck & investing multiplane, myofascial structures

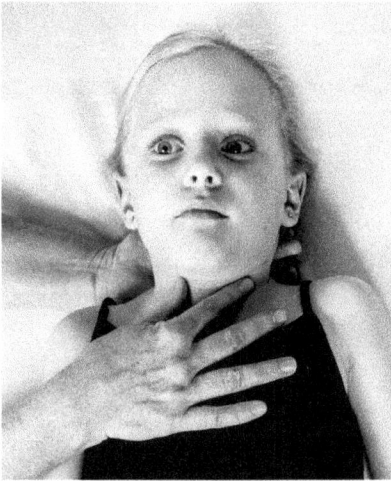

Hand One: Place index finger & thumb on the greater cornu (lateral edges) of hyoid bone.

Hand Two: Support back of cervical spine

Exquisite gentleness is required because of vast number of structures attached to or passing by hyoid. Any firm pinching could induce fascia distortion of the suspending tissue of the segments. Test gently for hyoid mobility and range of motion, & assess the relationship of investing myofascial field in which hyoid is suspended. Five paired directions of freedom of movement are optimal: three linear planes and two rotational. Treat in direction of ease with no-force, yet bring structures to any fascia barrier and apply Responsive Hold. Await tissue response. Keep in mind that the fetal segments of the child's hyoid still may not be fully mature.

Implications with history of tongue-tie, oral and tongue movements (gag response and tasting) along with emotional regulation and head/core control are related to structures in the anterior neck.

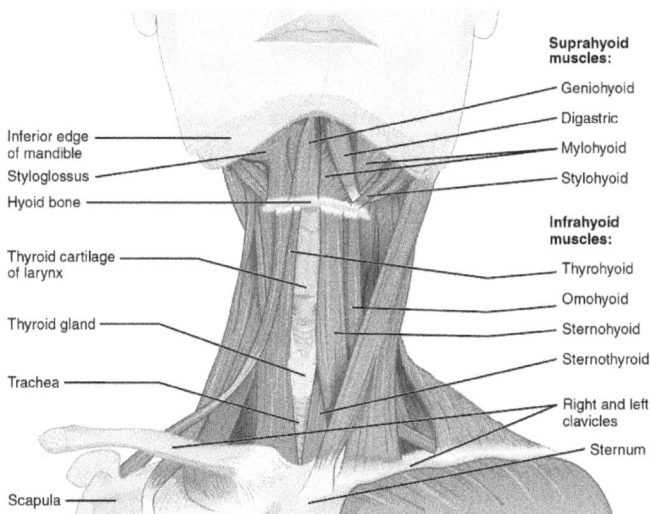

OpenStax, CC BY 4.0 <https://creativecommons.org/licenses/by/4.0>, via Wikimedia Commons

Global neck and shoulder; investing fascia of muscle terrain

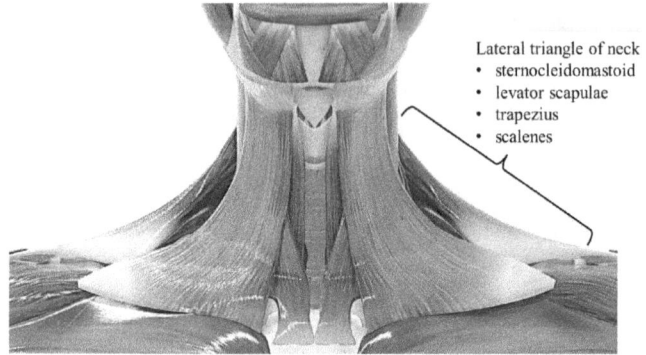

Lateral triangle of neck
- sternocleidomastoid
- levator scapulae
- trapezius
- scalenes

Both Hands: Resting over shoulder girdle make contact with span of anterior and superior surface with palms and fingers. Thumbs in contact with posterior surface touching superior angle/spine/border of scapula. Index finger along the lateral horn of hyoid, along with contact with the margin of the mandible. While in this position index finger can also contact the floor of the tongue attached to mandible curve. Assess investing fascia to structures attached to hyoid, mandible, tongue floor, manubrium and clavicular attachments, upper ribs and superior aspect of lungs (& pleura). Carotid & subclavian arteries traveling with primary branching of vagal nerve through lateral neck structures contribute to neck tone. Start globally with Responsive Hold & remain blended with tissues for as long as it takes for full self-corrective response. Use lines-of-pull to work with specificity for any myofascial tensions that manifest.

Horizontal fascia field at Occipital – Atlas junction (Cranial Base) [6]

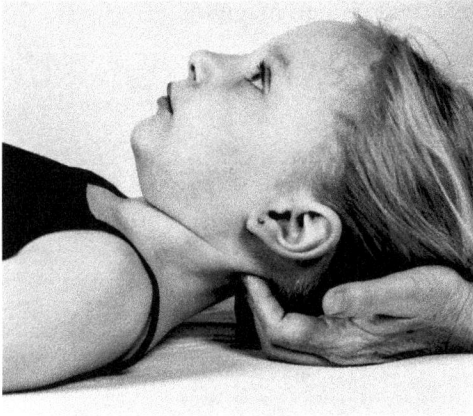

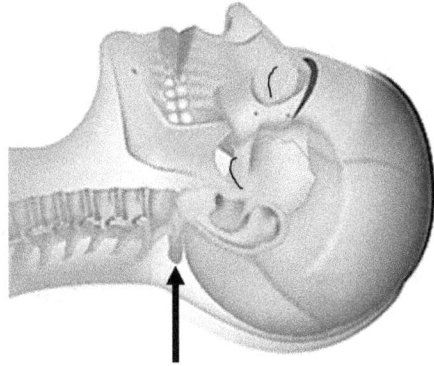

Both Hands: Two age-dependent methods

8+ years (maturation of occiput segment fusion). Align all fingers under occipital ridge. Fingertips rest at posterior arch of atlas (C1) with finger pads contacting occiput condyles. Head rests its full weight into palms of therapist's hands (not the fingertips)! This position gives the head an opportunity to slide posteriorly, stretching and soften the myofascial structures of suboccipital area, especially reaching the Myodural Bridge.

Less than 8 years of age: SKIP THIS METHOD. Instead, simply cup and cradle the occiput, as shown on page 213, without pressure towards atlas. Hold and wait for full self-corrective release with tissue-guided traction through upper cervical spine and creating space at occiput and atlas junction.

DO NOT perform this method on a child younger than 8 years of age

6 Should only be performed by manual therapist or osteopath with professional expertise

Tissues surrounding foramen magnum, transverse tissues at O-A junction

Fascial field suspending occiput segments which form the foramen magnum

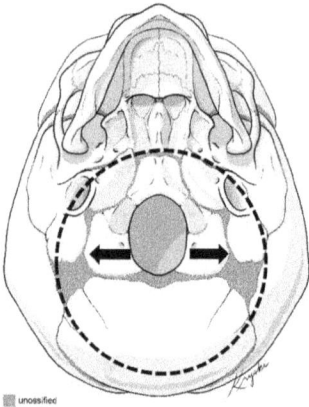

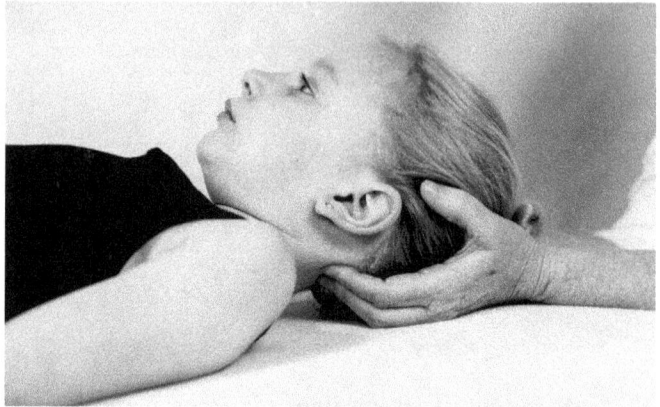

Illustration: Diana Kryski,
modified by author

Both Hands: Finger pads over lateral occipital condyles to balance fascia suspension around foramen magnum. Responsive Hold with a gentle stretch laterally on both sides. Retained compressions from birthing is commonly found here.

Spinous Processes

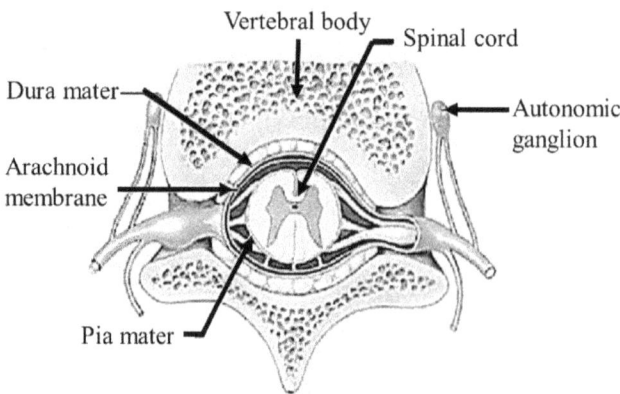

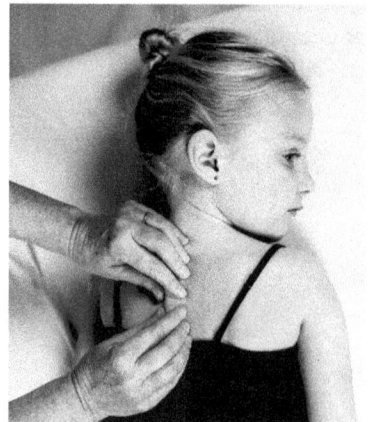

Both Hands: Gently pinch two adjacent vertebrae, apply Responsive Hold & assess tension patterns in segment's region. Blend generally through posterior fascia band lines that suspend spine. Then tune in more specifically to lines-of-pull that follow nerve pathways, revealing restrictions. Follow lines-of-pull to adjacent structures (often intercostal space to isolated fascia/lymph/nerve pathway or innervated site of muscle or organ. Sustain the gentle pinch, which engages fascia, and hold long enough for complete resolution.

Longitudinal bands

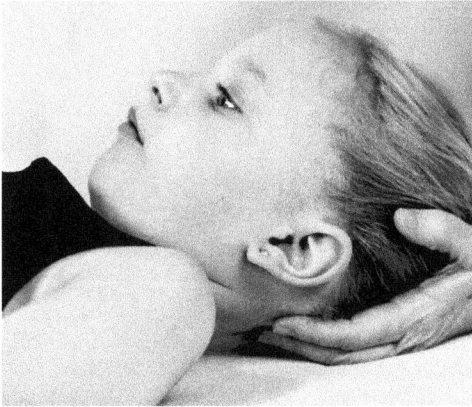

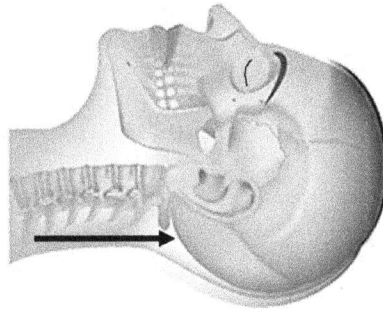

Responsive Hold direction of taking up
slack of self-expanding soft tissues
Do Not Traction with force

Both Hands: Cradle the cranium with weight of head comfortable in palms. Apply Responsive Hold with the intention of listening to the complexity of myofascial, tendons, ligaments, vascular and fluids in the neck. Wait for the natural tendency for the neck to lengthen superiorly with a "self-tractioning" effect. This is especially true in a child. Assist the posterior longitudinal bands (dural tube, paraspinal tissues, investing fascia of paraspinal myofascial structures). Soft tissues will slowly expand, hydrate, and lengthen with this supportive hold.

> **This is NOT spinal mobilization. No force is imposed to traction head or cervical vertebrae, despite a common perception that this technique is traction-based.**

Deep front line fascia band and associated structures

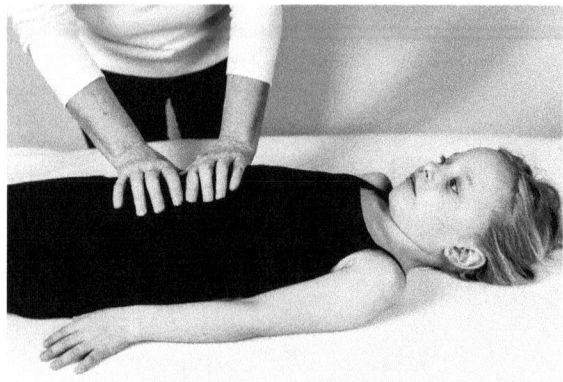

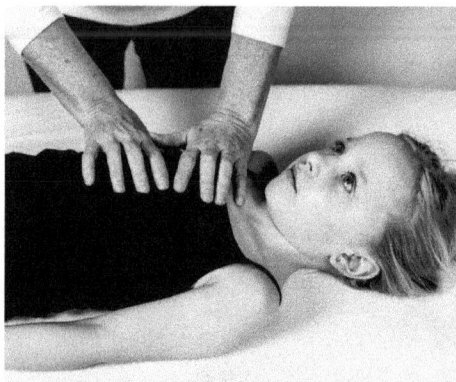

Relates to tongue flexibility, organ suspensions, nerve plexuses, and self-regulation. Arising from independent learning of work with hundreds of clients, this particular technique gained clinical favor providing bodywork pre- and post-surgical releases for tongue-ties. This fascia band connects the head to the toes, with central attachments of the middle of the tongue to the coccyx. An important anatomy train suspending the front core only recently has it gathered attention for structure and function.

Both Hands: Treat the Deep Front Line in two halves.

Fingertips pan the upper half from floor of tongue down over front of throat, investing fascia over hyoid, towards manubrium & sternum, resting at xyphoid process. (Photo does not depict upper hand making contact with anterior and suprahyoid muscles to tongue floor, but hands can shift for that placement.) Rib cage & thoracic organs are influenced by this band. Apply Responsive Hold & await self-correction. Lines-of pull technique are commonly needed.

Then fingertips pan the lower half from xyphoid process, over the solar (celiac) plexus and core of abdominal viscera, extending through the middle of the mesentery, to the pubic bone. Organs, blood vessels, & myofascial structures are suspended off this band. Apply Responsive Hold & await self-corrective response. Lines-of pull technique are commonly needed.

This technique can coincide when applying thoracic cavity, or extended from falciform ligament technique. (See pp. 175-177, 181)

Deep posterior bands and core fascia

Relates to the posterior column of fascia that surround the spinal column as well as connecting with the deeper craniosacral system of the brain and spinal cord. The craniosacral system (dura mater, arachnoid, and pia mater) is comprised of the meningeal layers and a "pumping" action that circulates cerebrospinal and glymphatic fluids. Any tension this deep directly affects the central nervous system as well as the peripheral nerve roots and branches. The dural tube (meningeal tube of spinal cord) is house below the lines of fascia that surround and suspend the entire spinal column.

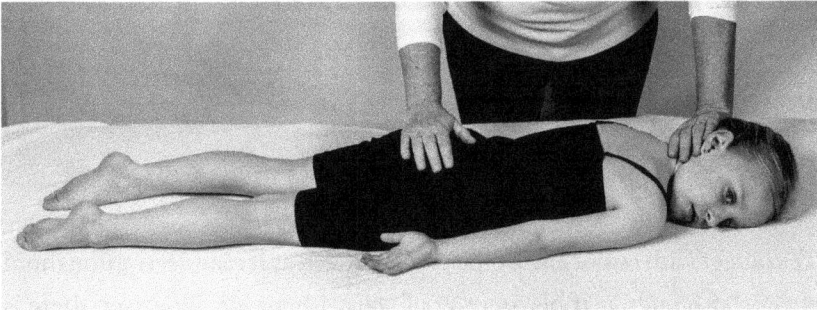

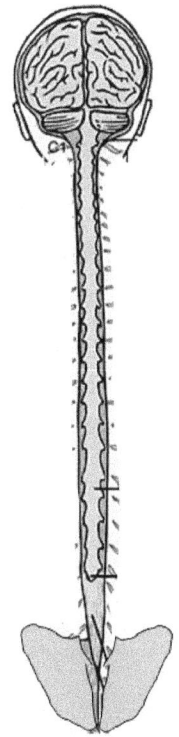

Both Hands: Can be done in supine or prone, place one hand on occiput and the other on the sacrum. This is a CST technique to monitor the craniosacral rhythm, but also an effective way to appreciate the posterior lines of fascia. Either hand can move from its starting point, (keeping the other in place), to treat any specific sites that may reveal tissue or organ tension. Begin with hands following the craniosacral rhythm, but hands can move into lines-of-pull or myofascial release techniques by what is revealed in the tissues. Posterior strains through rib cage are commonly discovered with this technique, but also dural tube tensions may have an adverse effect upon dorsal column transmission of somatosensory information.

Distinct strain patterns: Lines-of-pull in fascia

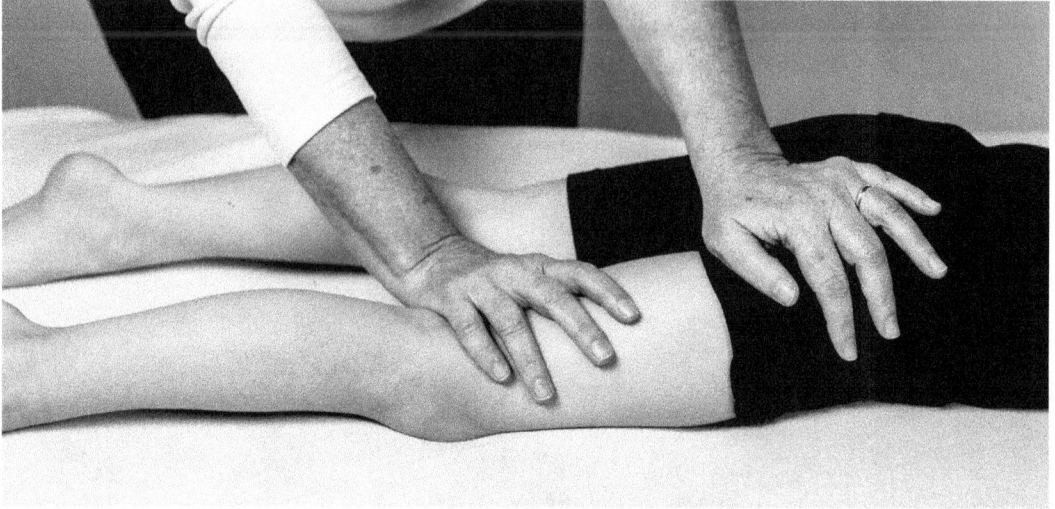

Both hands: Make contact with either end of the strain pattern as it manifests throughout the session. Gently engage both ends as if playing tug of war with a rope. However, **there is to be no tugging and warring.** This is different from a myofascial method of pulling. Instead, apply a gentle stretch, apply Responsive Hold, and sustain this stretch until you're sure of a complete resolution. Follow the tissues' self-corrective actions under the gentle stretch, follow the direction of any tissue expansion or movement,. The line is another clue of what the tissues/organs/vasculature are about to reveal. When completed, the line-of-pull should have dissipated. Recheck for any residual tension.

Lines-of-pull evolved from the traditional versions of myofascial release and strain-counter strain (for pain, joint, and posture rehabilitation). In *Bodywork for Sensory Integration*, we hold constant awareness of vasculature, nerve plexuses, & organs, along with other investing fascia attaching to muscle mass and bone. All types of connective tissue restrictions that can reveal during the therapy process can be so much more than just the musculoskeletal system these two mentioned methods refer to. A new description was needed to differentiate approaching these distinct fascia pulls.

Any direction or vectors, any soft tissue or vessels, can be involved. The fascia bands, a nerve pathway, a hyperactivated innervation site, a facilitated vertebral column segment, an organ with peritoneal restrictions, or muscle adhered to the periosteum of bone shafts can inform the hands with clear tension lines. It is optional to know exactly what structures are involved, but it is more important to wait to see where the line-of-pull takes the treatment process. Lines-of-pull aligns more with the concepts of CST for the following of tissues, but with the additional concept of treating for specificity. Lines-of-pull gives you permission to veer from a protocol you're comfortable with.

Our therapeutic hands must remain within the realm of non-invasive touch to not alarm the ANS and set off protective retraction, enhancing restrictions. Most commonly, these types of restrictions reveal themselves after superficial work. This can happen even in muscles, so too much pressure or work on the therapist's part is counterproductive. The lines seem to pull us to an epicenter or ground zero of an issue, but if we only focus on that line, we miss the messages coming from other (connected) tissues. Hold the ends and wait for the rest of the story. Something else is always revealed in the vicinity.

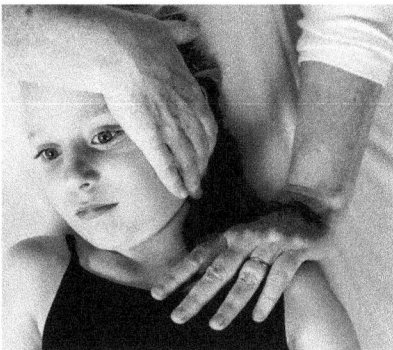

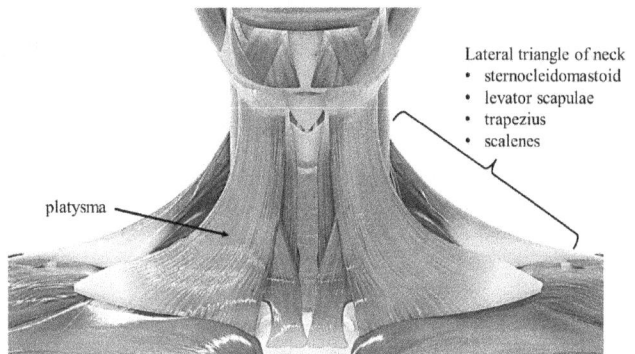

Lateral triangle of neck
• sternocleidomastoid
• levator scapulae
• trapezius
• scalenes

platysma

Hand One: On side of head with fingertips around ramus of mandible and palm in full contact of face & skull

Hand Two: Covering the shoulder complex from clavicle to scapula, humeral head and acromyun process, to the fold of the neck. Apply Responsive Hold to assess the tension and tone of investing fascia and mass of lateral triangle of neck musculature. Index finger can contact the hyoid to monitor ANS tone. Tune into the layers of structures beginning with the platysma (attached from jaw line to length of clavicle). Treat globally, allowing adequate time for protective retraction to self-correct. Follow tension to edge of restriction. Apply line-of-pull to any specific restriction that surfaces.

Treat the other side.

occiput
atlas (C1)
mastoid process
middle scalene
sternocleido-mastoid
anterior scalene
posterior scalene

tension sites that immobilize 1st rib
(raising tension in lungs & pleura)

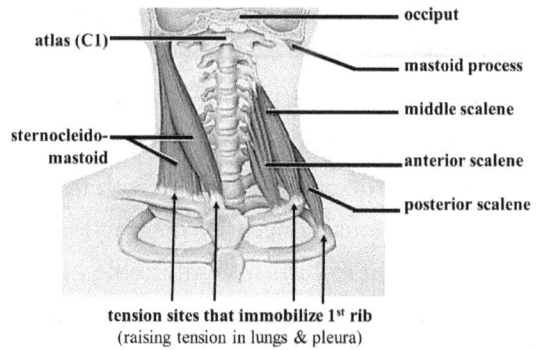

Both Hands: Remain in same position as above. Tune into the first rib by the curve of index finger and thumb. Fingertip locates rib at manubrium attachment and thumb at C6 attachment. Assess the flexibility of the rib; it should wiggle slightly like a loose tooth. Apply Responsive Hold, followed by an inferior line-of-pull to mobilize not only the upper rib but the intersection connective tissues of the thoracic cavity (mediastinum, pleura to the lungs). Wait adequate time for a full self-correction for the amount of invested fascia in the region. Completion is indicated when rib slides inferiorly and laterally wtihout much effort on your part. The secret is in holding it long enough. Too much invasion, the lung & vessels alarm.

Treat other side.

Cranium

Key Objective: Using bony canister as handles to reach the fascia container surrounding brain and spinal cord, and the vasculature structures housed below dura mater. Decompress osseous plates, using bones as handles in impose the gentlest of stretches, to promote soft tissue expansion and fluid exchange. Minute traction of spongey, expansive layer of arachnoid villa (and possibly a fourth meninge layer between arachnoid and pia) to promote cleansing. Decompress cranial nerve outlets and pathways. Sensory integration function is optimized when cranial nerves are free of compression, well-hydrated, and well-nourished.

Adequate science exists supporting the practice of CST, and the clinical judgment of the practitioner is paramount for successful outcomes. Cranial sutures are not fixed as has been documented extensively by Upledger's clinical research. It has been demonstrated on cadaver tissue of intact craniosacral system tissues that intracranial membranes can be mechanically affected by traction on various cranial bones. Structural compression of cranial bones were proven to alter human neurological behaviors by William Sutherland. Upledger's work expanded the science of this truth with his CST methods

Holding bones long enough lift bones to increase the tiniest of spaces. (This could be studied using ultra sound to measure changes in intrameningeal spaces.) Bones are handles to reach membranes. Membranes are handles to reach the vasculature of arteries, veins, and lymphatic (glymphatic pathways). Only miniscule displacement is needed for self-correction of structures.

There are little to no published records of adverse effects of CST applied. Though there is frequent confusion in comparison, cranial osteopathy and Upledger-CST are not the same thing. Two different premises serve each theoretical foundation. In *Bodywork for Sensory Integration*, applying Upledger's CST techniques ensures safety for the client because these methods were designed for that goal intentionally. Optimal physiology to the child's neurocranium and facial structure for cranial nerves and brain stem/spinal cord is the objective.

Cranium regions organized for clinical application

Neurocranium:

- **Bony Container** – neurocranial plates both fused and unfused, in articulation development
- **Fascia Container** – dura mater, arachnoid layer, pia mater (frontal, parietal, temporal)
- **Intracranial membranes** – reciprocal tension system giving brain and skull tensegrity shape longer hold to get to it through layers, sphenoid directions reflect

Viscerocranium:

- **Face structures** – eye orbit, nose, cheeks: osseous framing & cranial nerve innervations ethmoid, zygoma, sphenoid, orbit structures, mandible
- **Mouth structures** – osseous framing & cranial nerve innervations maxilla/vomer/palatine

Bony Container

Skull bone is divided into 2 divisions: neurocranium (head) and viscerocranium (face). Neurocranium bones proliferate in growth via membrane ossification.

Neurocranium:

Frontal: begins as two halves with the metopic fontanelle (some refer to is as a suture). Complete fusion of the midline point of the forehead is generally by the 4th year, but does not fuse on everybody. CST releases eye orbits, olfactory bulbs and pathway, falx cerebri, cerebral frontal lobe where executive skills develop, and behind the frontal lobe the anterior limbic system. Compression here can translate mechanical forces inwards towards the amygdala, hippocampus, thalamus, hypothalamus, basal ganglia, and cingulate gyrus.

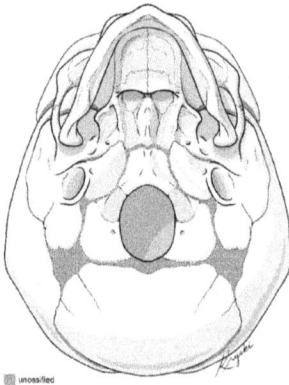

Baby cranial base. Heads at birth have more immature bone segments than an adult head. Here the four segments of the occiput bone reveal posterior soft spots (where mechanical forces can be retained, imposing structural stress on neurocranium).

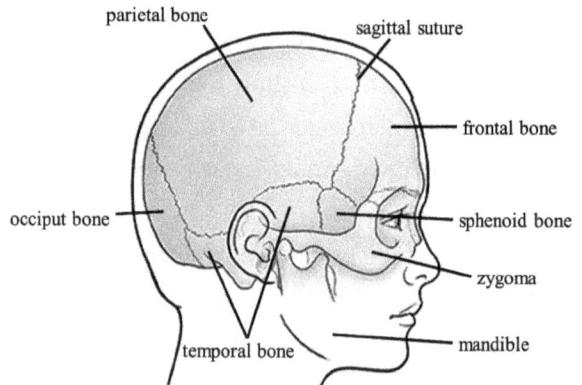

Child cranium. Membrane ossification to fuse bone segments can take several years to be completed. This fact makes the infant's and child's head moldable and responsive to gentle manual therapies such as CST.

Parietal (2): join at the sagittal suture where little movement is available. Ossification of plates progresses radially from bone center (arising from fascia cells specialized to ossify). Tentorium & falx cerebri attachments to this bone create essentially two pseudo cavities in the skull. Perceptual and language centers, sensory, motor and premotor cortexes at or near the sagittal sutures.

Temporal (2): Arise from four segments (squamous, petrous, tympanic, styloid). First 3 segments are fused via intramembranous ossification before the first year, the styloid takes into adulthood. Houses inner ear bony mechanisms and forms Eustachian tubes.

Occiput: At birth is four segments (squamous, basilar, 2 lateral condyles) form foramen magnum and cranial base. The squamous ossifies from two radiating centers taking 4 years to mature; all four segments take up to eight years to fully ossify; maturing into a single bone. The sphenobasilar junction does not fuse, remains a dynamic joint.

Sphenoid: Most complex bone of the body, forms from 7 segments. (2 each: lesser, greater wings, & pterygoid processes, situated around a central body). Connects with all other bones of neurocranium, providing foramina for neurovascular pathways. It also connects the structures of neurocranium and viscerocranium.

Ethmoid: Oddly shaped; the most anterior of the neurocranial bones. Houses cribriform & perpendicular plates and 2 labyrinths. Ossifies by replacing fetal cartilage with calcified bone. Fully formed by age two.

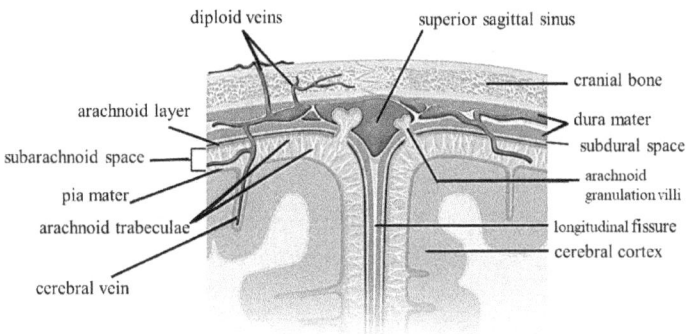

In CST, bones are handles to reach membranes.

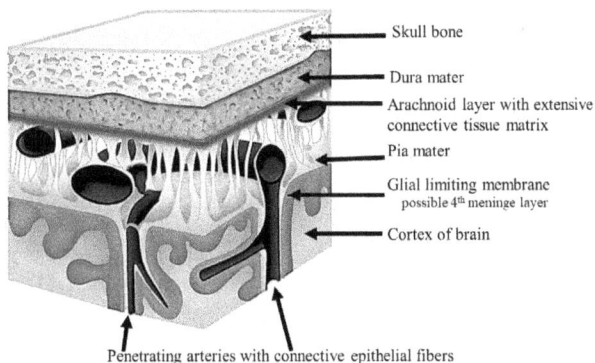

Craniosacral therapy treats the structural container to reach entry points of sensory information

In CST, membranes are handles to reach the vasculature of arteries, veins, lymphatic, and glymphatic pathways.

Intracranial membranes:

The fascia container of the head houses the growing bony container of skull. Within the cranium there are both longitudinal and horizontal bands of fascia, known as the intracranial membranes. They are two-layered extensions of the surrounding dura mater. The tentorium cerebelli is transverse plane extended over the posterior cranial fossa. This tough membrane separates the occipital and temporal cerebral hemispheres from the cerebellum and brainstem. The tentorium cerebelli is the second-largest dural reflection that extends over the posterior cranial fossa. It separates the occipital and temporal lobes of the cerebrum from the underlying cerebellum and brainstem, and divides the cranial cavity into supratentorial and infratentorial spaces. The other intracranial membranes (falx cerebri and tentorium cerebelli) continuation bands from the fascia container (meninges), under the bony container of the cranium. These three intracranial membranes constrain brain motion as well as form the suspension containers the different hemispheres of brain. These surrounding fascial layers have shock absorbing properties and are related to cranial shape distortions (plagiocephaly).

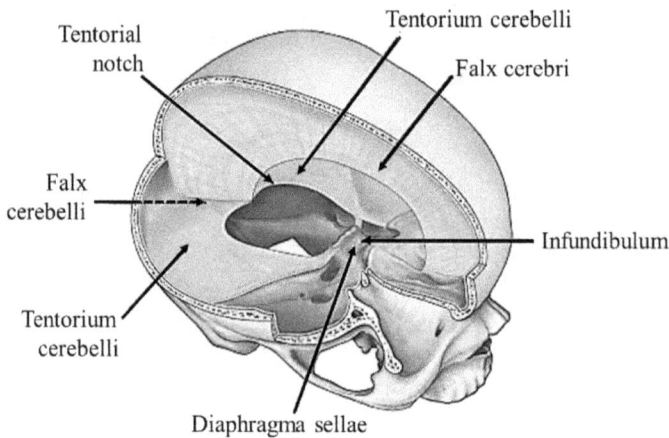

Intracranial membranes are continuation of the dura mater.

The cranial vault's squamous bones (paired frontal & parietal bones; squamous parts of the temporal bone; and interparietal part of occipital bone) are all formed by intramembranous ossification. This process is means that from the basic embryonic connective tissues surrounding all structures, specialized cells (osteoblasts) group into clusters forming an ossification center. As baby grows, these cells proliferate radially and condense to form the harder bone structure. Squamous portions meet other neurocranial bones at sutures. The sutures are the remnants of the original sheaths of connective tissue, and as Upledger proved, remain actively expansive and contractive through life for the service of the craniosacral rhythm.

Viscerocranium: (face and mouth)

Viscerocranium bones generally grow is size first, and then ossify through endochondral ossification. This means that bone cells replace cartilage by ossifying within the formed structure formed. (Contrasted to intermembranous ossification where proliferating osteoblasts form bony plates within the structure of fascia). Viscerocranium structures remain flexible and can be influenced by gentle, sustained stretch. The orthodontia profession has proven that with the science of braces for teeth.

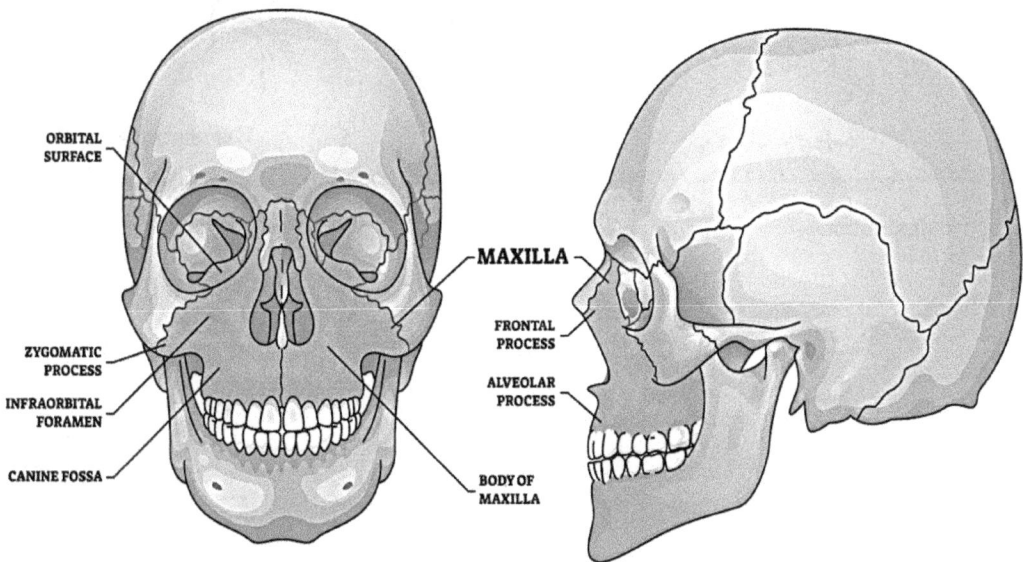

Face structures:

The structures of the viscerocranium forming the face include: maxilla, vomer, palatine, mandible, zygoma (2), lacrimals (2), and nasal conchae bones. These bones frame the eye orbit, base for the cartilage of the nose, cheek and lip stability frame .

Mouth structures:

Extensions of the bony viscerocranium, but forming the oral cavity including the hard and soft palate, the inner curve of the mandible as floor base for the tongue.

The key point for the neurocranium and viscerocranium in *Bodywork for Sensory Integration* is to utilize methods that decompress cranial nerve and neurovascular foramen and pathways.

Theoretically, reducing any structural bony compressions (retained from birthing or acquired from injury) helps self-corrective soft tissue expansion and promotes fluid exchange under the fascia container. The twenty-four pathways of the **cranial nerves** are thus optimized for the job of feeding the sensory integration process. Without structural stress, the nervous system is more likely to receive and process sensory signals in parasympathetic state.

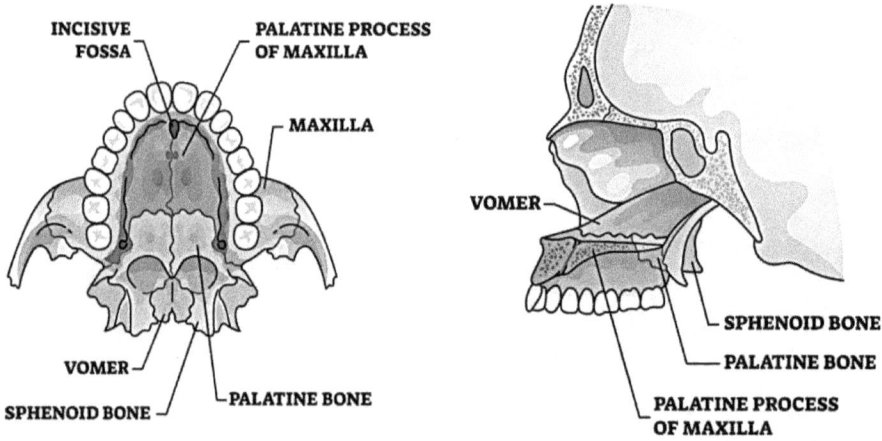

Cranial Nerves, Nerve nuclei and fluid pathways

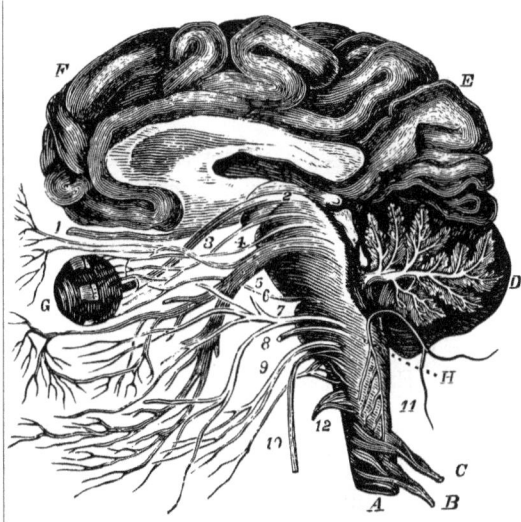

Cranial nerve in situ of the cranial vault. Visualize each nerve with its adjacent arteries, veins, and lymphatic vessels. It's a crowded field exiting bony foramen and sutures.

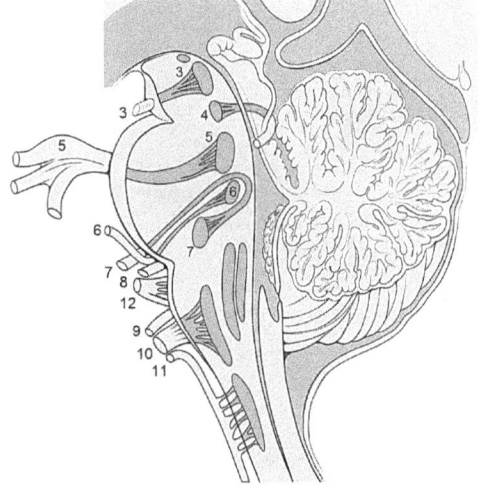

Patrick Lynch @ Creative Commons. Lateral view of cranial nerve nuclei distribution through the pons & medulla of brain stem.

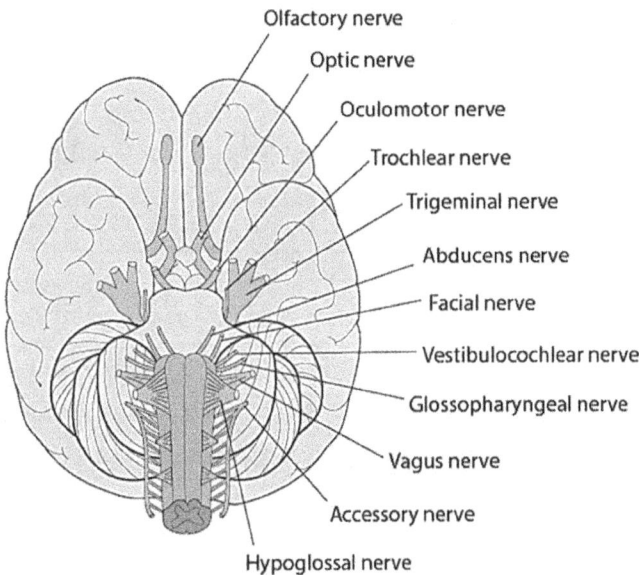

Olfactory nerve
Optic nerve
Oculomotor nerve
Trochlear nerve
Trigeminal nerve
Abducens nerve
Facial nerve
Vestibulocochlear nerve
Glossopharyngeal nerve
Vagus nerve
Accessory nerve
Hypoglossal nerve

Inferior view of cranial nerve location. Patrick Lynch Creative Commons. The central concentration of cranial nerve nuclei makes cranial nerves vulnerable to cranial-cervical compressions translated into brain stem and midbrain. CST decompresses the upper cervical (dorsal column and spinal thalamic tracts) and the cranial nerve centers for optimal ANS tone and sensory integration to have a foundation of working.

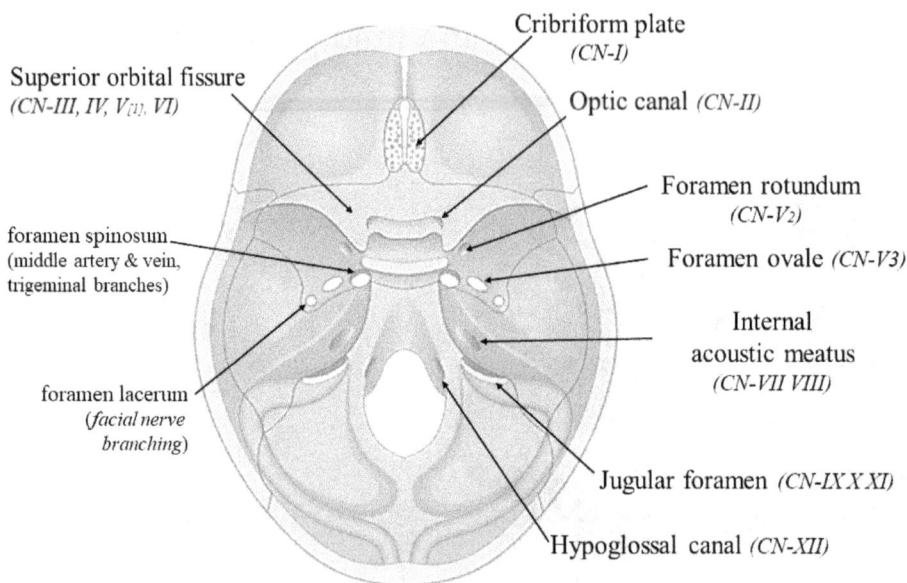

Superior view cranial bowl base, and locations for cranial nerve foramen exits. Proximity to sutures and fontanelles, with the possibility of bone migration through compression or meningeal torsion, make these openings and the structures passing through vulnerable for chronic sympathetic tone or dysfunction. Image: Shutterstock (modified).

Upledger CST mixed with other techniques assists each cranial nerve.

CN-I: Olfaction – frontal, maxilla/vomer, ethmoid decompressions; falx cerebri release

CN-II: Optic - frontal, parietal, sphenoid decompression; balance sphenoid membranes

CN-III: Ocular motor muscles – frontal, sphenoid, maxilla, ethmoid decompression; zygoma lift

CN-IV: Auditory/Vestibular – temporal, parietal, sphenoid decompression; tentorium release; occipital condyle balance with other segments, cranial base release (O-A) decompression

CN-V & CN-VI: Facial and trigeminal – mandible decompression (2 directions), sphenoid segments balanced and decompression, spheno-basilar decompression if needed, temporal decompression

CN-IX: Glossopharyngeal – cranial base release, cervical traction (for brain stem decompression); falx cerebelli release

CN-X: Vagal – temporal decompression, cranial base release, dural tube traction cervical & brain stem, balance occipital segments in immature sutures not yet fused; falx cerebelli release

CN-XI: Accessory – temporal decompression, cranial base release, dural tube traction cervical & brain stem, balance occipital segments in immature sutures not yet fused; falx cerebelli release

CN-XII: Hypoglossal – cranial base release and dural tube release (cervical and spinal meninge release), brain stem decompression, occiput segment balancing in fascia membrane at foramen magnum; falx cerebelli release

Bony & Fascia Container Surrounding Brain

Frontal bone & anterior cerebral
tissues and fluid pathways

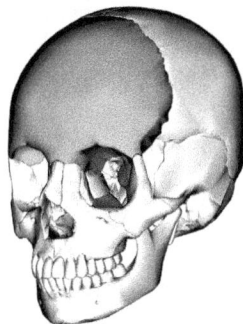

Frontal bone. Anatomography
at Wikimedia Commons

Both Hands: Draping contact over frontal bone, adding a slight upward stretch is enough to engage bone against dura mater, lifting to facilitate fluid exchange in subdural & subarachnoid spaces and the arachnoid vasculature layer. Always with Responsive Hold as your guide, blend with tissues and wait for self-correction of a hydroplane sensation of "self-rising" of the bone from the soft structures that lies beneath. Be sure to wait long enough for a thorough response for completion of tissue adjusting.

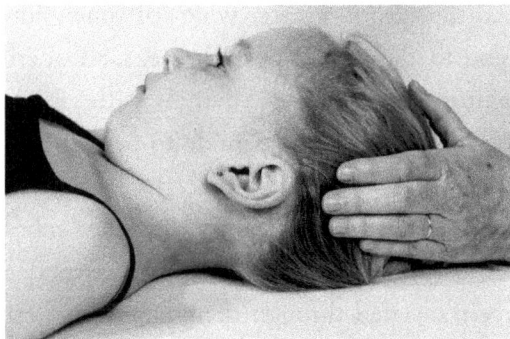

Parietal bone & lateral cerebral
tissues and fluid pathways

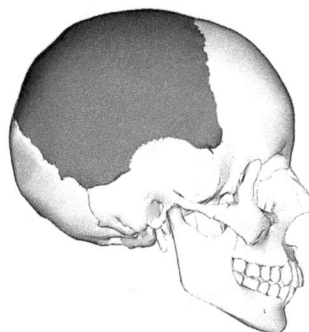

Parietal Bone. Anatomography
at Wikimedia

Both Hands: Assuming the person being treated is under 8 years of age, contact bone on the posterior aspect of the parietal bone, index finger at or slight behind the ear. (**No compressive technique should be done on anyone under 8 years as bones are not all fully fused**). A slight upward stretch of both parietal plates is enough to engage bone against dura mater, lifting to facilitate fluid exchange in subdural & subarachnoid spaces and the arachnoid vasculature layer. Blend with tissues and wait for self-correction of a hydroplane sensation from the bone. Wait for completion. Those trained in Upledger CST knows that a slight compression is first applied for those older than 8 years of age.

Temporal region, tentorium cerebelli, and inner ear structures

Both Hands: Gently grab ears and place thumbs close to the external auditory canal (deep in concha). Apply an acupressure "pinch" - engaging as much of ear cartilage as possible. Wait & assess tensions from fascia and muscular attachments projecting to skull. This long-practiced technique is used by several professions, with some variances in approaches. CST approach is a gentle yet direct fascia traction & tissue pull at a specific angle (based on age), and learned in formal training.

Clinically, this is also a temporal bone decompression technique and has been shown to resolve and prevent ear infections, routinely assisting in the clearing or drying out effusion of fluids. This ear-pull technique is also fantastic to relieve discomfort the change of inner ear pressure on airplanes.

Sphenoid (not fully fused until age 8 years of age!)

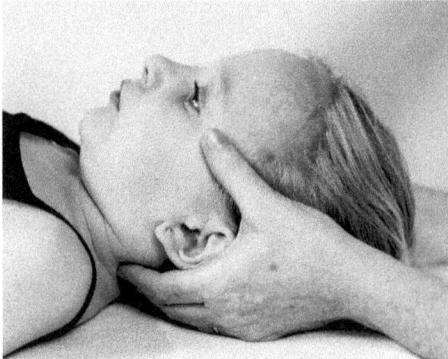

Both Hands: Cradle head in hands and gently place thumbs on greater wings of sphenoid; just lateral to the corner of each eye. CST methods monitors craniosacral rhythm with this and other cranial holds as a guide to performing the technique safely. Stabilizing the occiput with fingers, the thumbs gently lift the sphenoid anteriorly. The sphenoid absorbs a great deal of force translated through the cranium in birthing and may still be retained. The sphenoid has direct contact with pons. This technqiue balances immature sphenoid segments by release to structural stress in the intracranial membrances, and decompresses the flexible sphenobasilar junction where sphenoid and occiput join.

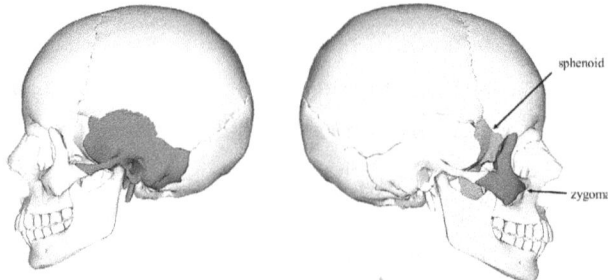

Temporal, sphenoid, and zygoma bones. Anatomography. Wikimedia Commons. 23303129

Face Structures

Ethmoid Decompression- from frontal/sphenoid: Two Directions

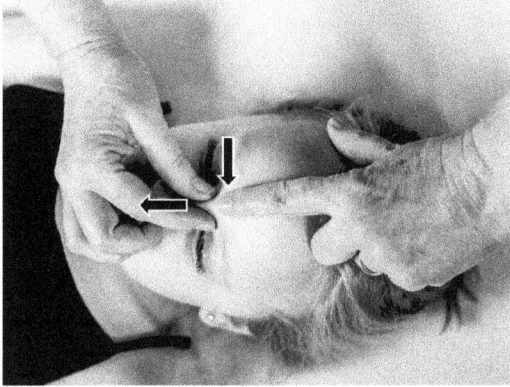

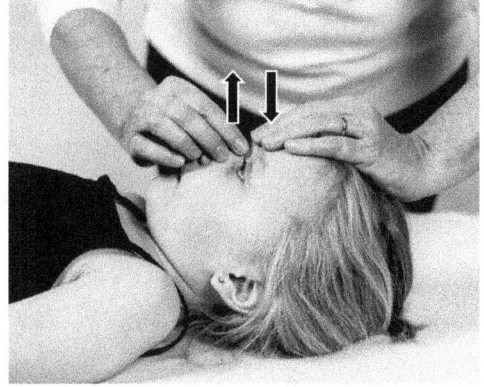

Hand One: Gently stabilize frontal bone just superior to frontonasal suture.

Hand Two: Gentle pinch on ethmoid and nasal bones, giving inferior stretch to decompress ethmoid bone from the frontal bone.

Second direction: **Hand Two** gently mobilizing anteriorly while **Hand One** continues stabilization.

Enhances olfaction by these two directions of pull by reducing structural stress or inflammatory stress into the cribriform plate and olfactory apparatus. Also, releases sinus congestion (related to inner ear health). Has been helpful in normalizing hypersensitivites to odor and restoration from illness and injury to the olfacotry nerve endings.

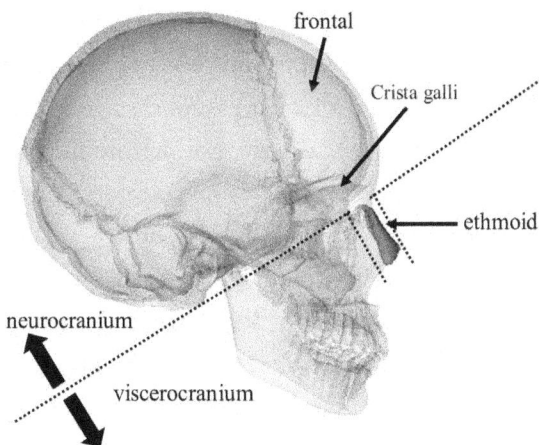

The ethmoid articulates with 13 bones: frontal, sphenoid, nasals, maxillae, lacrimals, palatines, inferior nasal conchae, and vomer. It's complexity and its relationship to many cranial nerves makes is a point of interest for sensory integration. Directly connected to the dura mater and extension of the falx cerebri is the Crista galli is on the superior side of the ethmoid bone and is a wedge-shaped vertical midline plate that serves as an attachment for the meninges of the brain.

Extra-ocular Muscles: Tension patterns during tracking teaming

The eyes literally are an extension of brain tissue. *Bodywork for Sensory Integration* offers clinical application of craniosacral therapy, lymphatic methods, acupressure, myofascial techniques to the ocular motor system. Because of the intimate connections and close proximity to meninge tissue behind the eye orbits, these methods are a perfect blend to explore supplemental treatments for problems such as strabismus and amblyopia. Clinical knowledge is outdated when only surgical options are offered to families. Progressive behavioral optometrists and ophthalmologist are also testifying to the advantages of combining routine behavioral optometry with sensory integration combined with manual therapies. Reports of changes include an immediate change in prescriptive lenses, and demonstrated support in optimizing ocular motor function, eye-teaming, and eye resting position. For facial traumas that may have contributed to ocular motor challenges, manual therapies are a very positive option. Retaining and regaining the fluid nature of the sclera, and balance of fascia network that suspends all extraocular muscles are the objectives. (See more at Postural-Ocular Dysfunction p. 310)

Global assessment & treatment: sclera & muscle attachment

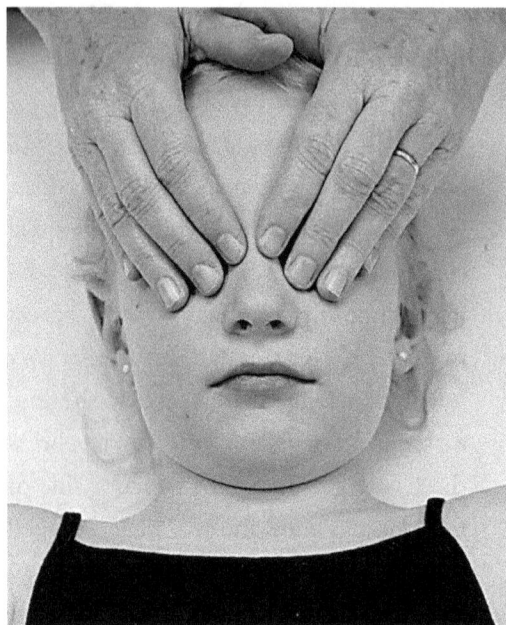

Both Hands: With eyelids shut, place finger pads exquisitely gentle upon closed eyelids, as a butterfly would land on a lily pad floating on a pond. Wait ever so patiently to be informed of tension patterns on all sides of the eyeball globe. Wait for any protective retraction to resolve or dissipate. Index finger(s) can then follow tissue response or tiny lines-of-pulls to specific muscles or fascia attachment restrictions. Apply only Responsive Hold at restriction sites. Hold microscopic space and wait for a self-correction reaction, happening quickly as tissues are thin and delicate. Check in at insertion sites of each extra-ocular muscle on eyeball (keeping eyelids closed).

Precise treatment of identified muscles holding tension

Index Fingers: With eyelids shut, place one fingertip on one or both paired eye muscles where tension is palpated. Ask the child to move eyes side-side, up-down, and in circles to assess any imbalance between all the extra-ocular muscle pairs. It is common to find bilateral medial muscles tense (over-convergence from screen time) and unilateral medial or lateral muscle tension with a lazy eye. Apply only exquisitely gentle Responsive Hold at any of these muscles and the attachments to the sclera. No pressure should be used. This is energetic acupressure style work. And it has been consistently effective for both children and adults in our clinic.

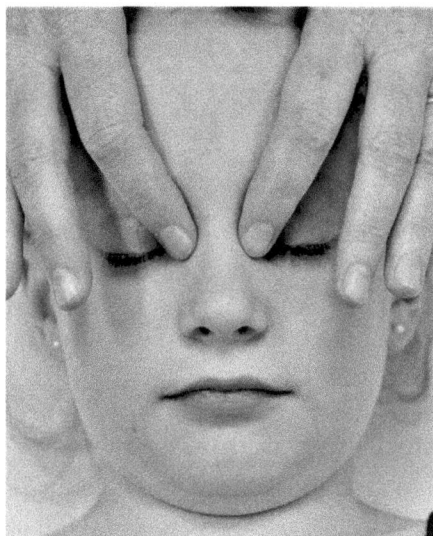

Additional thoughts during eye treatment:

The surrounding sclera of the eye holds the gelatinous vitreous and is the attachments for muscles. Extraocular muscles and their investing fascia extend to more vitreous tissue within the socket. Tune into common tendinous ring in posterior socket which supports and suspends all muscles and nerves projecting towards the eye. Have eyes move in different directions as a team (tracking) while hands are palpating to determine tissue tension and symmetry of eye movement.

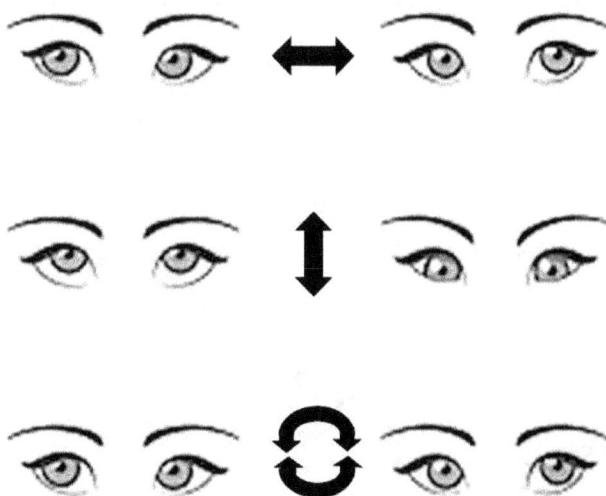

Eye-teaming movements during assessment of tension

Imbalances of eye-teaming muscles can be palpated. Too invasive and you can easily set off an extreme protective reaction and sympathetic response. Match the microscopic resistance if present. Feel for equal or lack of lines-of-pull which will express themselves to gently-placed fingers. ONLY use Responsive Hold, and trust that self-correction can occur with patient fingertips. Through closed eyelids, assess even the eyelid muscles (top and bottom).

Eye teaming and tension:

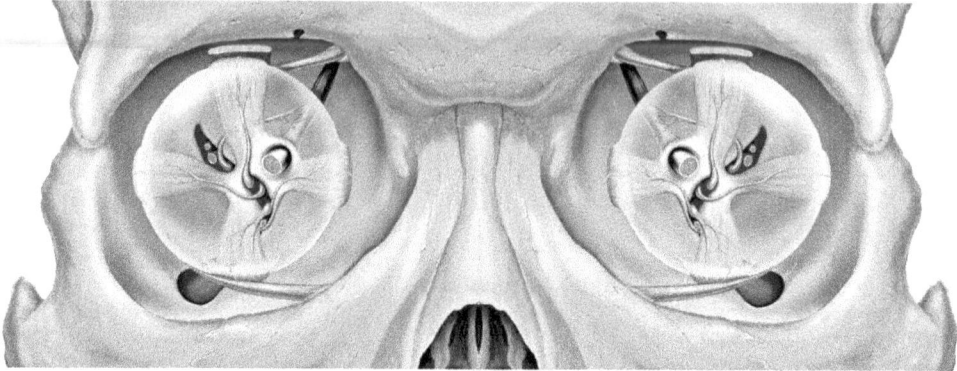

There is no better way to hone skills of non-invasive touch and patience for Responsive Hold than to work on a person's eyes. If you trust your manual skills, begin to trust that you also have the delicacy to not alarm and not harm the visual system. Facilitate a cellular and micro fascia self-correction under fluid movement phenomenon.

If a client is too sensitive or threatened, save this for a later session when the trust relationship deepens, or treat over the child's own hands placed over the eyes. And be sure your hands are extraordinarily clean but without perfumes or harmful chemicals from hand lotion.

Do not work on eyes until you trust your touch skills

Mandibular decompression: relief to trigeminal & facial nerves; de-retract innervation to face

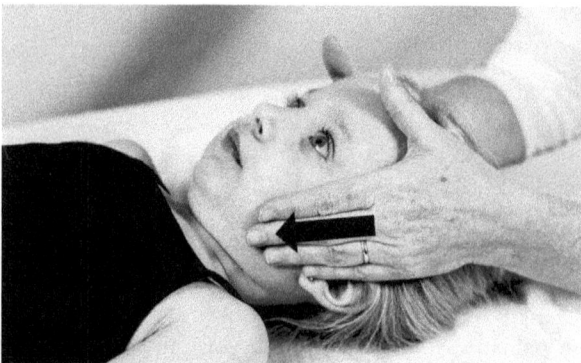

Both Hands: Place on both verticle ramus of mandible. Fingertips engage with inferior ridge of body of the jaw. First apply Responsive Hold to allow time release of any protective retraction that may contribute to jaw tension. Remaining blended and responsive to tissue give, impose a gentle **inferior stretch** to the mandible, which decompresses TMJ capsule ligaments, relieves pressure & stretches the sphenomandibular & stylomandibular ligaments. Adequate time holding the stretch should be given to ensure a complete release of compressions to the articular disk.

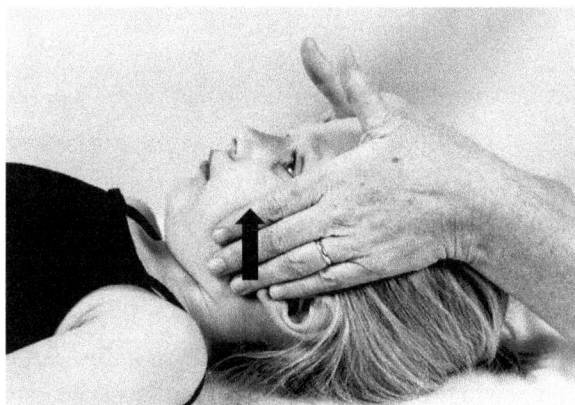

Both Hands: Keep hands over both sides of mandible but change direction of pull a gentle **anterior stretch**. Ring & middle fingers can use the lateral ramus edge for leverage. Repeat the same style of tissue treatment as the inferior stretch. Adequate time holding provides an accupressure to masseter insertion margion. In addition, any retained jaw compression from infancy is assisted into complete release. The main nerve assisted by these two techniques is the trigeminal pathway after it exits cranium at the foramen ovale (hole in posterior sphenoid), and travels just anterior to the joint capsule of the TMJ. Sympathetic activation of the TMJ can lead to jaw clinching and teeth grinding. TMJ issues can also be related to vestibular-auditory processing issues, either structural etiology or sympathetic reactionary.

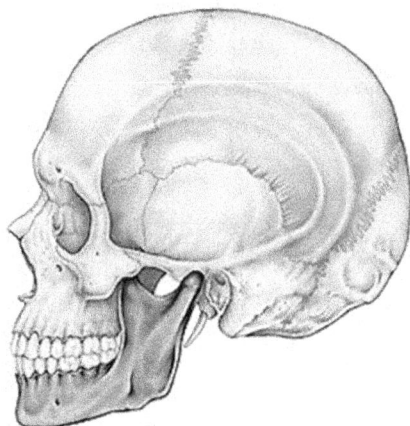

Mandible position on skull

Temporalis & Massester

The temporalis muscle is innervated by three branches of the mandibular nerve from the trigeminal nerve (CN-V) Movements of jaw mastication is this muscles primary function. To continue to down regulate the ANS, treating all structures innervated by the trigeminal nerve is helpful. The masseter muscle is responsible for elevating the mandible, with some protraction action as well. Motor innervation of the mandibular branch of CN-V can lead to jaw tension from clinching as a fight-flight-freeze response from an over-active sympathetic system. Suffice to say that both muscles are also helpful in activation the parasympathetic rest & digest state by activating peristalsis from chewing (on food or non-food items). Sucking thumbs, chewing on pencils or fingernails, or eating when not hungry are all behavioral indicators of self-directed ANS regulation.

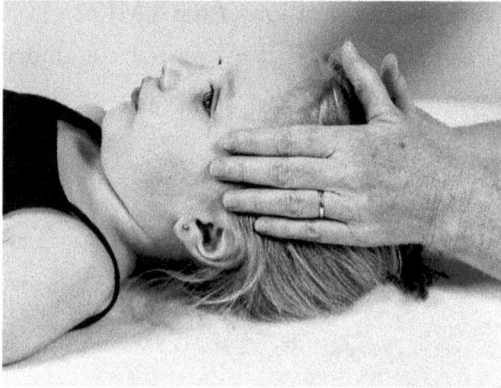

Responsive Hold, Acupressure, and line-of-pull to Temporalis muscle. Reduces tension at innervation sites of both trigeminal and facial nerves, inducing an immediate PNS response.

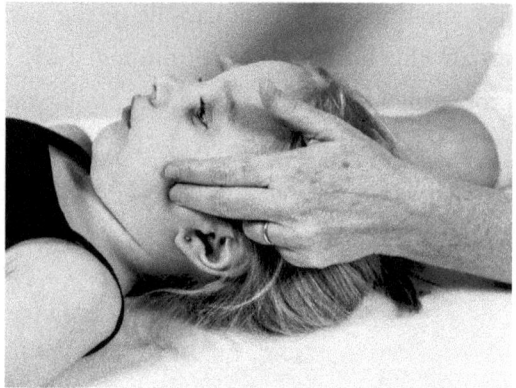

Acupressure into masseter muscle shown in photo. Apply Responsive Hold, line-of-pull, CST, or myofascia release techniques. Reducing tension here has proven helpful for teeth grinding, sound sensitivies (in some people), and general stress abatemen

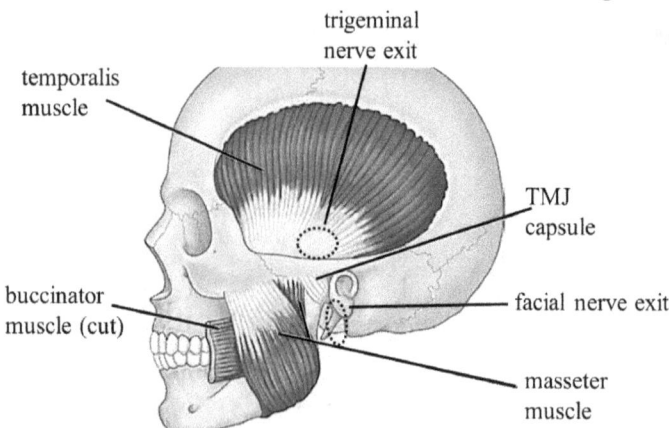

temporalis muscle

trigeminal nerve exit

TMJ capsule

buccinator muscle (cut)

facial nerve exit

masseter muscle

Temporalis & masseter muscles impact the ANS of the face and jaw because of vicinity to trigeminal and facial nerve exit sites. Image modified: Shutterstock

Zygoma decompression: mobilize soft tissues around orbit, towards Eustachian Tubes

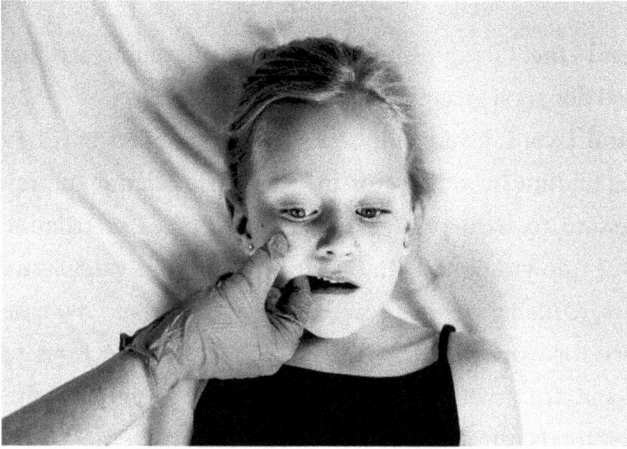

Mechanical forces can compress the cheek bone such as birthing or trauma to the face, and structural compromise can translate inward toward the middle & inner ear and the Eustachian tubes. Compression at other sites of the trigeminal or facial nerve can cause upper lip and facial retraction and muscular tension, such as temporal or mandible compression. John E. Upledger, D.O. described several phenomenon in fresh cadaver dissections while applying CST techniques to unembalmed cranial bones. One such observation was a zygoma decompression at the recommended 45° lateral angle broadened the floor of the eye orbit, freeing up the palatine bone of the hard palate. This technique is helpful in relaxing maxillary & zygomatic branches of the trigeminal nerve.

Hand One: Gloved, very gently slide index finger slightly under the inferior margin of the cheek bone; thumb makes contact externally with the body of the zygoma. Be extremely careful of the amount of pressure the thumb exerts as it is precisely over the foramen where the zygomatic branch of the trigeminal nerve exits bone. A typical adult finger should be able to be able to "lift" the zygoma bone, needing only microns of distance to decompress its position. The direction is approximately a 45° angle, but it is more important to move in the direction of ease with the tissues than to worry about a precise angle. This technique should cause no pain for the child; stop treating if pain is elicited. Many an old face injury has been discovered with this technique.

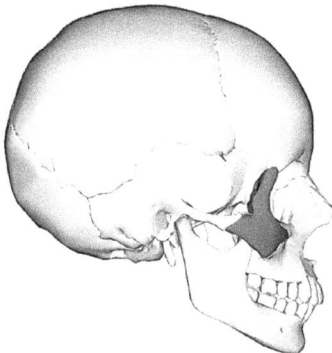

Human zygoma bone.
Source: Wiki Commons

Maxilla/Vomer Complex, Palatine bones, Hard & Soft Palate

For the young child under age 8:

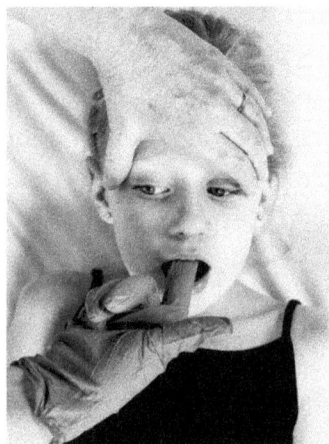

Hand One: Gently stabilize sphenoid with thumb and first or second finger on greater wings.

Hand Two: Gloved, index finger placed on intermaxillary suture at midline (avoiding crossing transverse palatine suture). Acupressure to hard palate at innervation sites of palatine nerves can down regulate the palatine nerve and vasculature branches, reducing a hyper-reactive gag response and increase tolerance for work on the palate. Wait for oral structures to accept and trust your therapeutic presence. Combine with CST methods of decompressing the maxilla/vomer from suture margin of sphenoid and palatine.

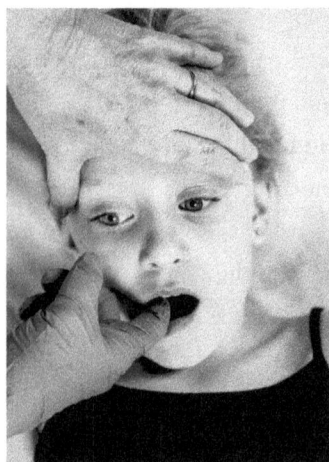

Hand One: Remain stabilizing the sphenoid

Hand Two: Move finger pad very carefully to curved surface of palatine bone, at margin of soft & hard palate. The palatine is the very posterior edge of the hard palate. Remaining within the scope of CST, mobilize the palatine in the way described by Upledger to reach the connective tissue ends of the Eustachian tubes. Be exquisitely gentle and mindful of palatine nerve and vasculature foramen and nerve pathways over hard palate surface.

These techniques assist self-correction of tissue transition of hard to soft palate, which can be compressed by a number of factors. Bubble palates (high arch) correlates with ankyloglossia (tongue tie) from lack of tongue pressure but also from genetic traits of narrowed palate formation. The foramen for palatine nerves and arteries are vulnerable to compression translated from the face (birthing; trauma). The trigeminal nerve serves the protective sensory aspect of roof of mouth.

Adhere to your scope of practice when considering oral therapies

Only a few examples of oral work was included in this book for sake of brevity

Any oral (mouth) work should be practiced ONLY after formal training

***Bodywork for Sensory Integration* aligns with mouth work methods**
of Upledger's craniosacral therapy

Part IV

Dysfunctional Behaviors

1. Sensory Processing Dysfunction (Neurodiversity & Sensory Wellness)
2. Oral motor and feeding aversions/Picky eater
3. Attention Deficit – Hyperactivity Disorder
4. Autism Spectrum Disorder
5. Sensory-Modulation and Self-Regulation Dysfunction
6. Tactile Defensiveness
7. Inner Ear Dysfunction
8. Posture/Ocular and Bilateral Coordination
    - Reflex integration for balance and postural righting
9. Dyspraxia
10. Toe Walking

1. Sensory Processing Dysfunction

A new narrative for neuro-diversity and sensory wellness

The following appendixes convey the historical concepts of sensory processing theory and practice. As all things evolve, new trends are moving forward using fewer terms of exclusion from a perceived norm. Sensory wellness is a term that embraces neurodiversity so that all people feel included within such descriptions. Every person has sensory needs and expressions. The intention thus becomes presenting the areas functional areas where sensory health can be explained so that all person's strengths and weaknesses have a voice.

Definition

Sensory Processing Disorder (SPD), historically called Sensory Integration Dysfunction, is described as the brain not efficiently synthesizing information. Based upon the work of A. Jean Ayres, sensory integration dysfunction was first recognized as a neurological-based problem. She realized sensory differences in children, otherwise of typical intelligence, yet struggled to perform routine tasks in school and daily life. Occupational therapists and researchers (like Dr. Ayres) expanded the science with greater behavioral and differential analysis clarification. Scientific strength exists in the classifications of sensory-based performance and behavioral challenges.

SPD lacks cross-professional recognition as behavioral perspectives attempt to merge opinions and theories. Nonetheless, pediatric therapists now routinely evaluate and treat children (and adults) where differences can be attributed to these underlying neurological (sensory) processes. SPD can be a stand-alone problem or co-occur with autism spectrum, learning disorders, attention deficit disorders, and psychological or mental health disorders. The severity of dysfunction depends upon how many senses are affected and to what degree the autonomic nervous system is affected by sensory responses. Performance differences are categorized into subtypes based on adaptive or maladaptive responses. The full scope of SPD can be challenging to understand and diagnose thoroughly without extensive training.

Behavioral manifestation

The person with SPD can feel bombarded by everyday stimuli when the nervous system does not quiet down (inhibit) input. There can also be the opposite tendency of being unaware of generally apparent sensory information and, therefore, not responding as others would. SPD's physical manifestations may include weakness, clumsiness, awkwardness, or delayed motor skill development. Social-emotional presentations can fall along a continuum of lethargy and apathy to irritability, restlessness, agitation, and explosiveness. Emotional instability as daily events can lead to less success and resiliency in educational and daily activities. Assumptions about a person's cognition have been made when reactions to sensations are not typical. Behavioral judgments and psychological labels have been assigned to people with SPD, furthering the difficulty of identifying underlying neurology.

As our societies become more sedentary, development trends change as learning environments change. Time will tell the effect of passive and technology-driven learning through a child's day affects the growing brain. In addition, lack of engagement in sensorimotor-rich activities is a new cultural observation and concern in some professional ranks. The body and brain need rich sensorimotor experiences to build a broad base of sensory processing foundation. For example, pediatric therapists have long understood that crawling is crucial for sensorimotor development for maximal physical performance. This phase strengthens the shoulders and hips, spines, and arches of the hand. However, we find ourselves in a new political environment where policymakers state the crawling stage is not necessarily required for maximal development. Crawling is one significant skill where neurological health is evaluated. Thus, lack of skills can be a combination of experiences and underlying neurological processes, even brain injury.

Sensory Processing Dysfunction Subtypes

- Sensory Modulation Disorder (can be applied to each sensory system)
  - Sensory Over-responsiveness
  - Sensory Under-responsiveness
  - Sensory Craving
- Sensory-based Motor Disorder
  - Postural Disorder
  - Bilateral Coordination (arms, legs, eyes)
  - Dyspraxia (with subtypes: movement execution and initiation, ideational,
  - constructional, sequencing, planning)
- Sensory Discrimination Disorder
  - Visual motor difficulties
  - Somatosensory and fine motor difficulties
  - Auditory processing difficulties

Sensory Modulation Dysfunction (See also pp. 251, 273, 288, 299)

The resulting behavioral and emotional expressions are hypersensitivity (over-responsiveness) or hyposensitivity (under-responsiveness) to sensory events. When hypersensitivity exists, sensory input causes a [sympathetic system] alarm. Fear, agitation, or avoidance can be coping behaviors that develop as a result. A degree of struggle in achieving and maintaining homeostasis of the autonomic nervous system can be an undercurrent. A standard indicator if this exists is the time necessary to recover their composure from the sensory alarm or the level of intensity of a reaction.

A certain information threshold must register within the nervous system to elicit a reaction. Sensory craving or sensory seeking is described as "never seeming to get enough input." It is suggested that a lack of full registration prevents the person from reaching complete neurological registration of sensory signals. Insatiable need to touch objects, unable to sit still, the body always moving, self-stimulation like picking at fingers or chewing on lips, taking too many risks, and being distracted by the movement of others. Conversely, a lack of response or oblivion, poor timing or coordination, and delayed reactions to sensory input are common expressions of hypo-responsiveness.

Another important aspect is a state of sensory overload as a way to cope with a continual onslaught of sensory information. When a person goes into a shut-down state, they may appear hypo-reactive to input. Distinguishing the difference between shut-down and hypo-responsiveness remains challenging to identify and live with.

Sensory-based Motor Dysfunction (See pp. 299, 310)

Postural control, coordination, and motor planning skills (praxis) are functional products of the vast sensorimotor processing of vestibular, proprioception, and kinesthesia information. The touch system and visual and auditory cueing also contribute to discrete body position in space and quick reactions/adjustments to physical challenges. Dysfunction occurs when the two sides of the body (brain hemisphere communication or synthesis of information in cerebellum and cortical centers) are ill-timed, or two-sided reactions are out of sync. Postural control implies core stability and integration of spinal and brain stem level reflexes to master gravity. Coordination dysfunction of limbs implies that core stability or timing of body side movements are out of sync. Praxis dysfunction (dyspraxia) involves much more complex brain processing, and Ayres' is best known for her work on identifying subtypes of dyspraxia. Dyspraxia is one of the most overlooked problems with independence in the child's maturation and is often misunderstood as behaviorally based. The subtypes include movement execution and initiation, ideational, constructional, sequencing, and planning. Motivation, emotional regulation, cognitive synthesis, and memory also contribute to the quality of praxis skills a person develops.

Sensory Discrimination Dysfunction (See also pp. 288, 299)

- Visual motor difficulties
- Somatosensory and fine motor difficulties
- Auditory processing difficulties

Visual motor difficulties are different from compromised eyesight (acuity). They are measured by convergence (eyes moving together), visual focusing, and visual endurance (visual attention). This area of human function is bilateral coordination and the success of both eyes moving and working together as a precise team.

Somatosensory receptors are located in all layers of the whole-body skin organ. There are six different types for detecting the sensations the skin will encounter, and each sends specific information to the brain. (See anatomy chapter with varying types of cells). Touch processing dysfunction can be a problem, either over-reacting to various touches or under-reacting to touch. Fine motor dexterity less than optimal writing and eating utensil use can result in touch processing difficulties. Intolerance of human interaction can also be a consequence.

Auditory processing difficulties may result in an inability to locate where sounds come from, hearing targeted sounds or voices in noisy backgrounds, or if spoken too quickly. Delayed reaction times and responses to directions are common. Frequent requests for repeating of instructions. Behaviors may be misinterpreted as hard of hearing or rude for lack of anticipated response to a social exchange.

Historical treatments in Sensory Integration

The therapeutic approach of sensory integration was initially developed by Dr. Ayres and is now formally known as Ayres Sensory Integration® (ASI; Ayres, 1989). ASI® includes the theory, precise assessment methods to differentiate each sensory function, and core treatment philosophies. Interventions are structured sessions for child-directed sensory-motor participation to work on weaker areas of function, yet well-designed as play-based learning and skill mastery. Tasks are designed to challenge the child towards more complex responses, with the therapist facilitating active control of the child's body. Specialized gym equipment to stimulate the hidden senses of vestibular and proprioception sessions are most often utilized.

The direct treatment environment of occupational therapists is where ASI is usually provided. Professional qualifications help the public recognize those certified to carry out the approach as intended by its originator and provide evidence-based practice. The outcomes of ASI have been documented as designed to assist improvements in sensory processing and adaptive responses, self-regulation, coordination and balance, and praxis skills. As a result, foundations of maturation, learning behaviors, social engagement, and skill competencies have improved.

Bodies communicate awareness and the reaction to sensory input. Motor responses are one way to measure this. Reflex integration is measured by the actual movement reaction to specific sensory events imposed upon the body. Retained reflexes have been shown by Dr. Ayres to be, in part, less than stellar primitive reflex integration. Her treatment routines demonstrated assistance for such integration and reduced over-reliance on them.

Complimentary approaches to Ayres' methods have evolved from this original beginning. These include but are not limited to astronaut training for vestibular system strengthening, deep pressure therapeutic brushing for touch desensitization, weighted vests and pressure garment wearing for sensory modulation, and individual 'sensory diets' for self-regulation. The occupational therapist scaffolds therapeutic activities to create the just right challenge so that areas of strength can bolster the weaker areas of performance and sensory comfort.

New discoveries

Data suggest 5-20% of children have their daily life events affected by sensory processing differences. Several factors have been implicated regarding SPD, such as prenatal and birthing consequences, sensory neglect in infancy, and genetic links. As with any developmental behavioral disorder, there can be links to environmental toxins or the lack of fundamentally solid bonding and interpersonal relationships.

Sensory processing differences are more prevalent than the incidence of autism spectrum and attention deficit hyperactivity disorder. SPD continues to lack recognition in medical and psychiatric diagnosis manuals. Relying solely upon behavioral measurements of sensory processing has left gaps and differences in interpretation. Brain science has confirmed the theories of sensory processing foundations of behavior and functional skill attainment. Physiological differences in brains have been shown to exist between children with SPD and typical controls. Many studies demonstrate quantifiable differences in the microstructure of white matter in comparing neurotypical children to those with SPD. Some studies show a correlation between the sensory and motor receptive control centers, aligning with Dr. Ayres' original theories of sensory integration processing.

White matter (fatty glial cells of myelin) wraps around axon branches of neurons and forms the tracts for projecting axons. White matter's role in neuron circuitry is maintaining neuronal connections and insulating pathways so energy is not lost from the tract. Long thought to be passive tissue, these cells are part of the modulation and regulation function of the nervous system, with a chief function of modulating the action potential of the electrical reaction of a nerve being stimulated. Brain studies of boys with SPD have shown decreased white matter connectivity, markedly in the parietal regions. Auditory and tactile processing differences have also been demonstrated in correlation to white matter microstructure.

Glial cells are also shown to be active players in the brain's immune system regulation and clearing function. Every single brain autopsy study on people with autism spectrum has revealed various degrees and distributions of inflammation. One study even showed evidence of the presence of heavy metals. Glial cells contribute to the lymphatic system within the brain and connection to external detoxification pathways from the CNS to the periphery.

A specific gene mutation has been linked to a reduced sense of proprioception, leading to clumsiness and less-than-optimal coordination. Neurophysiological studies have shown that the brains of children with SPD have different responses to control subjects. Sensory gating, the brain's ability to "tune out" a stimulus with repeated occurrences, is less in children with SPD. Sensory gating may be related to neurochemical differences.

Bodywork Observations – Functional and behavioral descriptions of each subtype paralleling practice-based Evidence of Bodywork techniques demonstrated as having a positive effect:

Sensory Modulation Disorder – difficulty regulating responses (too much or too little)

Sub-type	SPD Behaviors	Bodywork Methods Offering Positive Effects
Over-responsive Also known as 'sensory defensiveness'	Tendency to respond too much, too soon, too long to sensory stimuli which most people find tolerable, or extinguish to attend to other information, often with emotional spikes. Fight-flight responses to sensations to overwhelming. Ex: Avoidance behaviors. Covers ears or nose. Gravitational insecurity from vestibular overload, approach-withdrawal from touch overload and refuses to participate in fine motor tasks; controlling behaviors with food and objects with fragrance.	• CST to cranial nerve nuclei, tracts, innervation sites. • CST to cranium sutures • LD to skin & fascia layers • Acup. to trigger points of Vagal & other nerves • Vis. organ & peritoneum tissue mobilization to modulate Enteric and ANS
Under-responsive Can be 'shut down' from overload but seems oblivious to input	Tendency to be unaware of relevant stimuli, or a delay in reacting; reactions are muted or not meeting the required intensity compared to peers' reactions. Comes off as passive or apathic. Or need much more input to reach a motivated state. Poor body awareness and clumsiness.	• CST to cranial base to decompress connective tissues surrounding spinal cord and brain stem • Same as over-responsivewhen "shut-down" issuspected.
Sensory Craving Never seems to get enough input	Tendency to rarely reach satisfaction from obtaining sensory input (never gets enough, never sits still). Getting more stimulation results in more disorganization nor satisfy the need for input. Touches, tastes, smells, experiences everything available, but does not become clam or focuses by these behaviors.	• CST to cranial base to decompress connective tissues surrounding spinal cord and brain stem. • Decompress occiput from upper cervical soft tissues (nerve and blood vessel exit sites from osseous structures) • Decompress cranial bones

Key:

- CST = Craniosacral Therapy
- LD = Lymphatic Drainage Massage
- Vis = Visceral organ/peritoneum methods
- MF = Myofascial method
- Acup = Acupressure and other tactile inputs

Sensory-Based Motor Disorder – difficulty with balance, coordination, postural stability, performance in novel or unpracticed tasks; practice does not make motor difficulties better

Sub-type	SPD Behaviors	Bodywork Methods Offering Positive Effects
Postural – Ocular Disorder Difficulty in body stabilization in standing and sitting; body doesn't support for eye-hand coordination	Impaired perception of body position; poorly developed movements that need core stability; lack of reflex integration as a result; body seems weak or has low endurance. Lack of enjoyment of movement activities. Visual motor dysfunction of focus and tracking. Indications for a need of vision therapy.	• CST to cranial base to decompress connective tissues surrounding spinal cord and brain stem • Decompress occiput from upper cervical soft tissues nerve & blood vessel foramen • CST mobilize dorsal columns of spinal cord • *CST to all cranial nerve nuclei, tracts, and innervation sites
Bilateral Coordination	Impaired timing and rhythm of left and right body sides moving in sync; less than stellar eye tracking; two handed tasks are difficult (tying shoes) are total body performance (bike riding).	• CST to cranial base to decompress connective tissues surrounding spinal cord and brain stem • Decompress occiput from upper cervical soft tissues nerve & blood vessel foramen • CST mobilize dorsal columns of spinal cord • CST to all cranial nerve nuclei, tracts, and innervation sites
Dyspraxia Results in difficulty with planning and carrying out new motor actions; poor organization of movements.	Formulating a movement response with a combined processing from pre-motor, motor, and cortical brain centers. Needs to spend more than typical amounts of time thinking and planning; gives up easily. Breaks toys and lack of success in sports; fine motor difficulties	• CST to brain stem, spinal cord, neurocranium and facial nerves; entire craniosacral system; glymphatics; intercranial membranes

Sensory Discrimination Disorder- difficulty processing and interpreting sensory input qualities

Sub-type	SPD Behaviors	Bodywork Methods Offering Positive Effects
Olfactory discrimination	Difficulty discerning good and bad smells	• CST to structures of forehead, face, and hard palate
Taste discrimination	Difficulty discerning good and bad tastes; less than stellar tongue control.	• CST for cranial base • Cervical decompression • Thoracic and cranial nerves
Vestibular discrimination	Difficulty in recognizing and responding to head placement (falling) or a delay in postural or balance reactions. Difficulty moving through space well (playground or calisthenics)	• CST to temporal structures and adjacent sutures. • Sphenoid balancing • Temporal bone decompression to reach inner ear and E. tubes
Auditory discrimination	Recognizing when being spoken to; locating source of sounds; telling difference between words; appropriate social response to when spoken to.	Same as vestibular processing
Visual discrimination	Difficulty with meaning of visual input differences; form and shape recognition; depth perception is balanced visual focus	• Frontal and sphenoid • Orbit symmetry of size and tension through oculomotor muscles.
Proprioception / kinesthesia discrimination	Feeling and understanding a meaning of up/down, in/out, left/right; feeling and executing the correct pressure of action to meet the task demands; kicking/gripping/pushing too hard or not hard enough	• Decompress peripheral nervous system. • CST to entire system of dural tube for dorsal column pathways to cerebellum through brainstem and midbrain.

Sub-type	SPD Behaviors	Bodywork Methods Offering Positive Effects
Tactile discrimination	Difficulty with discrete fine motor tasks such as writing, buttoning, scissoring and other tool use; manipulation skills	• Lymphatic drainage to mobilize fascia structure of dermis and clear effects of inflammation. Mobilize tightness of dermis layers. MFR to promote C cell healing.
Interoception discrimination	Recognizing when both hungry and satiated. Recognizing tensions in abdominal organs. Recognizing when bladder and bowels are full and need to move/empty. Delays in toileting independence during the daytime. Night time enuresis is believed to be a different etiology. Or the converse, frequent stomach aches & urination.	• Visceral organ techniques to peritoneum and mesentery to promote PNS and vasodilation, promote peristalsis. • Tx fascia of entire visceral structures. Breath expansion. Adrenals and kidneys, stomach, liver, intestine branches of vagal nerve pathway to innervation sites.

Additional observations of positive effects of bodywork upon the subtypes of SPD are covered in the following sections.

Oral motor and feeding aversions

Definition

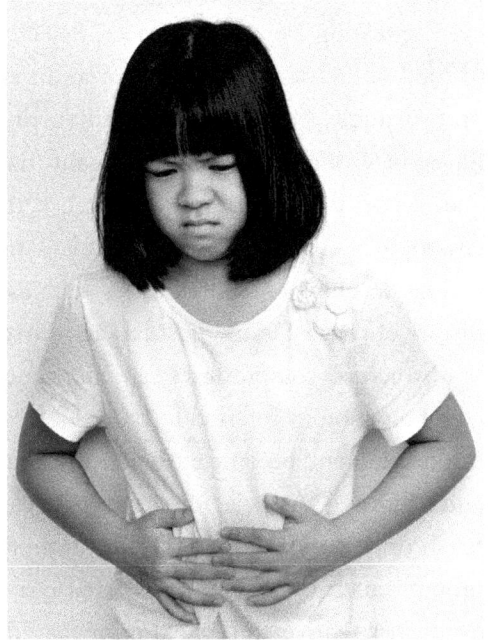

One of the most common issues pediatric therapists work with is the spectrum of sensory-based feeding difficulties. A continuum of picky eating restricted eating, and the more severe Avoidant Restrictive Food Intake Disorder (recognized within the diagnosable eating disorders). Negative associations with food mark the gamut with varying degrees of controlling rigidity and behavioral dysregulation around meals and snack times. A child with sensory-based feeding challenges has an elevated demand for control over what enters personal space or approaches the mouth—ranging from the run-of-the-mill picky eater to extreme behaviors of eating refusal. Refusals are often based on how food looks, tastes, smells and feels.

There may be underlying medical issues that need to be ruled out by a qualified professional, such as inflammatory conditions, delayed motility, organ emptying, dyspepsia, etc. These conditions can negatively affect a desire or tolerance for eating due to restrictive ease in swallowing or how the digestive tract feels once food begins digestion. Nutrition, weight gain, mood, energy endurance, and family stress levels depend on the child's ability to eat various foods without severe behaviors impacting mealtimes. It is the highly astute parent and therapist acting as detectives using deductive reasoning that can discern the underlying factors. Picky eaters may have sensory difficulties, or they may have very discerning tastes.

The difference and often the factor that propels families into seeking therapy is the degree to which food and mealtimes are negatively affected.

It is estimated that 20-25% of children and infants under eight struggles with various feeding challenges. Aversive behaviors of a child becoming upset with nonpreferred foods placed near them, refusing offered foods, choking and gagging with foods in the mouth, and even retreating from eating space. Selective eating can be defined as eating small amounts of food or restricting foods eaten to an extremely narrow selection of sometimes only one or two items. Sensory-based feeding disorders are often related to prematurity, chronic illness, neurological issues, unpleasant oral-tactile experiences, invasive medical procedures, and delays in oral motor function. Other associations include gastroesophageal reflux, eosinophilic esophagitis, food allergies, and triggers. Food aversions are common features of various developmental disabilities, such as autism spectrum disorder, anxiety, ADHD, and a history of torticollis and postural asymmetry.

Structural compromises can contribute to digestive tract resistance when swallowing. A common etiology referred to but seldom explored in greater depth is "GI issues." Behaviors can develop and become entrenched when root causes of feeding aversions are not identified and addressed. Associated stressors develop when adults' approach to the child's feeding is harsh or demands reminders and prompts. Repeated gagging is an SNS behavior that can prompt entrenched protective defense mechanisms that shape personalities. Past negative experiences with eating, choking, or gagging may contribute to this particular aspect related to eating. The underlying common denominator is that eating and ease of digestion all function within the state of being within the parasympathetic nervous system activation. Controlled by a sympathetic state of over-arousal or lack of modulation perpetuates a lack of readiness to engage in oral intake and hinders peristalsis.

Eyes, ears, and nose reinforce associations with food tastes, smells, and appearances. The oral structures give tactile and taste reinforcement. The mouth is the entrance to the GI tract, and the digestive system is one system that plays a primary role in PNS activation for getting to and staying in the rest and digest state. Personal success and performance in food intake reflect the balance of the person's autonomic nervous system and the peristalsis actions of the digestive system. The presentation of a meal to a child has cultural but also sensory ramifications for foods presented in appearance, odors, textures, and tastes. Conversations persistent of other sensations can affect the tolerance of a meal.

A sense of control vs. being forced to eat by others can affect the sympathetic nervous system. Other sensory aversions can also contribute to (sound, auditory processing – the sound of chewing), Oral motor, postural control, and motor planning of jaw and tongue with swallow and chew/swallow/breath synchronicity. Postural control of body symmetry supports both ease in swallowing (midline structures) and gives stability for hands to manipulate utensils and food pieces. Postural stress can create midline tension and recruitment, hindering ease of the upper esophagus in the swallowing process and adding resistance. Lack of sensory regulation may emerge as avoidance of certain vital foods and nutrients. It is essential to know the history of feeding and eating to uncover the structural and enteric system impact of current comfort levels with eating.

Behavioral manifestation

- Tantrums around mealtime
- Choking, gagging, vomiting when eating
- Trouble accepting new or mixed textures
- Refusal or solids or liquids
- Failure to thrive, poor weight gain
- Refusal to remain seated during mealtime
- Negative behaviors at mealtime
- Insistent on how foods are prepared

Sensory Modulation Features (hypo- or hyper-reactivity to sensations)

- Over or under reaction to strong odors
- Over or under reaction to differences or mixed textures of food on tongue
- Over or under reaction to other sensations in the vicinity of dining space
- Under reaction leads to poor suck and lip closure, inefficient tongue movements to move food
- Under reaction leads to over stuffing and pocketing food

Anatomy of the digestive system
Wiki Commons

Sensory-Based Motor Features

- Lacks in adequate oral closure during eating
- Lacks skills of tongue lateralization of food to molars
- Lacks tongue movement to initiate and the control food bolus posteriorly into oral cavity

Sensory Discrimination Features

- Lack of localization of food items on lips or tongue
- Messy eater, food or liquid falls from mouth during meal

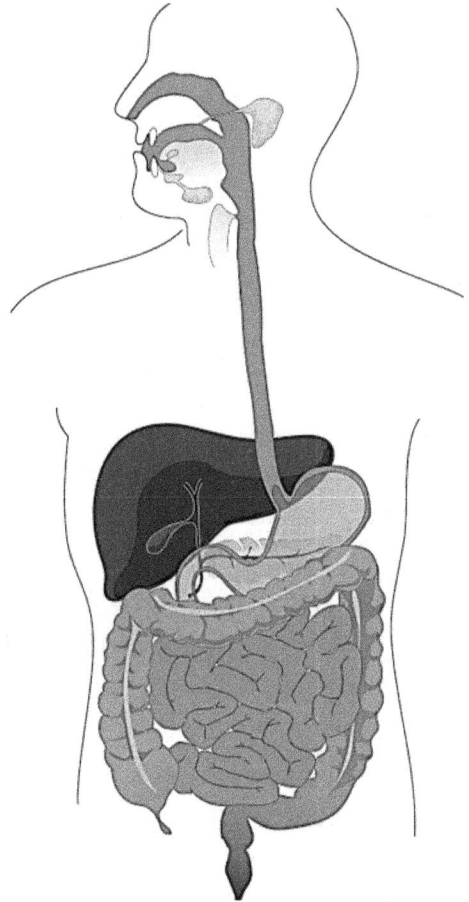

Historical treatments for oral motor and feeding aversions

Feeding dysfunction is more likely to be associated with infants and young children, whereas actual eating disorders are more likely to be associated with teens and adults. In all situations, underlying sensory processing can be at play, but left untreated can contribute to a myriad of maladaptive responses to food and events and people associated with mealtime. Personal success and performance in food intake reflect the balance of the person's autonomic nervous system and the peristalsis actions of the digestive system. The presentation of a meal to a child has cultural but also sensory ramifications for foods presented in appearance, odors, textures, and tastes. Conversations persistent of other sensations can affect the tolerance of a meal. A sense of control vs. being forced to eat by others can affect the sympathetic nervous system.

Speech-language pathologists and occupational therapists have historically addressed feeding therapy for sensory-based feeding and oral-motor difficulties. Focus can be solely on oral motor and swallowing function or expanded to include sensory processing. Chewing/swallowing and modulating odors, sound, and other inputs must be modulated without the dominance of sympathetic tone or state (which tends to tighten digestive structures). Sensory-based feeding problems can be diagnosed through two avenues and points of view: a clinical swallow study by a qualified speech-language pathologist and radiologist and a thorough sensory-motor history and clinical observations.

Behavioral approaches consider the cultural and interpersonal factors that can contribute to feeding performance issues. Meal times are often replicated, and observes parent-child reactions for various behavioral patterns of both to child's response to food and the parent's reaction to the child. Gradual comfort level increase with target foods through exposure and positive experiences (behavioral shaping) to offending sensory channels.

Sensory integration methods to promote self-regulation of the ANS have historically rested in the occupational therapy world (as a specialty area of practice). Sensory-motor oral play with non-food substances involves deep breathing and promoting the parasympathetic state of deep breathing in play. Straws blowing cotton balls, whistles and bubble blowing, and musical instruments desensitize the skin around and inside. Deep pressure into the face (wrestling, crashing into big pillows, linear acceleration to desensitize the face moving through space.

Such sensory integration activities must be child-directed and child-control. Feeding therapists should never be forced as the therapist who gets to work with the child after someone tried force-feeding has to deal with an entrenched sympathetic state and escalation of control behaviors. Overall, best practice management first considers any underlying neurological function that drives eating behaviors, then family and cultural factors that reinforce coping behaviors.

New discoveries

The mouth is the entrance to the digestive system. Movement of the tongue and jaw with sucking, chewing, and swallowing prompts the activation of the vagal nerve (by jaw movement in its proximity). Along with the mix of saliva secretions, other chemical and physical prompts, the PNS state turns on peristalsis. Optimal organ motility happens with reduced resistance through the GI tract to move food, chyme, and fecal matter. Tight soft tissues reflect increased sympathetic tone in the body and viscera.

Pediatric gastroenterologists are exploring eosinophilic esophagitis as a common source of dyspepsia or motility issues. Caused by substances triggering allergic inflammation, the esophagus walls can't contract properly during peristalsis. Pyloric stenosis or constriction is a common co-occurring symptom with esophagitis (in small children). Perhaps the body is protecting itself from absorbing intolerable food by constricting itself to limit flow. Outward symptoms have been a feeling of choking or gagging, reflux or regurgitation, motility delays, and even pain. Pain (of the esophagus or other organ) will always keep the ANS in a state of sympathetic tone.

A little explored aspect in therapy is how the digestive organs might react to the ingested food. The GI tract's scanning and interpreting any cellular threat from food substances can set off a neuroimmune response, activating sympathetic organ reactions.

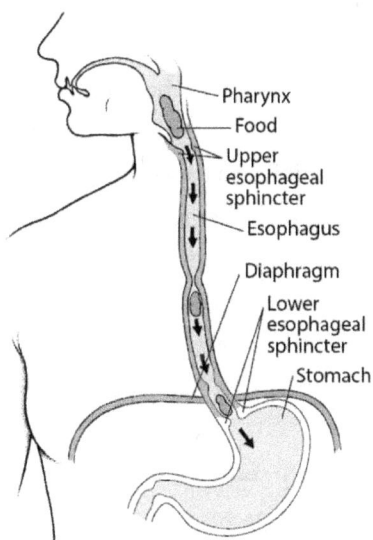

Source: Merck Manual.
Peristalsis in the upper GI tract.

Tension around esophagus dictates ease of peristalsis. Can be affected by pericardium and pleura tissues, chains of sympathetic ganglion in thoracic, restricted fascia bands in any direction, and respiratory diaphragm/rib cage tension.

Enteric nervous system response to food can be palpated by ANS tone. Treating organs, and specifically the digestive sphincters, have provided us to communicate directly with the "behavioral response of organs"

Craniosacral therapy is evidenced-based practice for the treatment of infantile colic and reflux, and has been shown safe in treatment application for premature infants. In one particular randomized control study, CST was demonstrated to be safe and the treatment group cried less, slept more, and reduce the severity of colic expression and intensity. Studies such as this support the use of bodywork methods in infancy. Much practice-based evidence exists on the rather quick and often spontaneous changes in feeding performance and sensory modulation from mixed methods of manual therapy.

Bodywork for Sensory Integration is a guide to address children with such historically issues that may not have had treatment or resolution. Therapists can begin to look for digestive issues in developmental history taking as red flags for autonomic nervous system structural stress. Bodywork to use in conjunction to rule out other medical issues. Can be used in differential diagnosis in close communication with gastroenterology professionals when concerns over nutrition intake and safety with eating are evident. With any question of safety with swallowing, a gastroenterologist or feeding specialist (speech or occupational therapist) should be consulted.

Bodywork Observations

- Evidence-based practice show that craniosacral therapy assists with improved breastfeeding. *Bodywork for Sensory Integration* is not a form of feeding therapy, nor should it replace such intervention when needed. However, *Bodywork* methods can assist in the correction of issues that might be interfering with a child's intuitive and natural ability to eat or be comfortable with eating.
- Eating is a whole-body process and assessment and treatment should reflect the entire digestive system, beginning to end. The organs of the gastrointestinal tract include the mouth, pharynx, esophagus, stomach, small intestine (three portions), and large intestine. A novel concept to consider for eating performance is the flexibility and state of the digestive sphincters. Seven in total, these constricted bands of smooth muscle, regulated by the vagal nerve system, play a critical role in peristaltic action of the tract. These sphincters offer a direct treatment to the vagal system, and when treated have consistently demonstrated positive changes to sensory modulation and self-regulation behaviors. From mastication of food in the mouth, from churning within the stomach for chemical decomposition, to movement to intestines for more degrading to molecular level and eventual capturing by absorbed into the mesentery blood stream. The common denominator of optimal digestion, is the state and flexibility of the sphincters by parasympathetic system activation.
- Feeding and eating performance is directly related to the tension state of all the structures in the digestive system.
- Whole-body myofascial, viscerofascia tissues, and deep fascia bands are related to conditions such as tongue-tie, a significant factor in oral motor performance. Treating the whole fascia field to gain flexibility and tissue expansion has consistently proven an indirect as well as a direct effect upon tongue and oral structures.
- Increasing flexibility and motility of the fascia that surrounds organs within a respective cavity improves the function of that organ (in the digestive process). For example: the esophagus penetrating through midline fascia structures, the respiratory diaphragm, and the two esophageal sphincters has improved ease in swallowing and reduced gag response, followed by activating a relaxed state for a meal.

- Structural compressions of the cranial nerves can have a negative effect upon the sensory processing of the mouth structures.
- Tone of breathing components. Compressions of rib cage or respiratory diaphragm that might have cause limited expansion of rib cage, or ease of expansion during breathing.
- The freedom of tongue movement (for both activating and modulating tastes pressing to the hard palate) reflects fascia organ flexibility as well as cranial nerve integrity of four cranial nerves.
- Hyoid and all the attachments on anterior throat and neck. General tone of components of GI tract In addition, Few if any oral motor therapy approaches address the hyoid attachments to tongue, clavicles, cervical vertebrae and the massive amount of investing fascia suspending these structures. Until explored through palpation, many false assumptions that a child's breathing is optimal in structural expansion, flexibility, and optimal ANS tone. The anatomy involved in chewing and swallowing are distal moving parts, supported by proximal structures providing stability.
- Historically from a sensory integration behavioral interpretation approach reactions are categorized as hypo- or hyper-reactive. Hypo-reactive responses have been observed related to compressed nerves to tongue and face (facial, trigeminal, hypoglossal or glossopharyngeal). Behaviorally this can be related to lack of food movement, placement, or tongue articulation. Compressions at the cranial base, occiput segments pushed anteriorly, fascial imbalance at foreman magnum, or mandibular with translating fascial distortions are but a few examples of osseous sites to evaluate. Hyper-reactive response expressed as behaviors of food refusal, controlling over how food is fixed or presented, rigid food choices and limited repertoire of foods and is generally related to compression into ganglion of sympathetic nerves.
- Tongue movement modulates taste, so lack of full tongue control over food in oral cavity can create a habitual hyper-protective state during eating. Tongue tie, compressed maxilla vomer palatine where soft and hard palate meet, restricted tissue in tongue (nerve in middle of tongue) taut can create a sympathetic tone state causing hyper-reflexive gag.

Bodywork for the digestive system frequently has an immediate positive effect on sensory modulation and self-regulation

Hill, Arden, and Sarah Halloran. "Diagnosis and Management of Sensory vs. Behaviorally-Based Feeding Disorders." *Children's Speech & Feeding Therapy, Inc.*, 1990.

Redle, Erin E. "Evidence-Based Intervention for Toddlers with Sensorimotor Feeding Disorders. EBP Briefs. Volume 7, Issue 4." *EBP Briefs (Evidence-based Practice Briefs)* (2012).

Toomey, Kay A., and Erin Sundseth Ross. "SOS approach to feeding." *Perspectives on Swallowing and Swallowing Disorders (Dysphagia)* 20, no. 3 (2011): 82-87.

Attention Deficit-Hyperactivity

Definition

Attention Deficit – Hyperactivity Disorder (ADHD) is a spectrum of difficulties with individual manifestations with two main features. With a predominantly inattentive presentation, distractions are easily elicited. It is difficult for the person to pay attention, organize and finish a task, or follow instructions and conversations. (Distractions are also a feature of sympathetic dominance of the autonomic nervous system). With a predominantly hyperactive-impulsive presentation, the person needs to move and fidget and finds it challenging to sit still. Restlessness and impulsivity can interrupt situations, speak at inappropriate times, or have great difficulty waiting a turn. The impulsivity feature raises the risk of accidents and injuries for a lack of judgment or taking time to assess any present danger.

People can have a combined presentation of both types equally. To reach an accurate diagnosis of ADHD, other physical or mental disorders should be ruled out. Symptoms of ADHD mimic other conditions. As with most developmental disorders, early diagnosis of ADHD increases the positivity of long-term outcomes. Exaggeration of under- or over-responsiveness sensitivity to everyday sensory events and input. Over-responsiveness is often observed as hypervigilance of sensory events with a common aversion effect, and the coping behavior is to avoid attempting to achieve ANS homeostasis.

Behavioral manifestation

The label of ADHD often needs further explanation of not being the inability to pay attention but rather a result of modulation and inhibition difficulty with less than the efficient ability to extinguish alerting stimuli. The person pays attention to the alerting and peripheral sensory input and has challenges tuning out the background. Shared attention and selective attention skills to remain on-task are also often challenged. It is within these realms that ADHD can become problematic. Sensory modulation and inhibition are the shared features with sensory processing dysfunction. It is important to remember that ADHD and SPD are two differing disorders. Different brain areas are involved leading to varying mechanisms for resulting behavior. There is differing evidence for effective treatments.

Impulsiveness and hyperactivity are two features, but not all with ADHD are impulsive bodies with excessive movement. Compressed occiput into an atlas and compressed or torqued upper cervical structures can compromise blood flow in the vertebral and other arteries. Movement may raise the pressure gradient to assist the movement of oxygenated blood through the space of structural resistance. Sensory seeking may be reinforced by such things as intense pressure or intense linear acceleration to also aid in improving blood flow. Vasoconstriction and its etiology in various sites of the body have never been considered in the same discussion of sensory processing theory and resultant behaviors.

Scientists have long studied risk factors and cause for the best management of ADHD. Though true causes are still unknown most substantial evidence is the genetic factor. In addition to genetics, risk factors include brain injury, environmental toxins (lead, mercury, aluminum), alcohol and tobacco exposure in utero, prematurity, and low birth weight. The body of research does not support the long-held views of a diet with too much sugar and unnatural addictive, too much television watching, parenting styles, and poverty or family chaos. Though not the leading cause, it is readily acknowledged that these things might make symptoms worse in some people.

The behaviors of ADHD and SPD are often considered the same, but there are distinct differences. A few particular features distinguish between the two. Impulsive behavior with ADHD generally cannot be stopped regardless of sensory input. Craves novelty and activity that is not necessarily related to a specific sensation. Sensory input does not tend to organize the thinking or focus, and the person calms or improves attention when getting constant novelty of information or ideas.

When behaviors are more SPD, behavior and cognition calm and focus when sensory input provides the proper modulation the nervous system needs. Conversely, dysregulation becomes more evident when provided with novel sensory or movement activities. Behaviors of modulation and dysregulation tend to follow a pattern (occurring during particular tasks or exposures or at particular times of day). Sensory seeking or sensory sensitivity behaviors do not tend to improve with ADHD medication.

It is typical for children to have trouble focusing and behaving at one time or another. However, children with ADHD do not just grow out of these behaviors. The symptoms continue, can be severe, and cause difficulty at school, home, or with friends. A child with ADHD might:

- Spend a lot of time daydreaming; becomes off-tasks often
- Forget or lose things most than most people
- Restlessness or fidgety
- Excessive talking, often at inappropriate times
- Makes careless mistakes; misses details
- Takes unnecessary risks
- Difficulty resisting temptation
- Difficulty taking turns or waiting
- Difficulty getting along with others, argumentative
- Tune out background noises and other sensations
- Difficulty turning off the alarm / alerting response

Historical treatments in sensory integration for ADHD

Standard treatment recommendations combine medication and behavioral interventions with individualized educational accommodations. In sensory integration evaluations, it is common for co-occurring with ADHD lack of core strength stability, head lag and anti-gravity postural control weaknesses, bilateral coordination issues, touch defensiveness (moves to prevent being or getting touched), living in a perpetual startle state or Moro reaction keeps the body in extensor muscle tone. Not uncommon to see toe walking or bouncing on feet during gait as reflexes are not inhibited. Primitive reflexes are often retained.

A nervous system in a chronic state of sympathetic dominance can have the same behavioral expression. Behaviors change with environmental changes. Differential diagnosis is imperative, especially if medication is being recommended. It is not a stretch to state that ADHD is over diagnosed and medications are given too readily before other therapies have been trialed.

New discoveries

Specific diets have been tried as a treatment for ADHD for decades, and there is scientific support that food substances may contribute to symptoms in individualized ways. A disrupted gut biome can cause inflamed intestinal walls, which are palpated as tense and restricted. Tight mesentery and vasoconstriction are also commonly palpated within this state. Conversely, a healthy gut is palpated as soft and pliable, with organs sliding with neighbors and free of restrictions for sphincter activation. A leaky gut can be the result of an unhealthy gut biome. (See more in Self-Regulation p. 273)

Intestinal wall tension can restrict the ease of the common bile duct drainage (contents of the liver, gall bladder, and pancreas). In bodywork, we observe constipation and tummy aches related to tension here at the sphincter of Oddi. The tight ileocecal valve can correlate to constipation. A vasoconstrictive state is common with a leaky gut presentation. Limbic system dominance can interfere with the inhibitory function of the cortex cells of the executive system. Chronic states of sympathetic dominant behaviors can create hyper-vigilance, similar to ADHD behaviors.

Science does not support diet as etiology (on a large scale) though there is plenty of evidence that food triggers can contribute to behavioral challenges. Old knowledge of the success of specific diets no longer shows effectiveness. Perhaps food substances have changed as an uncontrolled variable. Though nutrition is not within the scope of most bodywork practices, the effect of food triggers, toxins in food, or other chemicals can be observed and noted. The Gut-Brain enteric nervous system connection can reveal measurable structural with resulting ANS behaviors. Add to that the knowledge that the gut biome communicating with various bacteria determines the amount and quality of the neurotransmitters. 50-90% of neurotransmitters arise in the enteric and gut lining; it is a safe assumption that gut health may be related to autonomic behaviors like ADHD. Indeed, a lot more study is needed. Visceral manipulation has proven valuable in gaining subject skills at assessing inflammation of organs and vasculature, possibly creating a helpful monitoring method. For example, one boy (age 5) regularly struggled with self-control at home and school. With sensory-based and visceral-based detective endeavors, matched with food eliminations, a clear correlation to his severe behavioral and emotional outbursts was related (in part) to eating cold cut meats with nitrates and other preservatives for his lunch.

Bodywork Observations

- Chronic protective retraction in multiple sites is a common observation, both abdominal and thoracic organs and mesentery vasculature
- Multiple sites of high tone, fascia tensions, and tight fascia bands are common
- Cranial base compression often from the birthing progress
- Frontal bone compressed posteriorly, projecting pressure towards diencephalon (limbic system)
- Head shape differences: ridges at suture lines suggesting cranial plate over ride from retained cranial compression; metopic fontanelle ridge suggests head remolding that did not fully complete, sagittal suture ridge suggests tension pattern over sensory homunculus

- Upper cervical compression, often compromising ease of vertebral and other vessel blood flow
- Array of gastrointestinal complaints: Constipation, irritable bowel, gastroparesis, history of prolonged colic or reflux, vasoconstriction

Clients obtaining bodywork routinely report improvements in calming, increased focus concentration, and attention span towards task completion. The body's reactions to a compressed cranial base can be analogous to a snake's reaction to having his head trapped body thrashing and wiggling (various degrees) to get his head loose. Varying degrees of self-stimulation (head banging) or sensory seeking (bungee jumpers, skydiving, and surfing) craving deep pressure input or constant motion has been routinely shown to have a positive effect with a reduction in the extreme or intense need for such input. Frequent parental reports include behavioral changes with "improved focus, increased concentration, reduced impulsiveness, increased executive function expression, and mood stabilizations. Behavioral changes suggest ANS self-corrections.

Autism Spectrum Disorder

Definition

Autism Spectrum Disorder (ASD) is a condition of brain connectivity. The "spectrum" is a contemporary psychiatric description indicating that a range or continuum of autistic features can be present. The continuum infers levels of severity, and the pervasiveness of several or all areas of function can be affected. A kaleidoscope of behavioral manifestations of a person with information processing or perception differences vaguely defines the range. Sensory disturbances are key features that often trigger people to explore a diagnostic workup. People with ASD most commonly have problems with social interactions, communication, and function is often restricted by repetitive behaviors or interests. There are often unexpected differences in learning, moving, and paying attention. The differences can be mild or can be extreme.

The prevalence of ASD is approximately 1 in 54, based upon a 2016 study in the United States, though the reported incidence has risen over the past several years. It affects boys four times more than girls, possibly related to a protective effect of estrogen upon the brain. Regardless of ethnicity, culture, or economic background, ASD is present in every country.

There is one exception: the American Amish communities, by comparison, have very few cases of ASD reported. It is important to note that people without ASD might also have some of these unique traits, but for people with ASD, these issues can be more intense, extreme, or inflexible, making one's quality of life challenging.

The causes of ASD have been studied methodically, and discoveries continue to be made. Given the complexity of the various dysfunctions observed, no single reason is known. There are probably many different etiologies, with genetics and environment playing a role. Several genes appear to be involved with an ASD diagnosis and the severity of expression, such as Rett syndrome, Fragile X syndrome, and various metabolic disorders. Some genetic mutations seen are inherited, while others occur spontaneously. Someone with ASD is likelier to have an immediate family member with ASD. A specific MTHFR gene mutation hinders the body's ability to transform folate into methyl folate, leading to low levels of glutathione and cysteine. These essential antioxidants are crucial for the body's ability to detoxify and chelate toxins and heavy metals. If not chelated, remaining toxins can enter the bloodstream, and some may cross the blood-brain barrier.

Advancements in neuroimaging now make it possible to move our knowledge base forward in exploring brain regions holding the cellular processes of sensory processing differences in ASD. Every brain tissue analysis of tissue donations from subjects with ASD are consistently indicating presence of brain inflammation and even cellular damages. The sensory processing issues of registration, modulation, and emotional motivation are issues addressed by those studying ASD and by those giving personal accounts of what it is like to have ASD.

The emerging brain science about brain and sensory processing function (especially in ASD) is consistent with some of Ayres' original hypotheses, though not all. Evolving science is revealing variances of the brain in the ASD and non-ASD populations. The future of brain imagery methods being used in ASD can serve as timely methods to finally study sensory integration therapy to its fullest extent.

Behavioral manifestation

Multiple causes, multiple brain sites, gut dysbiosis, underlying genetic and metabolic function can all determine and influence the expression of autistic-like behaviors. ASD is not one cluster of the same behavioral features. The identification of phenotypes of sensory processing differences in ASD can be useful in better definition of behavioral features, leading to specific and more effective interventions. Generalizations of behaviors are impossible to quantify and categorize. But what is apparent is what was once labeled as sensory integration dysfunction is comingling with ASD manifestations in clinical and school settings.

The list of autistic-like behaviors can range from mild to extreme and include, but not be limited to:
- Lack of, or delayed, social skill development (which can be sensory-based learning)
- Difficulty with expressive and receptive communication
- Presence of restrictive (rigid) and repetitive behaviors
- Avoids eye contact (which can be sensory-based disturbance or oculomotor dysfunction)
- Delayed or absent speech and communication skills
- Over-reliance on rules and routines
- Becomes upset by minor changes to routine or unexpected events (can be sensory-based intolerances)
- Difficulty understanding other people's emotions.
- Pain and significant neurological discomfort, sounds (can be inflammation based; meningitis; neuritis)
- Self-abuse may be an extreme coping mechanism to inescapable pain, onslaught on input that is not modulated or inhibited. (may be a large-scale neuralgia or neuritis)

- Unexpected reactions to sounds, tastes, sights, touch and smells, (not adapting to sensory overload)
  - Sound: fingers or hands covering ears, extreme behaviors with loud noises, intolerance of crowds, teeth grinding or jaw clinching, intolerance of haircuts or face grooming
  - Smells: food intolerances, poor cooperation during meal times, smells everything
  - Touch: self-abuse of biting or hitting, aggression especially to those getting into personal space, skin picking or hair pulling, intolerance of clothing types or tags or seams
  - Visual: hyperfocus on a particular, often minute, target, hyper acute visualization, photographic memory, conversely can become overwhelmed with too many things presented, inability to sort overwhelmed by complexity

Historical treatments for sensory integration in ASD

Many different professions work with children and adults on the autism spectrum, offering treatments or support that align with their professional training and philosophies. From a neurological framework, Neuro-developmental, sensory integration, and sensory-based interventions are some of the longest-utilized approaches in early learning and educational therapy environments. They are primarily utilized by physical, occupational, and speech therapists. Neuroplasticity is proposed as the mechanism of change in the outcomes of the sensory-based intervention, such as Dr. Ayres' work of Lorna Jean King and other similar approaches. Based upon neuroscience parallels through the decades, Ayres' sensory integration has provided a framework for understanding reasons for behaviors and coping patterns of autistic features, even though her original approach focused on a population described as having 'minimal brain dysfunction.'

From the behavioral sciences, Applied Behavior Analysis (ABA) is a therapy based on the science of learning and behavior. Positive reinforcement is one of the main strategies used in ABA, and the therapist decides or controls the reinforcement. (In contrast, the neurological approaches assume the successful execution and repetition of child-directed motor action in play-based gross and fine motor play are self-reinforcing that build initiative, idea formation, and sequentially carry out tasks without the need for cues.) The treatment objectives generally are to increase behaviors that are helpful to a child's involvement within the family and school and decrease behaviors that are harmful or affect learning.

There is little comparative evidence of these two primary treatment philosophies for ASD. The goals and objectives of the neurological verse the behavioral approaches to treating autism are often at odds, as the focus can be on different aspects of development. It is sometimes complicated for families to engage fully in either approach.

Rote skill, locus of control, carry over and generalization, initiative, and dependency upon cues compared to drill learning of skills verse child-directed engagement in sensory-motor activities that address cerebellar function and overall praxis skills. Based upon positive rewards encourage the person to continue practicing the response and using the skill. Over time, this [may] lead to meaningful behavior change.

Many children with true ASD have some degree of dyspraxia. Sensory integration approaches activate and synthesize information through active engagement of the vestibular, proprioception, and tactile systems. When kids can't do something, they probably can't feel their bodies well enough or sequence motor tasks. Sensory integration is one of the most highly utilized interventions in occupational therapy (for autism and other populations). Based on criteria standards established by the Council for Exceptional Children, Ayres' Sensory Integration is now considered an evidence-based practice for children with autism.

Relationship-based, interactive interventions have grown from recognizing the inherent sensory bases to communication and mutually satisfying human interactions. The child receives intervention to assist that goal without causing alarm from an unmodulated sensory system. Social cognitive skill training is combined with child-directed sensory preferences and interests.

All the above are supported both by research evidence and practice-based evidence. Parents of children with ASD have the ultimate decision upon which interventions work for their child. Sadly, not all families or regions can access various methods and styles equally, plus policies for reimbursement have been different in different US states.

For decades, children with ASD were assigned to behavioral sciences without adequate investigation into biological and neurological functioning. That has finally changed as more awareness of holistic treatment approaches has developed. This fact alone has been driven by parents seeking and demanding answers from the medical and behavioral science communities. The needs can be short-term but most likely long-term, and for the majority, a lifetime of support and remediation may be necessary.

Treatments and views of autism have changed as more personal accounts, insights, and perspectives of people with ASD have been shared with the world. Neurodivergent is becoming a preferred descriptive term, rather than being autistic or having autism. It speaks to authority and the impact of diagnosis labeling. Remediation with the sole purpose of "fixing" or changing to conform to the normality of function is also being rejected by the community of people with ASD. Those with ASD may have more or fewer differences than the person making the behavioral interpretation. Different yet but not incapable embraces the fact that all humans have challenges to some degree.

New discoveries

Post-mortem studies currently consist of small sample sizes, but several remarkably consistent with findings. Cortical layering is largely undisturbed, but there are consistent findings of reduced neuron column numbers and aberrant myelination. Though a great deal more research is needed, brain science studies have shown the following:

- Abnormalities or differences in the cerebellum
- Abnormalities in the cingulate gyrus and frontal cortex
- Neuro-immune modulation with impaired synaptic, metabolic, proliferation, apoptosis, and immune pathways are repeatedly implicated
- A lack of neuronal pruning at expected developmental stages
- Brainstem involvement
- Autopsies have discovered heavy metals or brain inflammation in every study
- Strong evidence of dysfunction in GABA, glutamatergic and glial cells
- Majority of studies revealed the presence of neuroinflammation with activation of astrocytes and microglia and abnormal levels of cytokines and chemokines

This is but one possible explanation of brain inflammation for ASD. Viral or bacterial infections can also create such a scenario. Co-occurring conditions with ASD are common:

- **Pediatric Acute-onset Neuropsychiatric Syndrome (PANS) and**
- **Pediatric Autoimmune Neuropsychiatric Disorder – Strep Infection (PANDAS)**

PANS is a clinical diagnosis based upon a dramatic, often overnight, onset of neuropsychiatric symptoms. Typical features are obsessions or compulsions or extreme food restrictions. Other neurological features can include: tics or jerky movements, anxiety, depression, mood swings, aggression, hyperactivity, regression in speaking, extreme fears, sudden loss of fine motor skills, difficulty with memory or academics, sleep disruptions, or regression in bedwetting or bowel control.

PANDAS is more of a gradual onset (within 2- to 3 days) of behavioral expression. Symptoms similar to PANS, it is classified as a subset of PANS. The marked difference is a **known** positive test of streptococcal infection such as strep throat or scarlet fever.

Causes of PANS and PANDAS are thought to be triggered by infections, spike in metabolic disturbance and decrease body detoxification, or other systemic inflammatory reactions. Regardless, both conditions recognize that brain inflammation is contributing to behavioral manifestations.

Obsessive-Compulsive Disorders

Obsessions can be described as inability to stop topic thinking or an excessive worry about an event or the sequence of events. Excessive talking about such things reflect fixated thought patterns. Compulsions are the acting out of such thought patterns through rigidly-held rituals such as closing doors, reopening them, and reclosing them or carrying out tasks in a specific order; or washing hands excessively and more than is practically required. When OCD of childhood appears, the National Institute of Health now recognizes this as a form of PANS or PANDAS.

Seizures

It is estimated between 11 and 39% of individuals with ASD have or will develop seizures. Research continues to explore if ASD (and its underlying biology) leads to seizures, or if underlying seizures leads to ASD. Therefore, seizures can be observed at all ages in people with ASD. Some seizures are less obvious than others (previously known as 'silent seizures'). The contemporary use of qEEG (quantitative electroencephalogram) in brain mapping has made evaluating for seizures less invasive and more readily available for the public. The qEEG provides a non-invasive look at brain functioning without the use of toxic contrast mediums. This scanning method, used in the practice of neurofeedback clinics, has helped clinicians identify source of neurological symptoms or behavioral dysregulation in the brain. (Such technology could be used to empirically study the effects of bodywork methods on brain function).

Gastro-intestinal dysfunction and Gut dysbiosis

It is well established that people with ASD commonly have gastrointestinal and gastroenteric problems. The most common symptoms are constipation, diarrhea, and abdominal pain. Less known or underreported symptoms include gastritis, esophagitis, and leaky gut syndrome (suggestive of inflammation to the inner walls of these organs).

Children with ASD have been shown to have simpler gut floras than neurotypical peers, and some observational studies report unusual bacteria species present in the gut biome. Furthermore, the flora of the gut not only involves bacterial species but yeasts (candida). Candida albicans are two times more abundant in toddlers with ASD than in neurotypical peers. Too great of a load of candida albicans can release ammonia and neurotoxins into the bloodstream. Levels of neuroactive compounds like dopamine, serotonin and its receptor sites (5-HT), and GABA are produced by good bacteria in the gut lining communicating with the enteric neuron mesh. Gamma-aminobutyric acid (GABA) is an amino acid that is the primary inhibitory neurotransmitter for the central nervous system. (Inhibition is believed to be the immediate functional problem of sensory modulation disorders). These compound levels can become altered due to gut dysbiosis. Microbial metabolites have been theorized to cross weakened intestinal walls (leaky gut), getting into the bloodstream where systemic inflammation can be triggered. These metabolites may then cross the blood-brain barrier setting off intracranial inflammation.

In one interesting study, mice developed autistic-like behaviors when their intestinal tracts were colonized by microbes extracted from the fecal matter of people with ASD. Though it did not prove that gut bacteria caused autism, it did highlight that the gut biome can contribute to some disorder features. A contrary view challenges the idea that gut bacteria influence brain development and suggests that restricted diets and picky eating drive changes in gut bacteria in ASD. The researchers make this opinion state only one species of bacteria out of 600 has shown a genuine association with ASD. They found picker eaters tended to have less diverse microbiome and more diarrhea. Their research did not support the view that the gut microbiome is the cause of autism. Much more research is demanded studying intestinal flora and treatments that alter brain behaviors and improve dysfunctional autistic features. Anecdotal treatments to rebuild the gut floral include the reintroduction of breastmilk colostrum (from cows, camels, and goats) and through fecal matter translates.

The term toxic encephalopathy is brain injury leading to dysfunction caused by exposure, inhalation, ingestion, or injection to toxic substances or chemicals. Symptoms of toxic encephalopathy range from mild subclinical deficits to obvious clinical disorders. Heavy metals, herbicides and pesticides, and paints are the most commonly documented forms of toxins.

Signs and symptoms of heavy metal poisoning are not unlike features of ASD: Tremors, insomnia, memory loss, mood changes, neuromuscular affects, headaches, cognitive delays, motor control and coordination dysfunction. Mercury exposure in children has been documents to show: impaired motor skill, cognitive problems, language comprehension and speech difficulties, hand-eye coordination, and perceptual motor problems (unaware of surroundings). Accumulating exposure will likely produce increasingly severe symptoms and behaviors. Aluminum exposure can cause muscle weakness, bones that hurt or break easily, seizures, speech disturbances, slow growth. Lead exposure is well documented danger to the nervous system causing grave affects upon the developing brain, leading to lowered IQ and cognitive functioning. A simple blood test can determine if a child's behaviors that may be problematic are related to lead exposure. Severe cases of lead exposure can cause anemia, seizures, and even coma.

The public should continue to become mindful that autism is not one thing. Treatments must be individualized and all avenues explored prior to assumptions that behavioral interventions are the best and most cost-effective avenues. Wide variation of individual variations. A thorough history and medical work up covering all areas of function is in the person with ASD's best care. Biological approaches help us determine active vs inactive sources of inflammation. Sympathetic C fibers were shown to be damaged in the skin of a study group of boys with ASD and touch hypersensitivity though the study did not identify or theorize the source of cellular injury. (Refer to Touch Defensiveness on p. 288)

Bodywork Observations

Responses to bodywork has been as unique as each individual client. The majority, however, presented with suspected or known biological underpinnings. Summarizing a collective of individual changes that can be attributed to the use of primarily craniosacral therapy and lymphatic drainage. Bodywork is difficult to reach a finite ending of treatment process in cases of active uncontrolled inflammation. Collaboration with biomedical or naturopathic practitioners have uncovered a host of biological correlates with clients seen in our clinic. They include, but are not limited to:

- Viral meningitis, subclinical without fever but children responded to anti-viral treatments
- Lyme (borrelia) and related co-infections
- Parasites, the greater the infection the more severe the symptoms.
- Indications of leaky gut, either through less-than nutritious diet or self-limiting food repertoire such as skin disorders, touch intolerance, other sensory intolerance, allergies and food triggers and bowel movement irregularities
- Bacterial infections
- Gut dysbiosis
- High levels toxic metals

Bodywork palpation findings have ranged from

- Tight, restrictive meningeal tissues
- Reduced craniosacral rhythm vitality
- Significant mesentery tightness and vasoconstriction
- Lymphatic stagnation, liver congestion, and engorged nodes systemically

Behaviors changes we routinely observe include but are not limited to:

- Reduction in self-abusive behaviors (head-banging, hair pulling, skin picking) but in conjunction with biomedical or naturopathic interventions to find source of inflammation.
- Reduction in aggressive or emotional outbursts (but also with biomedical or naturopathic care combined)

- Increase in communication intent and even improved vocalization attempts in non-verbal children
- Increased awareness to surroundings and other people
- Increased engagement in social or interpersonal encounters
- Improved sleep regulation
- Decreased sensory hyper-sensitivities; individual variations
- Improved social engagement; individual variations
- Improve cognitive performance; individual variations
- Results would much more positive in younger clients than in adults, suggesting entrenched coping or behavioral patterns may play a role in engaging or investing in the treatment process

Not everyone with ASD tolerated treatment, while some request treatment regularly. For a person with neurodiversity, or for a parent of a child with ASD the advice we give is the *Bodywork for Sensory Integration* is an experiential treatment. When methods are performed as designed, all these methods carry little to no risk of harm.

Do not Harm. Do not Alarm.

That is the motto our clinic follows when offering bodywork to those with nervous systems that have fickle sensory reactions.

Sensory Modulation & Self-Regulation Dysfunction

Definition

Sensory modulation expresses the resiliency of the autonomic nervous system continuum. Sensory modulation disorder or dysfunction is a lack of optimal resiliency of the ANS. The degree of any sympathetic response (from alerting and localizing - to alarming or fight-flight-freeze reaction) and the speed and quality of the parasympathetic recovery return to homeostasis. Too much alarm causes an amygdala response, sending a distress signal to the adrenal glands. The hormone epinephrine (adrenaline) is then pumped into the bloodstream. Vasoconstriction (various degrees) is the result. Chronic tension can excite the network of interoception towards a sustained state of stress. A sustained alarm that can't be inhibited or recovered sets the stage for sensory modulation dysfunction. Too much PNS, or not enough alerting signals, we would be lethargic and sloth-like. Analysis of behavioral challenges in a child should also include the overall sensory processing function of every system. Various manual therapies can directly affect both ends of this continuum. Too much SNS, and we are in a constant state of distress. Key players in the body's regulation include the vagus nerve system, the enteric nervous system surrounding viscera, the circulatory systems, and organs innervated by sympathetic and parasympathetic nerves. Vagal tone is a term recognized to measure or screen for behaviors of the ANS continuum. The continuum is influenced by:

- Chronic stress (maternal or family environment)
- Adverse Childhood Experiences (ACE scores)
- Head injuries
- Poor breathing mechanics
- Chronic or latent infections such as Lyme disease or strep

- Toxin load and micronutrient depletion
- Blood sugar imbalances
- Gut biome imbalance, contributing to leaky gut and systemic inflammation

Behavioral manifestation

Lack of self-regulation means having challenges understanding and managing one's behaviors and reactions within cultural norms. Struggling for self-control over sensory responses and the autonomic continuum between SNS and PNS recovery can interfere with optimal learning, being well-behaved, cooperation and getting along with others, and formulating one's goals for the future. Self-regulation should begin to be demonstrated around 18-24 months, with several years to mature fully. Unmodulated or challenged modulation leads to negative, manipulative, and less than socially acceptable or attractive behaviors.

Additional problems can persist or emerge at all stages of life. Impulsivity and impatience, lack of control over frustration, and becoming a target for bullying. Or become the bully from a developed sense of need to protect oneself. These skills are highly interrelated to other people in a child's family and local social network. The locus of regulatory problems involves the transactions and reactions of those around the child. Experiences of adult responses to the child's reactions to early caregiving that are positive, supportive, and encouraging are critical elements of the quality and style of self-regulatory behaviors. Forming enduring perceptions and patterns of interaction is a family-system foundation. Problems in one area can undermine development and maturation in other areas. Well-regulated behavior of a child is easy to recognize and be around but is much harder to define and remediate when behavior is not so well-regulated.

Self-regulation involves the ability to pause between a feeling and an action. Initiated by the child to think things through, make a plan, and wait patiently. Children need to develop these through parental guidance; some need more help than others. Internal and external factors contribute to this complex skill development. The body's physiology at play may help us understand why some children need more help than others to gain self-control and modulation.

> **Bodywork for Sensory Integration** offers a hands-on way to address the question "Is it Sensory or is it Behavior?"

Sensations are often the triggers of the above challenges. Not being aware of reactions to the hidden sensation or subconscious or nonconscious reactions can interfere with healthy relationships, social engagement, and quality of emotional stability. Sounds, smells, tastes, and vision not inhibited can set off a sympathetic reaction. Movement input not modulated well for balance and coordination can set off a fear of falling response. Touch not tolerated can be misconstrued as an assault "every time mom hugs me, it hurts my skin." Every family gathering is loud and noisy, and clusters of "sounds hurt my ears." These physiological reactions can lead a child to believe, "I hate family because they make me feel bad."

It is estimated to affect 5-16% of the general pediatric population. Subtypes of SMD have been explored and identified. Dr. Ayres was the first to define sensory defensive behaviors of touch and sound. Further research has further elaborated the neuroscience of these and other systems, now known as sensory modulation dysfunction. Since sensory input, present-day, or memories may cause the same reaction, separating the behaviors from the sensory responses can be difficult. An age-old conundrum for therapists is to discern if a child's issues are sensory-based or other behaviors. Whatever the problem behavior is, it is undoubtedly a child stuck in sympathetic dominance of the central nervous system. *Bodywork for Sensory Integration* offers a method to understand the underpinnings of a child with self-control challenges.

A real-life example of such insights can be theorized as the basis for colic in babies. When there is a history of colic, bodywork for babies has taught us that a great deal of tension can exist from retained in-utero positioning, remnant pulls from the umbilical cord towards organs & vasculature, and deep fascia band tension around other visceral structures and nerve plexuses. There has been a long history of assumptions that babies grow out of colic, with multiple references citing a range of nine to eighteen months. It is a false assumption that a child grows out of this stage. They may grow into it and adapt or maladapt. When we begin to see patterns of colic in infancy and frequently reported stomachaches, constipation, picky eating, irritability, and other mood issues in childhood, we have explored this connection with bodywork through the years. Months on end is a long time for families to wait out and try to cope with colic. No scientific evidence supports the age-old advice that [all] children work out their visceral tensions through motor activities, time, or growth.

Sensory modulation disorder affects sensory processing and self-control or self-comfort behaviors. A single sensory channel, or multiple systems, can be involved. Over-responsivity or Under-responsivity are defining behavioral features. However, an often-overlooked behavior of «shut-down» is a coping strategy that may appear as an under-responding nervous system but is the inability to cope with an onslaught of sensory input.

Categories of Sensory Modulation Dysfunction

Over-reactive sensory systems

- Hands over ears with loud(er) noises
- Behavioral problems at meal times
- Picky eater (degrees of pickiness)
- Hates baths, hair washing, grooming tasks
- Avoids hugs and other human contact
- Bothered by clothing tags, seams, textures
- Often complains of stomachaches, headaches
- Frequent constipation
- Overeats, or undereats for age and activity level
- Fussiness, irritability, controlling
- High-maintenance to appease
- Shuts down & appears under-reactive

Under-reactive sensory systems

- Bumps into objects or furniture during play
- Falling or crashing their bodies into obstacles
- Seems unaware of body
- Hugs too hard or pushes children when playing
- Clumsy; quirky body movements
- Doesn't respond to verbal directives; needs
- multiple cues
- Misses details on paper, clothing, objects in room

Human Limbic System

The limbic system is a deep, yet central collaborative functioning between the hypothalamus, anterior thalamic nucleus, cingulate gyrus, the hippocampus and their interconnections. Emotional expression and regulation, self-control, and memory storage and retrieval are managed by this complex mechanism. In addition, this system links the cerebral cortex with the autonomic brain stem functions, so that intellectual and autonomic functions communicate.

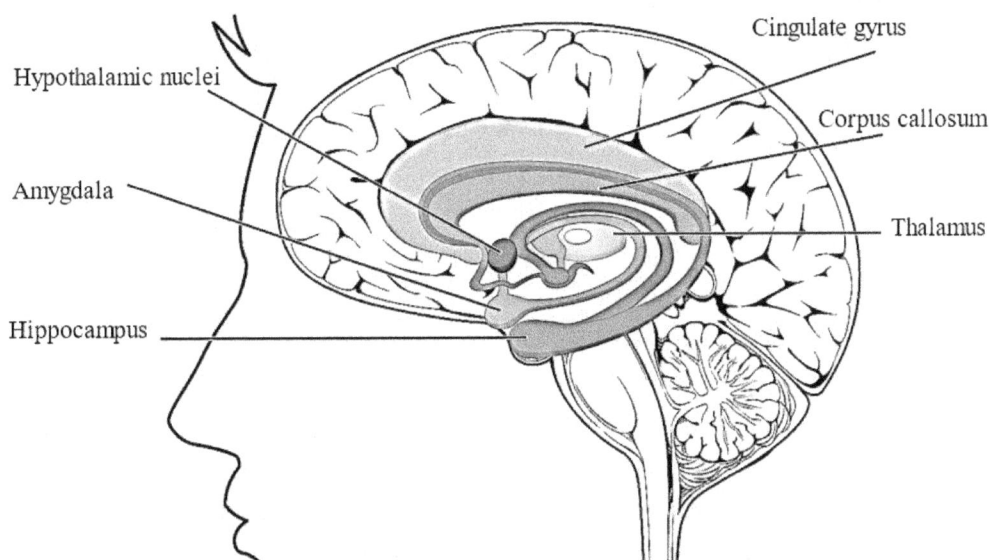

Human Limbic System: OpenStax College A&P Connexions: Wiki Commons. 2013 30148029

Historical treatments for sensory-modulation and self-regulation

Teaching strategies for self-control over emotions, impulse control, and attention skills are key concepts in occupational therapy, psychology, and professional counseling. Behavioral therapies offer options using positive or negative reinforcements, parent coaching, and behavioral contracts with the child. Sensory integration therapies have assisted children in activating sensory channels that inhibit the senses that alarm. For example, proprioception and deep pressure touch are incorporated into treatment to "cool" sensory input that is too fast or too loud. Interventions from a neuroscience-based viewpoint assist in activating higher cortical brain centers to modulate the limbic system's functions and behaviors.

Increasing self-awareness towards self-comfort and self-control within a sensory integration framework has been used for decades. Complimentary curricula such as "Zones of Regulation" and "The Alert Program: How Does Your Engine Run" are constructed in beneficial ways for ease of use and are common additions to a sensory integration practice. A certain level of insight into one's reactions to others and the world is necessary for the successful outcomes of these methods. Somewhere around the developmental age of seven years, this level of insight begins to emerge and mature.

However, none of the existing literature or curriculums suggest that the child's body structure may be a direct avenue to [re]gain regulatory function and gain it quickly. Little research exists to guide parents, teachers, and therapists in helping babies and young children achieve self-control by treating the organs and structures most closely to the functions of the ANS.

Perhaps the nervous system that has trouble regulating itself started life in a sympathetic state. During years of working with sensory integration dysfunction patients, many parents reported knowing something was wrong in their child's infancy. When babies and children are in situations that cause sensory overload, they often cannot (or might find it difficult to) self-regulate. Their emotions and behaviors go unchecked, even when unaware and overwhelmed. More information is needed to understand this lack of modulation. Bodywork proposes expanding the understanding of what has been learned through actually handling and palpating structures within the body directly affected by the ANS way before behaviors become maladaptive and entrenched.

Stress management in cognitive-behavioral interventions teaches cortical ways to 'relax.' When receiving bodywork, deep structural change can occur when expansion and flexibility are gained between the interconnected body regions. When this happens, the patient frequently exclaims, "I've never felt that relaxed." A child does not understand the abstract concepts of "relax," "calm down," or "sit still" until they feel parasympathetic tone deep in their muscles, tissues, and bones. A change towards ultimate freedom from structural tension is the education of feeling a complete and absolute parasympathetic state. ANS continuum is a matter of degree, not unlike the feeling of graded in other sensory systems. Organ and vasculature grading might be more like it.

The body senses autonomic changes immediately at the level of cells, vessels, and organs; long before the brain interprets it

The body senses ANS changes immediately at the level of cells, vessels, and organs; long before the brain interprets it. The sensory receptors within the fascia have already reacted. The enteric innervation sites within the gut-brain have responded before the cerebral brain formulates an adaptive response. One cannot feel relaxed until the deep structures, and cellular matrix are structurally flexible and tension is absent—relaxed, without structural stress. The balance of structural tensegrity represents balanced ANS and enteric systems, which can be palpated and directly treated by therapeutic hands.

Terms used to discipline and teach children to behave and control their emotions are often "calm down," "wait," "relax," and "be good." All of these words are abstract concepts for a young child. One cannot feel "calm down" or "be good" through words, but a child can feel relaxed. After structural tension is released, the words "calm down" now have a more profound meaning because one feels calm once in the parasympathetic mode. The neurochemicals released ensure that feeling. Words do not have much meaning unless associated with something specific or tangible. Abstract words used to control and shape behaviors have an even more challenging purpose because of their abstract nature. Concrete meaning to words is gained by somatic awareness and understanding of the difference between tight and loose, calm and upset, and relaxed or tense. Cognitive-based training methods include identifying specific emotions, goal-setting to gain self-regulation, teaching options for reacting, and practicing strategies for self-awareness. Mind over matter. The missing piece of these approaches is an assumption that the mind can release fascia restrictions. That mental capacity gained maximal vasodilation or maximal rib cage and diaphragm expansion. Or they released a compressed neighboring organ from another—structural reasons to keep the body in a state of sympathetic behavior.

Individualized treatment can meet the needs of a unique constellation of sensory challenges and coping behaviors. Structural therapeutic activities, often in a richly appointed sensory gym, activate other brain centers away from limbic system dominance, assisting in modulation. Remaining within the occupational science that neurological maturation and development occur within the concept of play-based participation. Sensory modulation treatment uses sensory-based activities designed to desensitize aversive sensory reactions or hyperreactivity, and use sensations recognized to inhibit the whole nervous system (deep pressure input, proprioception, linear and other calming senses.

Much of the science about self-regulation treatments is focused on the psychological end-point of behaviors. Infants and young children cannot regulate their emotions on their own; they need loving adults in their lives to help them immediately regulate their emotions and behaviors and to learn skills to do so independently. With practice and support, young children can learn skills to help them cope and manage daily stressors. Even when neurological processing endeavors, such as sensory processing theory, are applied, the field of study on self-regulation analyzes the consequences of adaptive versus maladaptive behaviors. Current interventions wait until the child has spent years either adapting to or maladapting to their body's tension state and the autonomic nervous system. *Bodywork for Sensory Integration* offers an option for families to help prevent mood and behavioral difficulties of childhood turning into adult difficulties. If maturation proceeds as expected, the child will develop self-control and progresses into higher executive functions of managing emotions and behavior. Personalities and coping strategies form and are reinforced around the level of tensions of the core autonomic nervous system.

New discoveries

Vagal behaviors

The Polyvagal Theory (Stephen Porges) describes three circuits of the ANS that seem related to branch-specific behaviors. Sympathetic spinal activation elicits the fight-flight response, and the dorsal vagal evokes the "freeze" state of fear. Behaviors of a neurological shutdown, powerlessness, or hopelessness seem related to the latter primitive vagal nerve branch. Other behaviors reflecting this pathway could be dissociation, self-isolation, and inhibition from homeostatic recovery. The converse of this theory is that the ventral vagal circuitry activates feelings of safety and comfort with a readiness to engage socially. Moving in and out of these states are normal human responses to all stress levels, but remaining stuck in any state is problematic for health and wellness. Traditional mental health therapy attends to the cognitive and emotional elements of traumatic experience; the somatic experience is often not considered. Self-regulation in childhood can become issues that need psychotherapy in adulthood (if left untreated).

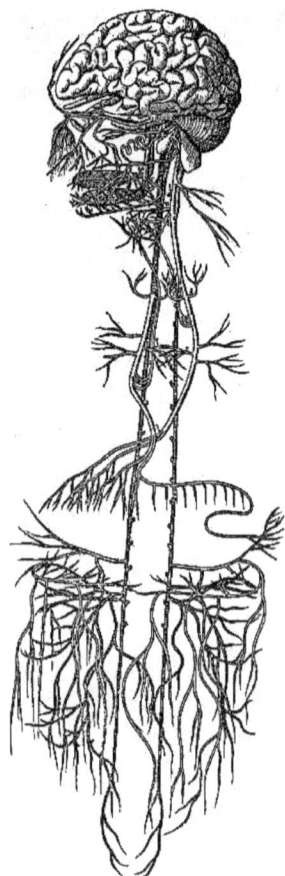

Various somatic experiencing treatments, marrying psychology with manual therapies, are emerging. They can share similarities but also distinct differences in addressing the somatization of stress. Somatic Experiencing® training, post-birth trauma release, and trauma-informed care are but a few examples. Eye movement desensitization and reprocessing (EMDR) therapy is a scientifically-supported method mental health professionals use to recalibrate associated neuronal pathways of post-traumatic stress.

The key in comparing and contrasting any method to treat the ANS is to consider if cognitive drives or physical insights (akin to biofeedback) are utilized. Physical insights can be cognitively guided (guided imagery), but manual therapies directly treat tension areas. The level of invasion to touch and the intention of such touch is imperative to understand. In his book, "The Body Keeps the Score: Brain, Mind, and Body in the Healing of Trauma," psychiatrist Bassel van der Kirk opinions that bodywork of any kind will be helpful to release trauma and energy from the body as long as the person trusts the practitioner.

Upledger's Somato-emotional Release® (SER) methods within his CranioSacral Therapy curriculum training are based upon his discoveries of how past traumas and dramas could be spontaneously expressed and [re]experienced during manual therapy sessions to address physical pains. Guided by the craniosacral system's rhythm, SER is a tool for the bodyworker to assist a client through any recalled event that might surface during a CST session. It often does. Upledger studied this phenomenon with Dr. Elmer Green (the father of modern biofeedback). The term SER was given as a description of the subconscious and the conscious mind having a conversation that started by exploring the depths of a person's body pain or dysfunctional pattern. Suppressed emotions trapped in the body can create defensive behaviors, which in turn can present the release of these energies. The SER training process was developed over the years to improve a bodyworker's palpation and evaluation skills to help locate areas of such trapped energy patterns while remaining within the scope of one's practice. SER is not psychotherapy, though it can be confused as the same. SER is a beautiful tool to gain a profound understanding of the physiological and psycho-emotional aspects of dysfunction and disease, as another layer of helping the client towards a self-corrective process.

Several structural sites have been recognized to assist in accessing and mobilizing what Upledger termed the "avenue of expression."

- soft tissue structures associated with the hyoid bone
- hard palate (and neuron columns projecting above the maxilla towards the limbic system
- hypoglossal tissues and nerve network of the tongue
- thoracic inlet

Mainly within the science derived from animal studies, signals from the gut organs trigger vagal and central neural circuits—behaviors that indicate avoidance, or conversely engaging, are either promoted or curbed by interoception. Interoceptors within the gut receive sensory information from neural, hormonal, cytokine, and microbial signals. This information creates a signaling network received by the central nervous system that ultimately, in reverse, modifies emotional states and behavioral reactions to events. The science is derived mainly from animal studies, and most conclusions are that these signaling primarily reduce approach behavior and increase avoidance behaviors.

It has been demonstrated that the microbiome of the gut organs has a significant influence over behaviors that motivate the animal. The vagal nerve system responds in kind. Much more study is needed to determine the mechanisms of interoceptive signaling to the central nervous system. Diet, stress, and other factors interact to alter the gut microbial population. The role of vagal sensory signaling from probiotics or dietary changes still needs to be determined, but there is an indication that these factors mediate stress responses and emotional regulation. Much more research is required to understand how interoceptive gut-to-brain signaling pathways affect human physiology, emotion, and behavior for emotional and physical health and wellness.

Bodywork differs from cognitive – behavioral methods because we have direct palpation access for a quick change in vagal behaviors. To aid in sensory modulation and self-regulation control, what is needed is for the deepest of structures (i.e., mesentery vasculature, organ peritoneum, and lymphatic pathways) to achieve optimal parasympathetic tone. *Bodywork for Sensory Integration* offers a subjective analysis and quick diagnostic treatment to the ANS continuum for a child who has yet to develop the maturity of insight (on which many cognitive or behavioral-based approaches rely). Use bodywork to assess the tension continuum between parasympathetic tone (loose and flexible) and sympathetic tone (tight and withdrawn medially). Typically, kids will report (often immediately after a session) that I feel «different.» We don›t ask for any other insights than that. An absence of (visceral) tension is all the indication we need that the parasympathetic state has been assisted into activation. When they feel the difference, further insight is gained on the ANS continuum. We theorize that neurochemistry release is directly related to the tone palpated in the organs. Interoceptors report to all channels that the rest and digest state has been reached. It ‹feels› good to have softer organs and flexible digestive sphincters and for peristalsis to be unimpeded.

Comparing SER and combined bodywork methods, we have observed that one cannot relax, learn to relax, or fully understand the meaning of relaxation until one FEELS the depth of relaxation in the visceral organs and massive vascular fields. We can only relax as far as the structures of our organs and blood vessels, and lung structures allow. We teach deep breathing as stress management, but breaths only expand within the tension of fascia surrounding and suspending the lungs and intercostal tissue-lasagna of the rib cage. The expansion and flexibility of the respiratory diaphragm set the tone of breathing and tone over the stomach and liver, the penetrating aorta, the vena cava, and other midline structures. The midline core-nerve plexus require space for expansion and flexibility, but if retained birthing issues such as retracted umbilical remnants (falciform ligament) exist, the SNS state is more paramount. If tension persists in organs, then organs don't drain or express natural motility movement against neighboring organs, adding structural friction.

As we explored in decades of bodywork for babies, we came to understand that unresolved birthing issues can grow into sensory modulation and self-regulation issues in childhood (and adulthood) In the bodywork field, hyperactive and restless children improve in attention, concentration, and calmness for more significant periods after their structural tensions are located and eased through structural tissue releases. There is no placebo effect with babies. They lack the cognitive and psychological aspects that create a placebo effect, and it is unknown at what age the placebo effect emerges. The experiences of bodywork propose that poor self-regulation can begin when a baby›s body has not truly decompressed and managed to release the physical and structural aspects stresses of the birth process. Prolonged protective retracted organs may have been present in infancy. In very young children, a history of adverse in utero or birth process or a medical condition leading to gut dysbiosis and leaky gut walls have responded very well to visceral techniques that facilitate mass vasodilation.

Structural remnants of mechanical stress can be caused by:

- Prolonged compression in the birth canal; adverse head molding unresolved
- Nuchal cord (cord wrapped around the neck or an extremity)
- Pulling on the cord
- Knotted umbilical cord
- Immediate umbilical cord clamping before the placenta ceases blood pumping
- Medical interventions that can be traumatic to tissue and physiology: Cesarean section, forceps or vacuum extraction

Adverse Childhood Experiences (ACE Test) external events can create internal turmoil. Left unresolved can linger and manifest later in life. Personalities develop around the sensory status as well as the emotional status of a child. It has been firmly established that personality traits are set by age five and are predictors of adult behavior. But what if these traits reflect the ANS structures, tension, and emotional traumas of childhood left unresolved? Bodywork should become part of the future studies of these established beliefs.

Vagal tone

Vagal tone indicates the ANS continuum with both sympathetic and parasympathetic systems—a sliding scale of sorts, indicating body tension fluctuations as well as behaviors of stress or relaxation. The fight-flight-freeze of vagal behaviors suggests the degree of vasoconstriction on this ANS continuum, the tension held in organs (both thoracic and abdominal), and the continuum of optimal rest and digest abilities from these organs. Vagal tone can be measured objectively by heart rate, heart rate variability, cortisol levels, and oxygen saturation. Subjective assessment of vagal tone indicates where tissues and organs are stuck or habituated in protective retraction. Subjective measurements can also reveal the sliding scale of ANS tone at these locations. *Bodywork for Sensory Integration* adds to our knowledge base through the collective palpation assessments of these various bodywork methods, yielding subjective readings of vagal tone in organs, vascular fields, and fascia pulls that surround and suspend these structures.

The more aggressive a treatment, or the more control taken away from the client by someone else guiding the imagery or dictating the treatment process, the more likely the therapist can set off and wrestle with the body's response to invasive touch or invasive interaction. Even lymphatic tension may be a protective retraction. The parasympathetic state activates lymph flow, and tissues require that state to help detoxify the body. Conversely, lymph is less than mobile and moves against more resistance through the system in degrees of sympathetic state and tone. Combing CST principles (arcing, blending, melding) with lymphatic drainage and visceral techniques created the described Responsive Hold. This fusion provided more clarity in treating the population of children and adults with self-regulation and sensory modulation challenges. Responsive Hold has proven the best bodywork method to effectively locate and address protective retraction.

Enteric / gut biome regulates vagal system

Visceroception,, enteroception, and interoception is information processing now believed to be a primary factor in some mental health disorders and the undertones of self-regulation difficulties. The combination of cut biome, food substances, and enteric nerve responses are also players in regulating the vagal system.

Bodywork observations

- Self-recognition of which state a person is in is called insight. The curriculums previously mentioned promote insight. However, what helps further develop insight is to feel the depths of those sympathetic states. Chronic stress, or a chronic state of "stuckness," creates or maintains vasoconstriction, blocking neurovascular circulation (like when a foot falls asleep). Vasoconstriction can be palpated as visceral organ tension or higher tone. It is common to witness that tight and restricted organs and fields of vasculature have led children [and adults] to feel pain in their guts or adapt to chronic tension by tuning these sensations out. Changing the tone or tension of visceral organs has consistently assisted our clients in expressing control over their internal state.
- Tensions in organs in older children and adults are similar to tension patterns of babies with umbilical cord tensions and visceral organ torsions in the mesentery and peritoneum that wrap around and house the enteric nervous system. Visceral modified as it reflects the ANS behaviors of visceral organs based upon tone and less upon the structural position within the visceral and thoracic cavity. The visceral terrain offers therapists a direct way to help the autonomic nervous system feel a change in sympathetic tone as if a stuck valve has become unstuck. Change is reflected in tone in the vicinity of all the digestive sphincters. Peristalsis immediately changes when sphincter tone is reduced. Often we don't have to treat each sphincter directly, instead the fascia that suspends them.
- Formulated through decades of practice, consistent positive affect upon mood, emotional regulation, and vagal tone (of organs and vasculature fields) occurs when the following structures are attended to in bodywork sessions:
  - Peritoneum and mesentery and investing fascia fields surrounding organs and vasculature
  - Vagus nerve branches and associated plexuses at innervation sites

- Key lymph nodes around each organ
- Sphincters directly reflect that state of vagal tone

- **The absence of deep pressure is best**. Sensory integration practitioners indirectly promote the parasympathetic state through various activities and methods. Based on decades of clinical evidence and scientific studies, deep pressure input (imposed proprioception) and tasks that fire deep receptors in joints and muscles (exercise or sensory integration activities) promote and maintain calm-alert-attentive states and facilitate emotional regulation. But we just now had a guide to treat ANS and Enteric structures directly. *Bodywork for Sensory Integration* gives a roadmap for a practitioner to work with a child›s viscera self-awareness, a manual form of biofeedback, and promote sensory wellness through insight into mood differences due to a change in viscera tone.

- Minimal spoken language is required to perform bodywork. However, social engagement is often part of the therapy session (to build trust, educate parents, and invite the child to learn how their body works). People appreciate relaxed organs' differences once they feel that difference moving from higher-toned organs and vessels to maximal dilation and physiological relaxation. Words used in counseling that guide progressive relaxation has been practiced for ages attempting to achieve maximal relaxation. But telling people to relax and breathe deeply through cognitive-based guided imagery has assumed that all the structures involved in relaxation and deep breathing can structurally expand and be free of tissue strains. *Bodywork for Sensory Integration* has proven that guidance words of relaxation can have a more profound meaning after organs are assisted into maximal motility and mobility by treating peritoneum and mesentery tissues. Perhaps even the interoception and neurochemistry by changing organ tone. Sometimes organs or major blood vessels are pulled or pushed into a neighboring organ. In typically healthy children without medical issues, this is commonly associated with remnants of tight vestiges of the umbilical cord drawn towards the liver or up into the celiac (solar) plexus. Structure begets function. People are only as relaxed as the organs they try to influence with their thoughts or guided imagery.

- Clinical comparison of bodywork to talk therapy, insight-oriented therapies, and sensory integration methods for self-regulation pale in the speed of change we have observed with modified visceral techniques. Waiting for the tone to change through therapeutic touch at specific anatomical sites. Visceral manipulation combined with craniosacral therapy has repeatedly shown activation of the vagal nerve system.

- Over the two decades of using these methods daily, I have often been told, 'I've never felt this relaxed in all my life.' Likewise, a dysregulated child can experience significant mood, sleep, and eating improvements after just one or two visceral treatment sessions.

- Spontaneous changes are often observed in toileting, hunger & satiation regulation, elimination patterns, and circadian rhythms. We have clients whose families observe improvements in sleep patterns and routinely book appointments when sleep performance regresses.
- The level of executive skills in the cortex is far from the source of (enteric) stress. Years of working with babies to address colic, reflux, digestive discomfort, constipation, irritability, and self-soothing problems have led to the discovery that bodywork methods treat the interoceptors and the enteric nervous system directly. One can teach someone how to relax through progressive relaxation verbal dialogue. Still, the body and the brain can only be as relaxed as the peritoneum tissues that house the enterics.
- Protective retraction is housed in the visceral organs and the deeper vascular system. Vasoconstriction is a hallmark of sympathetic tone, but behavioral approaches have not been proven to maximize vasodilation through talk therapy or even behavioral strategies. Manual treatment can find retained retractions in organs and vessels. The mesentery houses a vast amount of blood vessels, one of the most common areas of tightness in emotional and sensory dysregulation.
- Fight, flight, or freeze isn't just happening in the cortical brain; it is happening within the visceral and thoracic organs and ALL the blood and lymphatic vessels. We ask children how they feel. Are they relaxed? Do they feel less stressed by our attempts to help? But until their visceral organs are addressed by remnants of birthing effects (cord cut too soon), umbilical cord pulls and distresses, gut inflammation, gut frozen in sympathetic retraction from traumas and dramas of life.
- Combining & modifying visceral methods with craniosacral therapy has been the most important discovery in our clinic's sensory integration practice for assisting self-correction of modulation and regulation behaviors.

Combining & modifying VM with CST has been the most important discovery in our clinic's sensory integration practice for assisting self-correction of modulation and regulation behaviors.

Berntson GG, Sarter M, Cacioppo JT. Ascending visceral regulation of cortical affective information processing. Eur J Neurosci 18: 2103–2109, 2003.

Critchley HD, Harrison NA. Visceral influences on brain and behavior. Neuron 77: 624–638, 2013.

Maniscalco, JW, & L Rinaman. "Vagal interoceptive modulation of motivated behavior." *Physiology* 33, no. 2 (2018): 151-167.

Bar-Shalita, Tami, Yelena Granovsky, Shula Parush, and Irit Weissman-Fogel. "Sensory modulation disorder (SMD) and pain: a new perspective." *Frontiers in integrative neuroscience* 13 (2019): 27.

James, Katherine, Lucy Jane Miller, Roseann Schaaf, Darci M. Nielsen, and Sarah A. Schoen. "Phenotypes within sensory modulation dysfunction." *Comprehensive psychiatry* 52, no. 6 (2011): 715-724.

Kuypers, Leah. "The zones of regulation." *San Jose: Think Social Publishing* (2011).

National Research Council. "From neurons to neighborhoods: The science of early childhood development." (2000).

Schaaf, Roseann C., Teal W. Benevides, Erna Blanche, Barbara A. Brett-Green, Janice Burke, Ellen Cohn, Jane Koomar et al. "Parasympathetic functions in children with sensory processing disorder." *Frontiers in integrative neuroscience* (2010): 4.

Vroland-Nordstrand, Kristina, Ann-Christin Eliasson, Helén Jacobsson, Ulla Johansson, and Lena Krumlinde-Sundholm. "Can children identify and achieve goals for intervention? A randomized trial comparing two goal-setting approaches." Developmental Medicine & Child Neurology 58, no. 6 (2016): 589-596.

Williams, Mary Sue, and Sherry Shellenberger. *How does your engine run?: A leader's guide to the alert program for self-regulation.* TherapyWorks, Inc., 1996.

Touch (Tactile) Defensiveness (TD)

Touch is the first sensation that starts evolving in the womb at five weeks. The early development of the touch (tactile) system provides an essential foundation for social and communicative behaviors (Cascio, 2010). To touch and to want to be touched is innate to all human beings. Mammals need healthy touch for acclimation, assimilation, growth, and health.

Definition

Tactile Defensiveness (TD) can be defined as an intolerance to touch input that most people would not find painful or might be oblivious to the contact. Touching the skin creates distress and discomfort when TD exists, and a distorted meaning to touch input are common problems. Based upon behavioral interpretations of a child's behavior, TD represents (mal)adaptive and coping reactions of the daily onslaught of touch input the brain receives from the skin's sensory receptors.

Behavioral manifestation

Behaviors of touch defensiveness reflect sympathetic (ANS) reactions to benign touch, the intensity in which the input was bothersome, and the time required for recovery from the touch stressor. The state of the ANS may influence the perception of touch that might be different than pain in the periphery. When habituating in a sympathetic dominance, sensory reception can elicit [the degree of the] protopathic reactions. Touch defensiveness can occur alone or as a constellation of other sensory processing difficulties. Like all sensory issues, TD can run from mild to severe, and is a feature co-occurring in disorders of connective tissue such as Ehrler-Danlos, ADHD, and other neurological and psychiatric conditions. Sensory aversions and TD, along with self-stimulatory behaviors, are frequently seem in in autism spectrum disorders (ASD), though TD can occur in people without autism. Self-stimulatory behaviors such as hand flapping, obsessively jumping, or skin picking are but a few examples. It has been hypothesized that self-stimming causes the release of beta-endorphins, leading to a feeling of anesthesia.

When someone has TD, clothing is usually a big issue related to fabric selection, weight of garment, and location of seams and zippers. Movement of fabric over the skin can set off alarm reactions. People with severe TD often cope by disrobing, all or some particular clothing. The person often spends extra time being vigilant of items or activities that might touch the skin. Haircuts and hair grooming can be particularly torturous for a person with TD, along with nail trimming, shaving, or other expected hygiene tasks. Hugs may be accepted if initiated and control by the person with TD, but unexpected contact may set off a sympathetic reaction (alarm, avoidance, aversion, or stress).

Behavioral and emotional responses can include escape or avoidance reactions to be hugged or kissed by family members, clothing changes are unpleasant events, clothing types can be intolerable against the skin. Coping behaviors can include hyperactivity (to avoid being touched and to activate diversionary sensory input), body rocking, hand flapping, controlling behaviors during bathing and when getting dressed. Negative reactions are out of proportion to the kind of input that is experienced. Self-controlled tactile experience are more calming and tolerated than when touch is attempted by another person. A child may not understand why their skin is uncomfortable and blame others for their discomfort

In cases of severe TD such chronic pain can be associated with extreme behaviors such as self-abuse (to skin or hair) or aggression towards others (especially when they veer into personal space). Conversely, sensory seeking behaviors can be rooted in TD where receiving diversionary sensory input can override the skin response. Aggression is a maladaptive way to get some relief from TD, and so is head banging and bungee jumping.

The sense of touch and touch processing holds a vital role in development of body awareness, motor planning (praxis), and visual perception (eye-hand coordination). Due to the atypical reception of the touch mechanism within the neural system, these children (who can carry TD into adulthood) are often in the state of high alert (sympathetic state dominance). They may react by lashing out (fighting), running away or avoiding (flight), or become clingy or complaining (fright). Avoiding social situation and escaping events that potentially force touch contact.

Behaviors Suggestive of Touch Defensiveness

- Intolerance of self-care activities: bathing, dressing, hygiene tasks, diaper changes during infancy
- Happier when alone in crib as a baby or prefers to sit away from groups
- Hyper-vigilant of personal space; can become [excessively] stressed if people enter personal space
- Withdraws from someone's attempt to hug or touch
- Excessive preference for no clothing, or conversely, upset when unclothed (exposing skin)

- Tense hands, feet, face as if withdrawing
- Frequency facial expressions of stress or discomfort
- Food texture intolerances
- Avoids messy tactile play: play dough, wet sand, sticky substances
- Avoids writing or drawing tasks; lack of maturation of hand dexterity
- Often becomes overwhelmed (as if always dealing with a hidden stressor)

Historical treatments for Touch Defensiveness

TD is one of the most common problems treated at pediatric sensory integration clinics. Occupational therapists understand the implications of skills of self-care and social emotional function from the touch system. Environmental modifications can assist someone with TD. Hugs, kisses, or any other type of contact should be done with forewarning and with full consent. This reduces anxiety and sensory arousal. Not respected, this can create disruptive social behavior. Curriculums exist that build awareness, insight and advocacy for TD (and other sensory defensiveness). Zones of Regulation and the Alert Program are but two examples of long-used standards. Behavioral treatment methods have shown little effect in changing coping behaviors of TD.

It is recommended that people struggling with sensory processing seek help from a qualified therapist trained in sensory integration methods. Ayres' Fidelity has been the gold standard for decades. Based upon clinical reasoning, therapeutically designed activities (by occupational therapists) are developed to desensitize the skin through child-direct engagement with sensory input directly into the skin. The therapeutic aspect of such activities is that touch input is not imposed and the child has control through the scaffolded activities (that urge motivation for fun and assist moving through the resistance caused by the historical discomfort). Most often therapeutic activities are NOT the activities the child has already associated discomfort with, but are designed to activate or inhibit the different levels of sensory receptors in skin layers. Application of deep pressure input through various means provides practice-based evidence of reducing stress and helping regulate and modulate reactions to the touch input of daily activities. The common-sense knowledge akin to pressing or rubbing skin over a typical injury supports this position.

Sensory diets, especially heavy work activities (to activate proprioception and kinesthesia to override tactile processing) have been consistently shown to be individually effective with daily coping and self-regulation. The use of a weighted blanket can be offered and used if the child (and adult) experience comfort and a change in autonomic nervous system state from the static compression it gives. Weighted vests have been used over the years and observed to inhibit negative reactions associated with TD. Studies have concluded that the static compression may need to be removed after 30-45 minutes as the nervous system acclimates. Ideal weight amount has not been established, nor has vulnerabilities to posture been studied from the use of weighted vests.

Compression clothing has gained favor in many circles as a different method to get constant deep pressure input into the skin. Various styles and densities of stretch fabric giving circumferential compression adds resistance with movement, but input is dynamic with the stretch of the fabric. Compression clothing, when tolerated, apply deep pressure as well as to prevents a fabric from moving over the skin surface. When clothing moves, it can fire C fibers, setting off constant protopathic alarms that something is crawling on the skin or even a pain response (neuropathy). Explains why weight vests, head caps and blankets are valuable in helping with touch defensiveness. Compression into the skin nullifies the excitation of ANS fibers by activating deeper receptors.

Therapeutic brushing has been a routine practice of SI practice. The theory behind decades of using Deep Pressure Brushing (Wilbarger Protocol) was based on the 'gate control theory" Gate control theory suggests that the spinal cord contains a neurological 'gate' that either blocks pain signals or allows them to continue on to the brain. This theory is often used to explain both phantom and chronic pain issues. It's a reasonable and plausible theory, to activate the deeper pressure receptors in the skin to override the sensory signals of pain and sense of skin irritation or unmodulated response to touch input. This theory proved effective for many children, but didn't work for all.

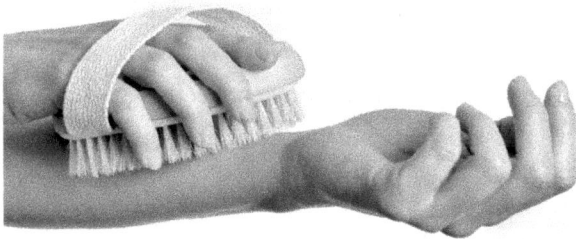

The Wilbarger Brushing program is one of the most commonly used methods to treat sensory disturbance of the skin. Anecdotal evidence suggests positive effects in reducing behaviors associated with TD. A lack of quality evidence to support the use of this protocol, and should only be used with caution after proper training. The method is applied in a very specific manner to stimulate deeper sensory receptors (proprioception), which are believed to release dopamine (a calming and regulating neurotransmitter).

New discoveries relevant to Touch Defensiveness

Dermatome mapping to evaluated sites of discomfort

Occupational therapists have needed fresh perspectives on the subject of TD. The knowledge came to us from our most reliable source of information: our young clients. We discovered that not all children seeking treatment for TD had an equal distribution of feeling uncomfortable. Using detective work for each individual, sometimes TD was only where socks touched, the edges of underwear, or sliding sleeves that felt uncomfortable. We observed some maps where only one side of the body was reported as feeling bad. It was also interesting to note the flavor of controlling behaviors each child had: mom could only hold a specific hand, or a haircut could only happen with three blankets draped over the whole body. Those differences suggest something more akin to neuralgia or neuritis, something we observe in conditions of known systemic inflammation or microvasculature compromise. The neurological training for occupational therapists was a foundation for this kind of detective work into the sensory terrain.

We started asking children to use a red or green crayon on a full-size outline of their body to indicate where touch didn't feel good and where it felt ok to increase their insight into the workings of their skin. As we compared more and more of these self-reported sensory maps, it became clear that sensory dermatomes were conveying a message. TD may not affect all areas. These body maps also because of quantifiable response to treatment measurements. If global discomfort is revealed, suspect whole-body ANS intervention. If localized, refer to the sensory dermatomes of the body and ponder the central core tissue strains that might exist at those peripheral nerve pathways. Craniosacral tissue strains and tensions can hinder cerebrospinal fluid movement, leading to soft neurological signs (routine chiropractic and osteopathic knowledge). Dermatome maps can suggest where spinal nerve roots might be compressed, and the craniosacral system has mechanical strains in the connective tissue. Body mapping revealed that not always is the whole skin irritated. It can be regions of sensory 'zones' or "hot spots," which suggest infections or toxin stagnation.

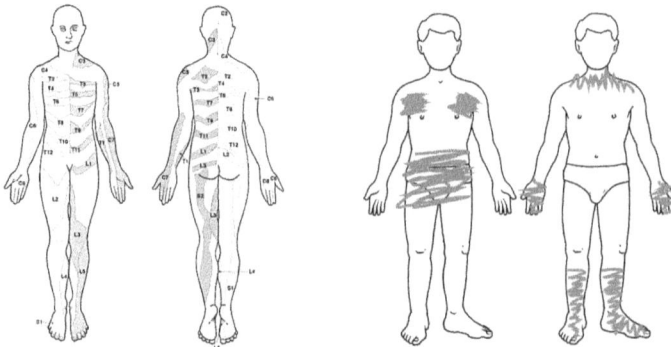

The standard dermatome map can be compared to a child's drawn, self-reported sensory map of where a they like and do not like to be touched, or where things feel prickly, or burning, or any such descriptives.

Providing sensory integration intervention to older children and adult populations has led to greater insights and recognition of triggers of discomfort. Adult clients often can verbalize their experiences in childhood with a history of TD. One particular woman who sought sensory treatment shared a growing understanding of her own sensory systems. "I always experienced hugs as painful and I always felt cornered attacked by those trying to hug me. Family gatherings were torture for me. It took me until adulthood to learn that hugs are gestures of love and affection. I didn't figure "me" all out, putting two and two together, until after my first divorce." This same woman taught the author about "sensory mapping" of how she figured out a survival method of telling people where it was OK to touch her during hugs and where best to avoid contact. A false assumption exists in the management of TD where global somatosensory involvement directs treatment endeavors. Keeping dermatome arrangement in mind helps discern if pathology of a related nerve root or segments related to somatic dysfunction.

We discovered there were pros and cons of the compression or weighted garments. Though comforting and calming for some childrne, they could feel restrictive and restraining, sometimes causing overheating as these garement tried to stimulate the deep pressure receptors. None of these deep input strategies helped mobilize tight tissues or fluid exchange of the superficial layers of connective tissue.

Children taught us to expand our vision regarding Touch Defensiveness

Consider the health of the Cutaneous Autonomic Nervous System

Inflammation causing stagnant lymphatics could be related to irritated or damaged C-fibers, leading to neuritis or neuralgia

Injury to C-fiber autonomic receptors in skin

In neurology, the densities of dermal nerve fibers are a major diagnostic target for small fiber neuropathy, but this part of anatomy has only begun to be considered in the behavioral aspects of sensory integration in children. The two classes of ANS sensory nerve fibers that transmit pain are unmyelinated C fibers (small and slow), A-delta, and A-beta fibers (both myelinated and fast). A- fibers also transmit touch information. C-fibers send diffuse pain signals in the dermis and other areas of the periphery to the spinal cord. A recent study of a small sample of boys with autism spectrum disorder who also manifested TD had dermal cell layers examined through skin biopsy and subsequent histology analysis. C-fibers were confirmed to be reduced in number or damaged. Part of the cutaneous autonomic system, these injured fibers were theorized to be a source of neuritis or neuropathy. Chronic pain and ANS dysregulation have a firmly established direct correlation. The study did not include the source, or any proposed source, of the cellular injuries. Other diseases affecting the cutaneous autonomic nervous system include pre-diabetes and diabetes, traumas and reconstructions of peripheral nerves. Diabetic neuropathy is the most frequent for autonomic neuropathies.

It has been described in other neurology studies that this pattern of autonomic sensory fiber damage accounts for functions of neuritis, hypoesthesia (under responsive to touch) and allodynia (extreme hyper-sensitivity to touch input). These observations suggest that individuals with autism and TD may have small fiber pathology, which may be associated with behavioral symptoms and coping traits, providing the beginning science of structural and physiologic evidence for the involvement of peripheral sensory nerves. This study is one of the first to be applied to the pediatric population with sensory differences.

Systemic inflammation and stagnant lymphatic system

Our clinic had the good fortune to network with several progressive functional and naturopathic physicians searching for underlying pathologies for autistic behaviors and other learning disorders. Pathogens and toxins were commonly discovered, which opened a new paradigm for analyzing sensory-related behaviors our clients were seeking treatment. Engorged lymph nodes and static lymphatic system (of waste removal) in the periphery became parallel discoveries. TD and systemic inflammation became one in the same thing as we observe children being more comfortable in their skin from a regime of bodywork experiences and physicians reducing the source of inflammation.

Comparing and contrasting dry brushing to Wilbarger's brushing was a natural segue. Dry brushing was embraced to help clear die-off pathogens and spike proteins stuck in superficial lymphatic vessels (Klinghardt's Lyme protocol). Dry brushing for treating chronic Lyme and co-infections is argued to sweep cellular die-off, which can be sticky and accumulate wastes in lymphatic vessels and interstitial spaces. Interstitial fluids give rise to lymphatic fluids, so die-off may block elimination flow, exacerbating chronic inflammation as immune cells continue to attack the stuck protein markers of the microbes. Wilbarger's brushing program probably serves lymphatic tissue and fluid mobilization.

Consider bloodstream health for the possibility that TD may be related to inflammation. A trifecta of lab tests detects and monitor inflammation, and these include C-reactive protein (CRP), erythrocyte sedimentation rate (ESR), and plasma viscosity (PV). These tests are now commonly used in primary care to monitor infections, autoimmune conditions, and cancers. These tests suggest toxins within the cellular terrain, either organic or inorganic. Particulates can irritate blood vessels and create protective vasoconstriction, even in the smallest vessels.

The health and state of the visceral organs should also be considered in matters of TD, as the ANS is intimately related to the visceral tissues and organs. The lymphatic node system surrounding organs should be treated when the goal is to help detoxify and clear systemic inflammation. This, in turn, helps regulate ANS responses to sensory input. Liver drainage clears the systemic effects of inflammation. A hidden or chronic biological issue may be at play that contributes to systemic inflammation. Mutually, a parasympathetic state facilitates lymphatic drainage and clears the effects of inflammation. Our young clients with TD also commonly have harmful bacteria overgrowth, viral infection, and even parasites.

A leaky gut can result from an environment with such issues as proteins from food not broken down for disruption of digestive juice secretion or harmful microbes out of balance, inflaming the gut walls, weakening them, and causing seeping into the bloodstream through the mesentery vasculature. The mesentery is a crucial site of systemic inflammation, as proven by a vast array of autoimmune conditions helped by healing the walls of a leaky gut. Inflamed cells, organs, and structures keep the nervous system in a relatively elevated state of sympathetic dominance. The presence of such an environment can be appreciated with palpation for tone at organs and the mesentery fascia and vascular field.

For the therapist is involved in a general practice that covers a wide variety of issues, begin to observe sensory similarities of TD in complex cases of systemic inflammation. Behavioral traits and adaptive coping for skin that is painful to the touch or chronic tight and taut has been observed in people with Ehrler-Danlos Syndrome, chronic Lyme and other infections, increased toxin load such as aluminum and mercury exposure, Chiari formation of brain stem/cerebellar tonsils. Other presentations suggestive of taut meningeal tissue and systemic inflammation (in both young and older children alike) can have these biological issues present with the behavioral presentation of TD. We can learn from those with acquired tactile defensiveness where neuritis or neuropathies have led to what the neurologist refer to as allodynia. It is possible that TD is an acquired condition, so age of onset of sensory (touch) intolerances is an important point to include in history taking. The skin is the entry point to the neurobiology of touch. Whatever is happening in TD is happening within the skin. Sensory integration approaches failed to consider the skin's biological terrain, the integrity of sensory receptors, the condition of microvasculature, or the lymphatic system. We need to consider the receptors' health, the lymphatic detox pathways, and the degree of dermal vasoconstriction. We use this knowledge to educate our clients and their families and urge them to work with their medical practitioners to investigate the full extent of biological underpinnings. In this way, we teach advocacy for themselves and their children and invite the thoughts that sensory wellness is obtainable.

A new hypothesis for Touch Defensiveness

To summarize, touch defensiveness has been shown to be directly related to sensory modulation and ANS imbalance. Gut health and systemic infections and inflammation have

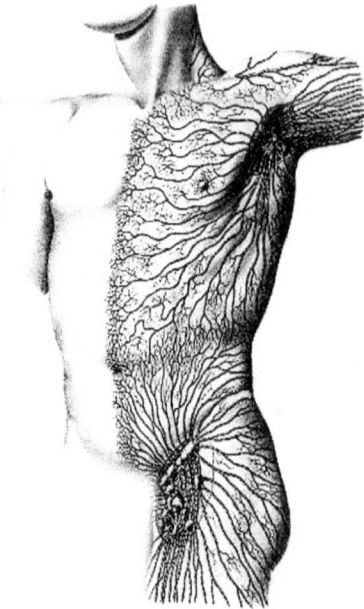

Superficial lymphatic anatomy

demonstrated to also be related to TD. Differentiation of subtypes of TD might be in order. This might explain why some children (people) don't respond as expected to deep pressure brushing, but others do. Some children have responded more favorably and quickly with lymphatic drainage where dermal layers were mobilized gently in counter-strain lines-of-pull, and stagnant fluids were evacuated and facilitated lymph and venous return. More study is needed into the relationship neurotoxins, infections, and the health of pain receptor fibers to the expression of TD. Could small fiber neuritis or neuropathy be the underlying etiology?

It is not acceptable to assume the existence of TD as simply a fact of neurodiversity without at least investigating the biological terrain of a person's skin. Out hypothesis is that pathogens and toxins stagnant within the microvasculature and lymph vessels in skin and surrounding organs may be a contributing factor leading to TD.

Bodywork Observations

- One branch of a new hypothesis for treating TD is that the effects of systemic inflammation end in the dermal layers, in the lymphatic vessels, and at the microvascular levels. Spike proteins and other pathogen wastes or toxins may not drain readily, becoming sticky and clog the system.. We have observed this consistently treating all ages for chronic Lyme disease, in the presence of MTHFR genetic mutation, and more recently in those with chronic symptoms of COVID-19 infection or vaccine reaction. Lymphatic drainage massage consistently reduces skin discomfort in these cases, treating both globally and locally where needed.

- Lymphatic drainage methods help identify local "hot spots" of discomfort to be added on the dermatome map (for documentation purposes of treatment effectiveness). These specific sites of discomfort, or maybe just a difference in the reaction, may be where localized infection or stagnant lymphatics reside. These sites of injury to the cutaneous ANS of A and C-fibers.

- Static, listening hands softly placed on skin can discern surface tension of dermal layers. Moving hands at the same level of non-invasive pressure can mobilize lymphatic fluids. This gentle lymph evacuation technique assists in reducing the tension and sympathetic tautness by assisting clearing and cleansing of the cellular terrain. Promoting the exchange of fluids in this way usually has an immediate reduction in tension levels in the connective tissues between dermal layers. There can be vasoconstriction in dermal microvasculature and it is this level of touch input (static or moving) the therapist can discern and influence without alarming the system.

- Lymphatic drainage massage (Chikly methods) is now our treatment of choice for tactile defensiveness and somatosensory difficulties. If may not always be effective, for it may take longer to clean the body in some people and address sources of active inflammation. But for the most part, positive effects are observed within 1-2 treatment sessions with global behavioral changes of touch tolerance and increased discriminative touch exploration, less approach-avoid-withdrawing and an incredible sense of wonder from parents as they sit and watch their child accept and tolerance the manual therapy.

- Lymphatic drainage massage matched with viscero-enteric fascia mobilization facilitates an active drainage of the liver (and kidneys). By enhancing the siphoning effect of the body's detoxification systems, parasympathetic tone in organs and vasculature results.

- CST communicates with all layers of the cellular network, from periphery to core structures. Our observation is that CST can alter the ANS state globally, in turn reducing protopathic perceptions to benign sensory input.

- One only has to try and give bodywork a chance. The longer a person maladapts to touch, the more avoidant behaviors can become entrenched. It is easier to try than to assume someone can't tolerate bodywork because of their TD. It is surprisingly consistent how well many children with TD have accepted and tolerated bodywork, when the touch is just right and with the right intention from the therapist.

Croy, Ilona, Helen Geide, Martin Paulus, Kerstin Weidner, and Håkan Olausson. "Affective touch awareness in mental health and disease relates to autistic traits–An explorative neurophysiological investigation." *Psychiatry Research* 245 (2016): 491-496.

Dunn, Winnie. "The sensations of everyday life: Empirical, theoretical, and pragmatic considerations." *The American Journal of Occupational Therapy* 55, no. 6 (2001): 608-620.

Foss-Feig, Jennifer H., Jessica L. Heacock, and Carissa J. Cascio. "Tactile responsiveness patterns and their association with core features in autism spectrum disorders." *Research in autism spectrum disorders* 6, no. 1 (2012): 337-344.

Fukuyama, Hiroshi, Shin-ichiro Kumagaya, Kosuke Asada, Satsuki Ayaya, and Masaharu Kato. "Autonomic versus perceptual accounts for tactile hypersensitivity in autism spectrum disorder." *Scientific reports* 7, no. 1 (2017): 1-12.

Glatte, Patrick, Sylvia J. Buchmann, Mido Max Hijazi, Ben Min-Woo Illigens, and Timo Siepmann. "Architecture of the cutaneous autonomic nervous system." *Frontiers in neurology* (2019): 970.

Jensen, Troels S., and Nanna B. Finnerup. "Allodynia and hyperalgesia in neuropathic pain: clinical manifestations and mechanisms." *The Lancet Neurology* 13, no. 9 (2014): 924-935.

Kinnealey, Moya, and Margo Fuiek. "The relationship between sensory defensiveness, anxiety, depression and perception of pain in adults." *Occupational Therapy International* 6, no. 3 (1999): 195-206.

Kuypers, Leah. "The zones of regulation." *San Jose: Think Social Publishing* (2011).

Orefice, Lauren L. "Peripheral somatosensory neuron dysfunction: emerging roles in ASD." *Neuroscience* 445 (2020): 120-129.

Schoen, Sarah A., Lucy J. Miller, Barbara A. Brett-Green, and Darci M. Nielsen. "Physiological and behavioral differences in sensory processing: A comparison of children with ASD and sensory modulation disorder." *Frontiers in Integrative Neuroscience* (2009): 29.

Silva, L, & M Schalock "1st skin biopsy report in children with ASD loss of C-tactile fibers." *J Neurol Disord* 4, no 262 (2016): 2.

Smirni, Daniela, Pietro Smirni, Marco Carotenuto, Lucia Parisi, Giuseppe Quatrosi, and Michele Roccella. "Noli me tangere: social touch, tactile defensiveness, and communication in neurodevelopmental disorders." *Brain Sciences* 9, no. 12 (2019): 368.

Williams, Mary Sue, and Sherry Shellenberger. *How does your engine run?: A leader's guide to the alert program for self-regulation.* TherapyWorks, Inc., 1996.

Inner Ear Dysfunctions

Modulation Dysfunction

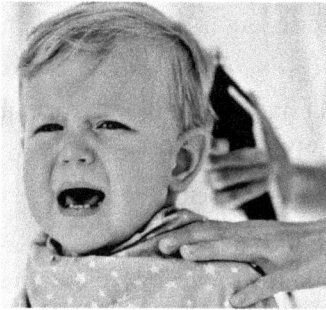

- Auditory Defensiveness or Hyperacusis
- Gravitational Insecurity
- Balance and Posture Control (See p. 310)

Discrimination Dysfunction

- [Central] Auditory Processing
- Bilateral Coordination (eyes, arms, legs)
- Balance and Posture Control (See p. 310)

Definition

Humans hear when sound waves enter the auricle, pass through the ear canal, and strike the fascia membrane of the eardrum. The eardrum vibrates as a result, and the three tiny bones absorb a transfer of this energy in the middle ear (malleus, incus, and stapes), which in turn transmits the sound waves to the inner ear toward the cochlea. The energy of sound within the air medium compresses and decompresses air molecules, but the eardrum is not the only receptor that senses and responds to these noise energies. The world's hullabaloo also stimulates the mechanoreceptors within the skin.

Sounds can have different physiological effects on a person, ranging from barely aware to powerful reactions. Hearing is a primary warning sense, and a sudden, unexpected sound can startle and alarm a sympathetic response. This will release cortisol to increase heart rate and breathing. Sound stress and noise pollution can significantly impact the overall health and wellness of all people. Auditory defensiveness is a modulation dysfunction, and the problem can either be within the ear structure itself or in the central processing centers of the cortex and other brain regions. Hyperacusis, a much more severe manifestation, can be a perpetual trauma reaction to painful input to the ears that cannot be inhibited, dampened, or modulated. Loud noise can injure or damage the delicate nerve endings within the cochlea that receives and transduces sound waves into electrical information.

This has been shown to relate to inflammatory reactions within the brain potentially. The health of the middle and inner ear is a critical aspect of sensory processing and should be considered. The structural health of inner ear components requires tensegrity balance of connective tissue. The suspension within the bony container of the temporal bones is a unique auditory and vestibular function perspective. The lateral sides of the skull are vulnerable to birthing compressions or even interventions such as forceps or vacuum extraction, where forces can translate inwards to the inner ear.

The brain receives vestibular and hearing information from a relatively short distance (compared to other cranial nerves)—the closeness to the brainstem and cerebellum for vestibular and auditory input. Each inner ear mechanism with a shared [split] nerve activates vestibular and auditory processing, but it is complicated by two equivalent structures that send information along parallel nerve branches. The crossing here is one of the first myelinated pathways in the CNS, probably because the vestibular system has constantly been stimulated through fetal development (via body movements of the mother). **The crossing of both sides sends two signals, yet the brain centers involved need to synthesize two channels to form one perception to prepare for a response.**

Humans' relationship with gravity is essential for feeling safe and secure with movement. Linked directly to the vestibular system, the proprioceptive, auditory, and visual systems provide feedback for checks and balances for body stability. The ear mechanisms hear, but the body is what listens. Whole-body listening responds to the frequency and vibrations of background sensations in proprioception, joint and tendon receptors, and even osseous receptors. The eardrum is not the only structure that receives the vibrations of sounds.

Dysfunction of vestibular processing can be both a modulation problem (gravitational insecurity) or a problem of discrimination, timing, and synthesis (balance, posture, and coordination). Auditory processing dysfunction can also be a modulation deficit (hyperacusis, excessive sound sensitivity) or a discrimination deficit (central auditory processing, depth perception and locating sound sources, social communication difficulties).

Behavioral manifestation

Gravitational insecurity results from multiple senses not processing optimally, resulting in fear of mildly challenging movement or when the body is moved with feet off the ground (sitting on a swing). Gravitational insecurity can be a significant source of chronic sympathetic firing and is considered one of the modulation disorder features. It is a common issue treated in sensory integration clinics and is often misinterpreted as behavioral problems, oppositional defiance, anxiety, or obstinance. We've learned from people who can verbalize their sensory experiences tinnitus and labyrinthitis can, as can dizziness or vertigo. Gravitational insecurity can impact a child's active participation in everyday contexts such as school, home, and community settings.

Vestibular input is caused by movement or a change in head position. This can include standing off the floor, bending to pick something up, crawling or walking over an uneven surface, or climbing up playground equipment. Organs in the inner ear detect movement and the pull of gravity and are responsible for perceived vestibular input. Signs of an over-reactive vestibular system which is also not modulated well with visual and proprioceptive feedback mechanisms, include an emotional response and refusal to do a backward roll, jump off a chair with eyes closed, get up after lying supine on an exercise ball, and step on. For the children who can verbally express themselves, they might verbalize a strong preference of sitting still, hating recess or gym class, despise elevator rides, or easily become car sick.

Bilateral coordination is an adaptive response and skill sets learned through experiences of playing and working with these sensory processing. Gravitation insecurity, or even a lack of exposure and participation can prevent the development of skills of balance, posture control, and smooth coordination. These skills are the purposeful use of the somatosensory information as the brain relies upon discriminative features. It's important to address vestibular processing challenges underlying gravitational insecurity to help prevent children from being chronically triggered into a state of fight or flight, as a prolonged elevation of cortisol levels associated with this sympathetic nervous system response isn't healthy for the brain.

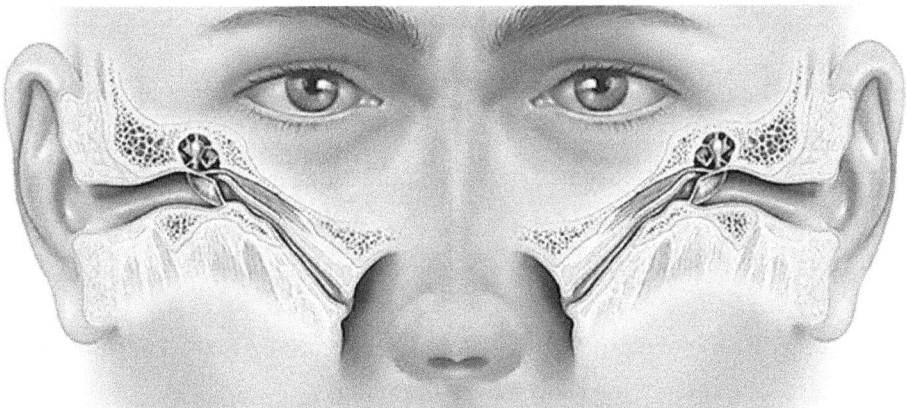

Children who feel gravitationally insecure or have balance and postural challenges may avoid going on swings, jumping, having someone pick them up, or sitting on a chair that is too high for their feet to touch the ground. Children with these sensory challenges can have difficulty walking on uneven surfaces such as beaches, lawns, or foam mats. Bending over or being picking up and held upside down can cause elevated distress. Fear or avoidance of elevators and escalators is common and tolerating car rides are frequently a common problem. Exaggerated emotional responses to movement and strong controlling behaviors through the day are reasons to rule out vestibular processing dysfunction leading to feeling insecure with changes of gravitational forces (into the inner ear). This condition is summarized as the child strongly insists their feet remain in contact with the ground/floor and any challenge elicits intense fear or resistance.

Auditory defensiveness is an over sensitivity to sounds in the environment. These children hear sounds louder than normal. They may have a flight or fright or freeze reaction to sudden sounds or loud sounds. Hyperacusis is a diagnosis (using given to adults who give accurate histories) with a hallmark symptom being reduced tolerance to benign, routine sounds. People with this condition often complain that their hypersensitivity to sound is living in a world where the sound volume has been left on too high a setting. People sound hypersensitivity often have non-integrated primitive reflexes such as the Fear Paralysis Reflex or Moro Reflex.

Auditory processing disorder is related to one of two scenarios: not perceiving or filtering target sounds from background and decreased processing time within the brain pathways. Functionally this can lead to slowed reactions, missing social cues from others, and less attention to details of the world. Auditory processing disorder have known causes: multiple ear infections, genetic mutation, head injuries, or birthing complications. Speech and language pathologists can treat these disruptions or differences in the way that the brain "understands" what the ears are hearing and sending information.

Modulation

Gravitational insecurity

- Strong preferences for sedentary activities; avoids physical activity
- Extreme fear of heights with exaggerated emotional reactions; extra time to recover
- Excessive fear or intolerance of riding elevators or escalators; extra time to recover
- Resists head tilted any direction (hair washing)
- Avoids having feet off the ground
- Feels queasy with movement activities; frequent or intense car sickness
- Delayed success at climbing stairs, clings to railing

- General avoidance of movement
- Head lags when crawling or when body moves between different positions; when pulled to sit
- Over-reactive ocular nystagmus in testing
- Imposed or force body movements causes tantrums

Auditory Defensiveness or Hyperacusis

- Difficulty in calming and soothing
- Easily irritated by sounds and physical activity
- Unsettled or distressed in groups of people talking
- Frequently cover ears to sound; in pain or discomfort
- Low intensity sounds seem too loud (refrigerator)
- Ordinary sounds (voices at conversational volume) are too loud or distorted
- Sudden, loud noise can be difficult to cope with
- Formal testing pinpoints volume level of discomfort
- Avoids activities that have loud environments such as parties, ballgames, indoor music and movies

Discrimination

Auditory Processing

- Can't locate directions of sounds
- Difficulty understanding spoken words with competing backgrounds
- Difficulty understanding too many steps to verbal instructions
- Longer response times in conversation
- Frequent requests for others to repeat words
- Inappropriate responding to spoken directives
- Difficulty learning verses, song lyrics, new languages
- Misunderstanding, such as detecting inflection changes that help to interpret sarcasm or jokes
- Poor musical and singing skills
- Difficulty paying attention
- Easily distracted by sounds
- Poor performance on speech and language or psychoeducational tests in the areas of auditory-related skills
- Associated reading, spelling, learning problems
- Early or frequent ear infections

Bilateral Coordination

- Trouble cutting with scissors
- Struggles with handwriting
- Difficulty tying shoes
- Hard time dressing themselves (pulling on socks, pants, and shoes)
- Trouble with fasteners: buttons, zips, snaps
- Clumsy movements
- Trouble catching a ball
- Often under-reactive ocular nystagmus in testing
- Difficulty with coordination of body, eyes, & visual tracking
- Facial asymmetries, especially cheekbones and ear placements on skull

Balance and Posture Control

- Persistent head lag beyond 3-4 months of age
- Prone extension / supine flexion
- Inability to remain seated for lengths of time
- Balance on one leg, eyes open and closed
- Lack of rotational of body with postural adjustments
- Retained primitive reflexes beyond age of integration

Historical treatments in sensory integration

Children with gravitational therapy can benefit from working with an occupational therapist trained in sensory integration to improve their vestibular processing and help them feel more secure during movement across various environments. Given well-designed activities and graded exercises for systematic desensitization, children can learn to feel more secure with a wider variety of movements and be more successful with their activities of daily living. Reflex integration activities, sensory calming strategies, a sensory diet plan, and accommodations to cope with a loud environment at school or the community have been a historical part of OT intervention.

Sound and vibrational therapies have been tried for years to treat auditory processing and hyperacusis. Various commercially available listening therapies are precisely controlled sensory information to entrain the sound spectrum. These music programs, grounded in science, are electronically modified to activate listening attention and synchronic bilateral auditory and vestibular processing, with the end goal of self-organizing the nervous system. Some systems have added bone conduction for sound to reach deeper structures of listening with the whole body. Therapeutic deep-pressure brushing and reflex integration programs have been shown anecdotally to reduce high-frequency sound defensiveness.

New discoveries

The complexity of vestibular-auditory anatomy and physiology demands equal respect in consideration of the functional consequences of structural compromise. Here are but a few points to consider. There are known causes of inner ear dysfunction: medications, infections, poor circulation, calcium debris in semicircular canals, and head injuries. Certain antibiotics are recognized as ototoxic: capreomycin, erythromycin, azithromycin, clarithromycin, and vancomycin. History taking of sensory challenges should consider medication history as these antibiotics are well-documented to cause hearing changes and balance disturbances. Nutrition plays a role in neurological and auditory hypersensitivity, such as deficiencies in magnesium, essential fatty acids (DHA-Omega 3's), and Vitamin A.

Most of the skull and the face's osseous structures directly or indirectly affect the encasement of the bony vestibular-auditory apparatus. The birthing process can impose mechanical and tissue restrictions in the cranial vault and the fascial structures that compress into the inner or middle ear, including the Eustachian tubes. The temporal bones and sphenoid segments can be compressed from multiple directions—posteriorly by the frontal bone, superiorly by the parietal bones, anteriorly by the occiput, and lateral compressions from sustained time in the birth canal. Dysfunction of the craniosacral system may result, which means normal physiological movements of osseous plates could be hindered, the dura membrane can be compressed or restricted, and fascial restriction can lead to asymmetrical temporal bone motion that can elicit vertigo or bilateral processing challenges. Temporal bone misalignment and motion asymmetry have been proposed as one cause of vertigo.

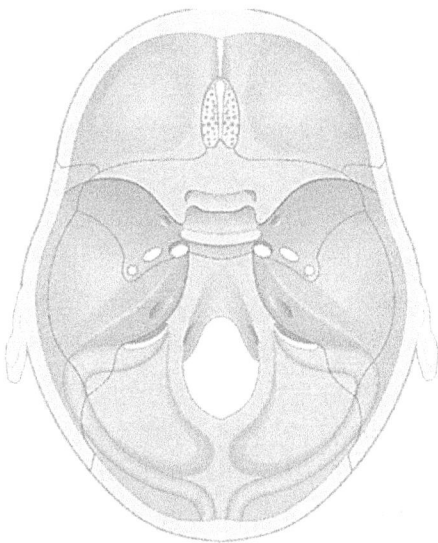

Cartilage of ears are handles to grab temporal bone; with gentle sustained stretch minute shifts occur to mobilize, lift, and balance fascia suspending temporal bone and internal structures. Balancing the fascia container of the cranium helps to balance connective tissue and osseous aspects of the inner ear. This is easiest to do before the ossification of the temporal bone encasement. Treating temporal bone balance through intracranial mobilization is a well-known technique in both the chiropractic and osteopathic fields.

Historically the anatomy and physiology of the Eustachian tubes have yet to be factored into sensory processing theory. The mechanical mobility of the Eustachian tube is a foundational success of middle and inner ear function. The developmental process and the mechanics of swallowing and tongue pressure put into the roof of the mouth naturally drain the Eustachian tubes as long as the tongue has a full range of motion and adequate strength for upward pressure. An expanded theory is that the Eustachian tubes formed at and above the pharyngeal recess act as an acoustic "tunnel" for internal body sound (one's voice, breath, chewing, heartbeat).

Lack of proper physiology of one or both Eustachian tubes can adversely affect the foundation of sound, movement, and perception sensory processing. Dysfunctional tubes can lead to abnormal ear pressure, persistent fullness and overreactive baroreceptors, sound distortion or hearing loss, and movement processing challenges. Accumulated fluids can occur from, and then perpetuate, infection and inflammation in the middle ear (otitis media). Nasal allergies can interfere with optimal sinus drainage leading to back-washed mucous. Esophageal reflux can cause digestive contents to flow to the lower opening of the tubes in the back of the throat.

If both ears cannot structurally transmit equal signaling, a distortion of vestibular signals can lead to balance and posture issues. A distortion in auditory processing can result from two vibratory signals transmitted to the auditory cortexes. A vestibular or auditory sensory processing challenge can interfere with midline orientation. In infancy, head tilting or having a side preference in breastfeeding and positioning can be misinterpreted as torticollis when it might be a coping posture from conflicting vestibular signals. Each nerve, surrounded by its sheath of pia mater, crosses the cistern of the cerebellopontine angle and is situated above the temporal–occipital junction. Compression from all directions can limit the temporal bone and the CN-VII pathway, short as it is to the primary sensory cortex.

The complementary effects of antagonist muscles enable modulating the transmission of sound and the actual awareness of bodily sounds. Dysfunction here can lead to decreased awareness or hypersensitivity to one's bodily sounds. Several reflexes require joint-reactive communication for complex reactions to movement, being moved, and stabilization of eyes for focusing, eye contact, gaze shifts, and visual tracking. The beginning of efficient sensory input to stimulate and integrate these complexities starts with the sensory organ free from mechanical compromise of retained compressions or torsions.

The thyroid regulates such things as muscle tension. There can be structural tightness in the throat, causing the hyoid bone to restrict the thyroid. Tight hyoid attachments relate to tight tongues. Hyoids tighten in a sympathetic state (hard to swallow when stressed) to tighten the upper esophageal sphincter and all the vast investing fascia in the anterior neck. If the hyoid bone in the throat still structures are restricted, there is a possible direct connection to the flexibility and movement of the Eustachian tube muscles. The zygoma and maxilla can be compromised by compressions or distortions within the cranial vault or pulls through the intracranial membrane system caused by the birthing process.

The bony labyrinth receives of the inner ear receives blood supply from three arteries, which also supply the surrounding temporal bone. The status of the temporal bone, its balance at fontanelles, and sutures can directly affect the middle and inner ear because of the softness of the calvarian bone. Considering the vascularization of the inner ear, the lymphatic vessels parallel suspension in the fascia. Blood compromise to internal arteries has been shown to cause deafness, with low-frequency sounds particularly vulnerable to vascular compromise. The vestibular system is more resistant to vascular challenges, but fluid dynamics may play a different role.

Endolymph fluid fills the inner ear's labyrinth and serves as a source of sensory input as the head moves. Vibrational waves transmitted with the fluid displacement convey information about sound, position, and balance if both sides function relatively equally. Endolymph fluid is strikingly different in chemical composition compared to nearby perilymph and cerebrospinal fluid. Perilymph is similar to extracellular fluids and spinal fluids. Conversely, endolymph has higher levels of potassium and a lack of sodium concentration. Endolymph is produced and secreted from specialized epithelial cells in one wall of the cochlea and sites within the membranous labyrinth, creating an active pumping action. Perilymph is similar to the cerebrospinal fluid and surrounds the compartment holding endolymph, so normally these fluids do not comingle. It is, however, anatomically possible for cerebrospinal fluid to enter the cochlea through perilymphatic ducts in the vicinity. Clinical experiences with the sensory consequences of concussions and head injuries, where applying lymphatic and glymphatic mobilization techniques to help the brain drain, raises questions about the health of structures that manage fluid movement and drainage in the inner ear. Vascular and lymph supply implications to vestibular-auditory apparatus have yet to be considered in children and for developmental issues.

Bodywork Observations

Bodywork methods treats the middle and inner ear directly addressing the structures, and is believed to help the structural balance and ease of tensegrity. The structural and physiological health of the inner ear structures contributes to the foundation of sensory processing of auditory and vestibular processing due to the complex intercommunication these systems have with other sensory channels. Because of the intimate proximity, both systems are treated simultaneously by reaching anatomical components of the middle and inner ear via the fascia suspending the surrounding bones.

- CST methods help decompress the osseous plates of the cranium (that may be retained from birthing or acquired head injury). The petrous portion of the temporal bone forms part of the base of the cranium, the base being a vulnerable site. The middle and inner ear structures are housed within the petrous, and a bony tube through the middle of the petrous forms the upper Eustachian tube. Ear infections are a common correlation to vestibular and/or auditory processing issues treated in sensory integration clinics.

- Connective tissue compression can create endolymph and lymphatic stagnation and hinder Eustachian tube drainage. Combining CST with lymphatic and glymphatic mobilization, and acupressure reflects the fusion of treatment options in *Bodywork for Sensory Integration*. It is common to observe a reduction in inner ear infections, especially in younger children.
- CST has been reported to reduce the effects of vertigo and similar disorienting conditions. Many children with vestibular-ocular dysfunction have responded just as favorably. Head lifting, rolling, and head stabilization as well as gaze stabilization are measures examples of evidence measures.
- CST (Upledger's methods) for oral and facial structures has also been reported to reduce Eustachian tube pressure and increase drainage; treating the inner ear structures in a direct way.
- Several acupressure points have proven helpful in individual responses to sensory integration challenges of the inner ear. By theorized release of tension in the Eustachian tubes and sinus pressure relief, reduce tinnitus, and help drain inner ear canals (See pp. 188-191)

Coelho, Carlos M., and Carey D. Balaban. "Visuo-vestibular contributions to anxiety and fear." *Neuroscience & Biobehavioral Reviews* 48 (2015): 148-159.

Cohen, Helen, Bernard Cohen, Theodore Raphan, and Walter Waespe. "Habituation and adaptation of the vestibuloocular reflex: a model of differential control by the vestibulocerebellum." *Experimental brain research* 90, no. 3 (1992): 526-538.

Christine, David C. "Temporal bone misalignment and motion asymmetry as a cause of vertigo: the craniosacral model." *Alternative therapies in health and medicine* 15, no. 6 (2009): 38-42.

Dai, Mingjia, Mikhail Kunin, Theodore Raphan, and Bernard Cohen. "The relation of motion sickness to the spatial–temporal properties of velocity storage." *Experimental brain research* 151, no. 2 (2003): 173-189.

Lopez, Christophe, Olaf Blanke, and F. W. Mast. "The human vestibular cortex revealed by coordinate-based activation likelihood estimation meta-analysis." *Neuroscience* 212 (2012): 159-179.

May-Benson, Teresa A., and Jane A. Koomar. "Identifying gravitational insecurity in children: A pilot study." *The American Journal of Occupational Therapy* 61, no. 2 (2007): 142-147.

Mudry, Albert, and Rinze A. Tange. "The vascularization of the human cochlea: its historical background." *Acta Oto-Laryngologica* 129, no. sup561 (2009): 3-16.

Sankar, Dr U. Ganapathy, and A. Prema. "Reliability of Short Gravitational Insecurity (SGI) Assessment among Indian children." *IOSR journal of Pharmacy and Biological sciences* 9, no. 4 (2014): 41-45.

Postural-Ocular with Bilateral Coordination and Balance

Definition

The outcomes of efficient sensorimotor integration culminate skillsets, grouped by performance categories. Postural-Ocular Integration or Postural Disorder; and Bilateral Coordination are classical sensory integration categories that reflect balance, posture, timing, and rhythm in the movement of both sides of the body. Multi-system processing of vision, vestibular, and kinesthetic senses help adjust for balance and coordination demands. Vision is used to guide and adapt movements to prevent falls. Auditory and tactile systems contribute to peripheral awareness, precision of limb placement, and eloquent reaction times. When not integrated well, movement is not regulated, expressed as ill-timed or clumsy. Posture is regulated by constant feedback and interrelated processing from all sensory systems, including interoception. Movement skills are adaptive responses to the synthesis of vestibular, proprioception, and tactile input matched with visual and auditory co-processing.

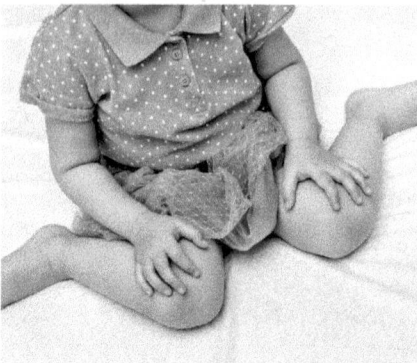

While Postural function is related to core and midline stability, with recovery when the body righting is challenged, bilateral coordination is more complex. Bilateral coordination is first an awareness of left and right body parts, reflecting the whole brain communication between centers. The cerebellum, just now gaining the brain research it rightfully deserves, has vast communication tracts between the cerebral cortex and sensorimotor centers.

Bilateral coordination is the complex use of left and right-sided body parts that move in sync with each; eyes, arms, legs, hands, and feet. Moving in and out of midline, maintaining balance and posture, and preventing falling and hitting the head, is the collaboration of postural control with bilateral coordination. Postural control can be measured by the child's ability to move into and hold postures such as spine flexion and prone extension with equal speed and balanced power. An example of bilateral coordination is eye teaming in visual tracking.

Primitive reflexes are involuntary motor responses originating in the brainstem and spinal cord that facilitate survival and the inherent need to move and ambulate. Most reflexes that facilitate active movement in a baby are inhibited (integrated) by the middle or end of the first year of life, replaced with voluntary control over movement, but may return with the presence of neurological disease. Or, as observed routinely in sensory integration practice, primitive reflexes may still be dominating movements that the child has not been able to gain control over, leading to incoordination, clumsiness, and postural challenges. Integration of primitive reflexes has a profound effect on efficient postural, balance control, and all coordination skills needed for human performance.

Some of the primitive reflexes most commonly addressed in sensory integration practices are:

- Moro Reflex
- Tonic Labyrinthine Reflex (TLR)
- Palmar Grasp Reflex
- Asymmetrical Tonic Neck Reflex (ATNR)
- Spinal Galant
- Symmetrical Tonic Neck Reflex (STNR)

In addition, ocular muscles and cranial nerves also have primitive reflexes addressed in the whole-body realm of sensory integration. They include but are not limited to pupillary reaction, Doll's eye reactions to passive head turning, eyelid reflexes, localization, fixation, pursuits, and gaze shift reflexes. Post-rotary nystagmus was a reflex addressed by Ayres, with the theory that this reflex activated lateral eye muscles in preparation for any pursuit or tracking reflexes. If vestibular processing is not registered or registered equally, the theory holds that eye adjustments to rotary input are not activated. This can lead to balance, posture, and coordination challenges. Ocular motor function is skills of bilateral coordination and reflect whole-system integration of visual, vestibular, and proprioceptive information.

Motor development begins with primitive reflexes that mature into automatic movement reactions to sensory input. Reflexes trigger the development of head control and balance muscle tone between antagonist groups. Primitive movement needs to gain freedom from reflexes to give rise to more mature reactions of righting, equilibrium, and postural reactions that remain as "background" feedback and support mechanisms for body protection. If reflexes are too dominant, the quality of movement and skill will be influenced.

Behavioral manifestation

Postural – Ocular Dysfunction

- Pull to sit with head lag
- (not normal to see after age 3 months)
- Stand / balance on one leg; eyes open & closed
- Sitting tolerance & endurance
- Crossing midline of arms and eyes
- without need to adjust posture or
- midline orientation
- Supine flexion of body is weak
- Prone extension of body is weak
- Focus and smoothly track with eyes
- because head is stable
- Motivation & interest to move
- Hold posture control on escalator
- Cross midline and maintain balance

Bilateral Coordination Dysfunction

- Understand "sidedness" of body
- Comprehend and follow navigational directions
- Understanding concepts of left-right,
- up-down, front-back
- Learning or managing bicycle
- Execute jumping jacks
- Learning basic dance steps
- Copying clapping patterns
- Learning to tie shoe
- Write legibly (writing is a 2-handed task:
- stabilize paper & posture – move pencil
- pencil & change direction of strokes)
- Opening jars and other containers
- Eye tracking (reading) difficulties

Visual motor control is the ability to coordinate perceived images (optic nerve) with motor output (ocular motor nerves). Forming perceptions of such visual information is complicated by if the object is stationary, moving, or has moved between stationary positions. Movement responses are adaptive reactions to all the visual information bombarding the eyes. More inhibition is needed to develop focal attention, or the visual system can be drove to distractions. Addressing the behavioral manifestation of ocular motor dysfunction, additional training is helpful in learning the neurological foundations for which both eyes can team in movement. These skills contribute to the overall quality of Postural – Ocular function and Bilateral Coordination performance.

These types of eye movements can be part of a sensory integration evaluation & measured for response to treatment:

- Fixations: eyes hold steady without moving off target; occurs monocularly and binocularly
- Saccades: eyes make accurate jumps as targets change; in horizontally, vertically, oblique planes
- Pursuits: eyes smoothly follow moving targets as a coordinated team

Skills of fixations include:

- Convergence: coordinated bilateral eye movements inward; focusing on object approaching or held near face
- Divergence: coordinated return from convergence to central position in orbit, still focusing on target
- Gaze stabilization: coordinated eye focused on a stable target while the head is moving in space/side to side

Ocular Accommodation:

- Gaze focus: the lens of the eye to change shape, allowing for a change in focus from far to near
- Refocus: lens rounds when objects are brought into focus and flattens to bring objects into focus

Historical treatments for posture, balance, coordination, & ocular-motor

Children who experience sensory processing challenges can have symptoms from several subtypes. A child with postural dysfunction may have more challenges than a child with bilateral coordination issues. Regardless, day-to-day function can mean that extra effort and energy expenditure is required for all activities. Stress indicators can rise as postural endurance waxes and wanes. Increasing a child's postural strength and control increases the capability of moving more efficiently in space, walk on uneven surfaces without tripping, balance getting on and off the bus, sit or stand upright, and navigate the school terrain. Gaining strength, poise, and endurance increases confidence and raises self-esteem, which in turn makes it easier to develop meaningful peer relationships, play sports or engage in healthy exercise, and trust that safety will prevail if engaged tasks that challenge postural, balance, and coordination.

Therapists trained in Ayres' Sensory Integration methods (most are occupational therapists) spend a dozen weeks learning theory, testing, and treatment certification. This approach raises the standard for reliability by dedicated scientists who train interested therapists in this specialty area. Core concepts of include therapist-structured, child-directed, play-based activities that stimulate and entrained weaker sensory channels. The whole nervous system needs are considered, to maximize a sense of confidence, mastery, and achievement. Tasks are designed to maintain appropriate levels of alertness, and tailored for the "just right" challenges for postural, ocular, and bilateral motor control. Reflex integration is promoted through increasing challenge levels toward complex motor patterns.

The tasks tap intrinsic motivation from the child, allowing activity choice but often within guided parameters. Social engagement intertwines with a therapeutic use of self (therapist) as a playmate within a therapeutic alliance for goal (skill mastery). Sensory integration therapy is not a list of exercises for the child, nor is it a free-for-all in a gym full of fun therapy equipment. The equipment has been designed specifically as therapy tools to address individual needs for postural, balance, and coordination. The therapeutic allegiance between therapist, child, and equipment (activity) makes for the "just right" challenges.

Sensory integration is now considered evidence-based practice within occupational therapy to help children increase success and sensory wellness in participation of daily activities. This approach is not rote-skill training. Instead, activities stimulate subconscious, lower brain centers (cerebellum, spinal pathways, hemisphere crossing sites) t and foundational

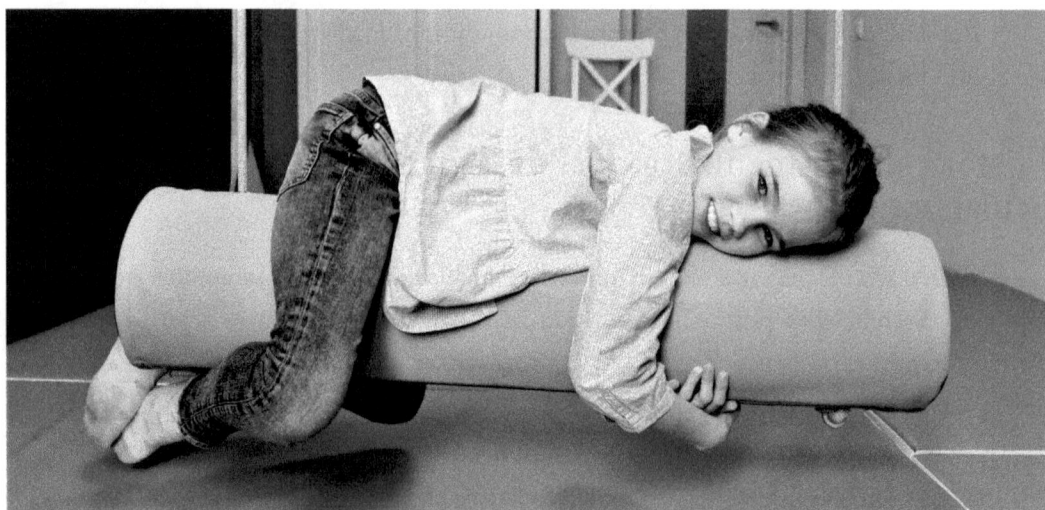

sensorimotor skills that support all activities. Neurological principles apply that facilitating maturation of the whole brain occurs through active participation by the child in tasks that stimulate relevant pathways.

Physical therapists may employ neurodevelopmental therapy (NDT) to address postural, balance, and coordination challenges. With a philosophy that a synchronicity between body stability and body mobility is not optimal. Working through various ways to refine weight shifting and weight bearing, facilitating rotational components of movement, and improve body control through gaining control over narrowing centers-of-gravity. Both OTs and PTs have combined NDT and SI approaches for eclectic and broader scope of therapeutic options in their practices. It's been a good marriage for many therapists.

Vision therapy (developmental or behavioral optometry) has evolved as a specific treatment option for oculomotor challenges. Comparing and contrasting vision therapy to sensory integration shows the approaches comes from different objectives. Sensory integration comes from the science of neurological development and the hierarchal levels of maturation of sensorimotor skills. Entraining the deeper neurological structures and multi-system stimulation to first activate eye muscles in preparation for tracking and visual pursuits is done by first activating, balance, and strengthening the system that activates eye movements: the vestibular system. Sensory integration takes into consideration the multiple systems that drive, support, and monitors each other. For example: baby's eyes begin to focus on a target with dark/light contrast, but it is tactile cues of a hand that grabs that toy with contrast the cues the hand to bring that toy towards the face. This hand-face (mouth) pattern is the beginning of convergence, therefore tactile cues of contact with toys and proprioceptive of the grasp on the toy, and the kinesthesia of the arm moving towards the face all guide the development of visual convergence.

Convergence insufficiency and other treatable vision problems can overlap with, or be misdiagnosed, as sensory processing dysfunction. Convergence insufficiency can be successfully treated by vision therapy but that skill does not develop in isolation from all the other sensory channels. Without sensory integration training, many therapists refer children with tracking (reading, writing behavioral) issues and defer all care to a development optometrist. Our experience is that sensory integration precedes the need for vision therapy, activating cerebellum and bilateral through the larger sensorimotor systems, not just the ocular motor muscles in isolation.

The objectives for sensory integration are to improve targeted areas of struggle such as gross and fine motor coordination, balance, tactile awareness, bilateral awareness, and hand-eye coordination to perform daily tasks optimally. Vision therapy objectives are to strengthen the eye-brain connection through entraining program of ocular motor skills teamed with balance challenge. The inner drive or intrinsic motivation of the child is not a usual factor in vision therapy. Strategies used in a variety of ways (testing, guided eye exercises, occlusion (patching), lenses and prisms) treat a range of visual problems such as tracking for better reading and amblyopia for eye balance. Vestibular entrainment and the triad of vestibular-visual-auditory processing with background feedback of proprioception and kinesthesia lays within the realm of sensory integration.

In the author's long career of working with children with postural-ocular challenges, some participated in vision therapy and some did not. Key postural measurements of dysfunction and measurable changes was the initiation and holding of anti-gravity postures of prone extension and supine flexion (10 seconds each posture). A common correlate was a history of head lag with pull-to-sit test, persisting into childhood. Within 6-8 sessions of sensory integration work we would consistently child would achieve these anti-gravity control milestones, benchmarks of postural readiness. Various sensory integration equipment used to engage in play-based challenge of core isometrics.

There is little research that compares and contrasts outcomes of occupational therapy utilizing sensory integration to that of vision therapy methods. One case study was published comparing two children with both sensory integration and ocular motor issues who received both treatments in isolation, and outcomes were compared.

New discoveries

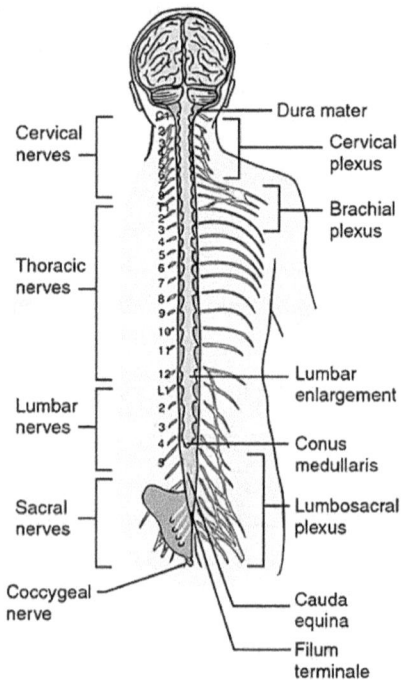

Fascia research suggests a "background" sensory function and structure mainframe for tensegrity of the human body. Tensegrity effects respond to gravitational forces constantly faster than the spinal tract can account for. The speed of sensory and motor nerve conduction does not explain the rapid reaction times bodies have adjusting to pain reception or other stimuli that alert or alarm all the nervous systems. Sensory information for postural control and bilateral coordination may have immediate control within the field of sensory receptors before signals reach the spinal cord.

Anatomy research has revealed that fascia suspension of organs and vascular fields in thoracic and abdominal cavities can affect posture. Body alignment, as well as the autonomic tone continuum, are reflected by peritoneum and mesentery flexibility.

Interoceptors in organs react to postural, balance, and coordination challenges. Emotional reactions to physical activities give feedback back to the viscera. Meninges that create the craniosacral system can be tight, compressed, torqued, or restricted. Most likely, the sources are retained from birthing or early traumas.

Small children can fall many times in the developmental process of learning to walk, run, jump, and climb. Fascia functions as a shock absorber, but it is a false assumption that all tissues release the mechanical stresses under local tissue release. Quite the opposite has been observed. Having had ample opportunity to treat people of all ages, we routinely treat "old injuries" of childhood in adults, often manifested as sites where headaches reoccur or necks that always act up since "I fell off the bike when I was ten."

Cerebellum tracts with Cerebral Cortex – The Upbrain and the Downbrain

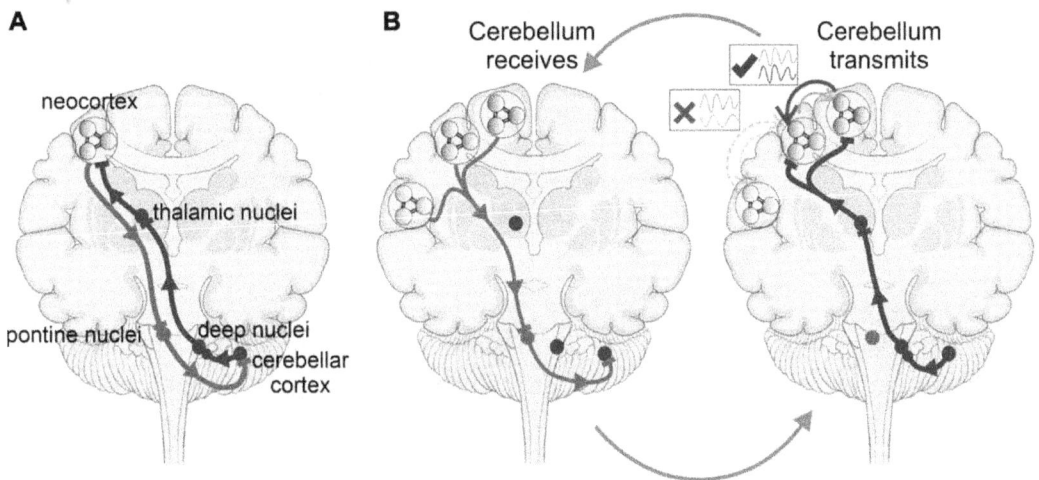

McAfee, Liu, Sillitoe, and Heck/Creative Commons Attribution License

Bilateral coordination reflects the ease of conduction in crossing sensory and motor information between the cerebral hemispheres, thalamus, and cerebellum via the midbrain. The cerebellum and brain stem are vulnerable to compression into the posterior skull and the cranial base. The posterior skull is also susceptible to stagnant occiput lymph nodes, a primary site for energy from whiplash to injure soft tissues. Chronic lymphatic stagnation can hinder fluid exchange.

The Myodural Bridge

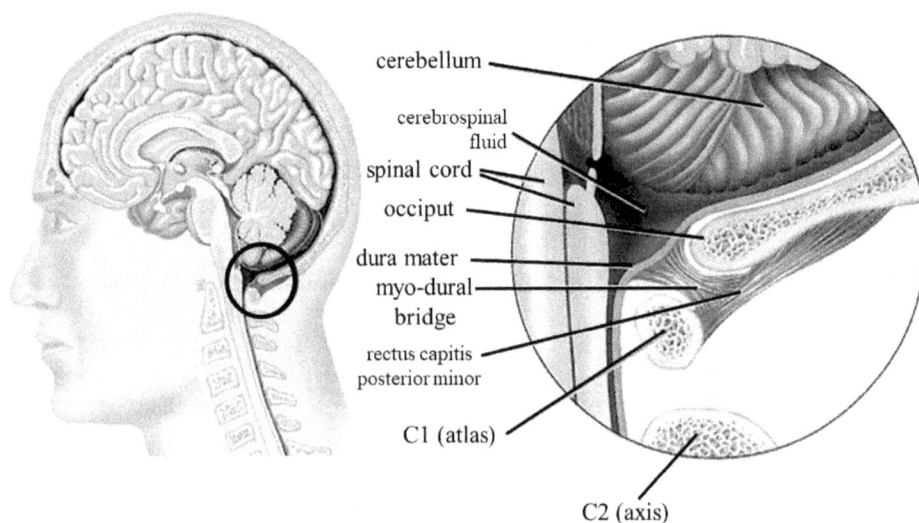

cerebellum

cerebrospinal
fluid

spinal cord

occiput

dura mater

myo-dural
bridge

rectus capitis
posterior minor

C1 (atlas)

C2 (axis)

Illustrator unknown. Extracted from published case study
by G. D. Hack and R.C. Hallgren.

An anatomical structure called the **Myodural Bridge** arises from connective tissue between suboccipital muscles, attaching directly to the dura mater in the upper cervical spine. Communicating through proprioception and kinesthesia, head movements are directly monitored by meningeal tissue at C1 and C2 vertebrae. This functional unit may be activated in visual tracking and stimulated through vestibular input, creating a triad balance system coordinating visual, vestibular, and proprioception directly with the dural meninges. This unit can be felt in active movement by gently placing your fingertips at the base of the occiput and the first cervical vertebrae at midline. Tracking eyes side to side, you should be able to feel active myofascial activation with this ocular motion. First identified in the 1990s, it has been theorized to be one source of cervicogenic headaches following whiplash and head injuries. It is safe to assume this structure plays a role in balance, posture, and coordination. Best treatments for the myodural bridge have yet to be determined, but CST is arguably one.

Physiological action of the rib cage

The ribs are under appreciated in their role in postural and movement integration. The bony constituents of the rib cage include twenty-four ribs, the midline structure (manubrium, sternum, xiphoid process), costal cartilage extensions, and a container of investing fascia. Breath is influenced by positioning and movement of all these parts, and so are posture, balance, and body-side coordination.

The 12 pairs of thin, curved rib bones are classified into three groups: true, false, and floating.

- True ribs (1-7): articulate with manubrium (1) and directly to the sternum (2-7) and vertebrae
- False ribs (8-10) articulate directly with vertebrae and indirectly with the inferior sternum by converging tough but flexible extensions of the seventh costal cartilage (costochondral joint).
- Floating ribs (11-12) articulate only with vertebrae.

Tradition knowledge informs us that the rib cage forms the thoracic cavity, protects vital organs, assists the diaphragm in respiration, and provides structural support for the scapula and upper extremities. Movement of the rib cage reflects rib position, flexibility in intercostal spaces, and movement axes at vertebral and sternal attachments. The ribs are staggered by layers of intercostal muscles and investing fascia.

During inspiration, the ribs elevate, and with expiration, they are depressed (returning to a neutral position). Each rib has an individual anterior-posterior motion with each breath. When the spine flexes, the whole cage displaces posteriorly and displaces anteriorly when the spine extends. Lateral flexion of the trunk of the rib cage also shifts the cage laterally in the opposite direction. In trunk and spinal column rotation, ribs move posteriorly-anteriorly in the transverse plane.

However, rib motion is more complicated than separately linear directions; instead, the cage is a three-dimensional dynamic canister with many parts affecting the total movement pattern. Each rib articulates posteriorly at three places: a tubercle meets the costovertebral joint, and two facets meet the thoracic vertebrae (forming a segment). One exception is that the first rib articulates only with the first thoracic vertebrae. The first ten ribs connect anteriorly with the stabilizing source of the manubrium-sternum with the semi-flexible costal cartilage extensions. The nature of all these articulations creates three functional portions of the rib cage, described by the section's action-in-motion:

- Upper ribs move with a "water pump-handle" motion, with a three-dimensional quality of the individual rib's anterior end shifting outward and upward from the center.
- Middle ribs have a "bucket-handle" swing motion as ribs shift laterally and upwardly.
- Lower ribs have a "caliper" movement where the entire length of the rib shifts laterally & anteriorly.

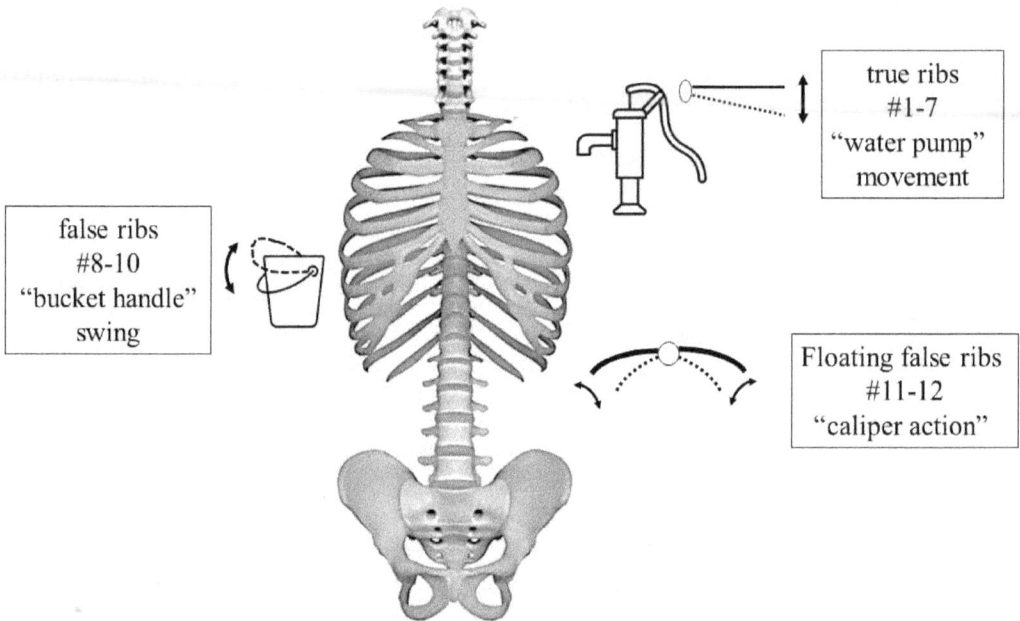

true ribs
#1-7
"water pump"
movement

false ribs
#8-10
"bucket handle"
swing

Floating false ribs
#11-12
"caliper action"

The perspective of three-dimensional movements layered with liner movements layered with individual rib movements makes therapeutic mobilization of ribs a serious endeavor. Traditional linear mobilization of ribs needs to be reconsidered. A more discerning look at the human rib cage offers better specificity in treatment.

As we breathe the ribs have to move in such a way to optimize thorax expansion.

Bodywork Observations

Multiple intersystem processing sensory channels are involved in supporting the functions of posture, balance, coordination, eye-hand dexterity, and visual motor skills. These skills reflect the mobility and flexibility of the whole-body tensegrity. Bodywork methods address the locations of cranial nerves and spinal column tracts, the sensory and motor cortex, and the cerebellum. All the methods addressed in this book hold potential to assist self-correction of the sensorimotor foundations (held in body tissues) for postural-ocular, balance, and coordination skills.

- Craniosacral system: spinal cord & dorsal columns, cranial base, and balancing occiput segments
  - The spinal column needs freedom from compression between vertebral segments to ensure the absence of restrictions of dorsal columns for the best transmission of somatosensory information. The cranial base needs freedom from compression at upper cervical and shoulder structures to ensure optimal tensegrity and ANS

tone through the spinal cord and brain stem enroute to the cerebellum. O-A compression is near the site of these tiny neck muscles. Myodural bridge here, too. Cranial base, occiput condyles rib mobilization on around core bands

- Cranium: craniosacral system, intracranial membrane, fascia container (meninges), brain stem
  - All of the CST that treats the cranial bones are relevant, highlighted by a few examples. The cerebellum has a massive number of tracts connecting with the cerebral hemispheres and thalamus, as well as the spinal cord. Via the peduncles attached to the brain stem, these tracts are vulnerable to mechanical pressures extending superiorly from the cranial base and upper cervical vertebrae, most likely retained compressions from infancy birthing.
  - The sensory, motor, and premotor cortex in the vicinity of the sagittal suture on the crown of the head can be compressed by the frontal bone from either birthing or early head and face injury. Segments of the sphenoid bone (which takes several years to fuse fully) have a structural relationship with the spheno-basilar junction with the occiput. These bones' placement within the fascia container can hold ever-a-slight retained compressions and alignment distortion. Treating bony segments and cranium sutures addresses the exit foramen of cranial nerves.
- The intracranial membranes: contributing to the spherical shape of the head, create tissue drainage pathways called sinuses. Bilateral coordination reflects information crossing between hemispheres and other sensory integration centers. Coordination skills synthesize all sensorimotor information put into practical use and adaptive responses. Areas of information crossing are of interest in the practice of CST as a vulnerable point of the young child's brain, including the corpus callosum, cerebellar tracts, thalamus, optic chiasm, auditory pathway, and spinal tracts (sensory and motor).
- Rib cage motion
  - The wholeness of the cage has strong links to postural control, balance, and movement precision. Mobilization in traditional methods is much too invasive for a young child, and deserving of attention through alternative means. The combination of CST guiding visceral techniques under the careful touch of Responsive Hold has shown better results than myofascial methods. All the intercostal muscles suspending the ribs, the diaphragm attachments, and the crus (posterior projection into lumbar vertebrae of the diaphragm) rest at a critical balance point between the upper and lower body. Restrictions here can affect the center of gravity control with breathing. All soft skeleton and fascial tissues detect balance as background sensory information. Retained reflexes may reflect tightness around the meninges and body fascia bands. Improvement from bodywork has been shown in one leg

balance with eyes open/closed, prone extension, and supine flexion; integration of asymmetrical and symmetrical tonic neck reflexes, head lag reaction, and overall balance and postural reactions. Treating the deep front line fascia band has global implications on rib cage expansion, organ motility, and postural balance between body sides and planes. Along with physical behavior improvements, we very often observe emotional behavior improvement.

- Peripheral fascia matrix and its connectedness to everything
  - In the peripheral body, the flexibility of horizontal and longitudinal fascia bands allows rotation and counter-rotation of the body for complex movement and balance reactions. The external myofascial tissues surrounding the rib cage play a role in thoracic flexibility for this function.
- Peritoneum and mesentery connections of all the thoracic and abdominal organs have a role in the body's internal tensegrity, affecting spinal flexibility and balanced postural reactions. Visceral organs can influence posture by way of peritoneum and mesentery fascia lines-of-pull. This can create off-center pulls and affect midline orientation. The residual effects of torticollis can also affect midline alignment and perceptual orientation.
- Structures of the face
  - For ocular motor coordination (binocularity), the orbit of the eye and the cranial bone constituents need freedom from compression into cranial nerve innervations to extraocular muscles. Visual balance and teaming Both eyes have a special monitoring role in postural and coordination functions. Dysfunctional ocular motor skills may reflect bilateral coordination of the whole body. Convergence insufficiency can result from structural compressions into the soft tissues related to the orbit. Suppose any of these small muscles or the investing fascia that suspends them are adversely affected by retained birthing or mechanical forces. In that case, the neck structures contribute to the success of visual motor skill mastery (tracking, gaze shifts, convergence and divergence, and accommodation).
  - Compressions can keep SNS tone elevated, contributing to pupils' dilation or constriction. Bodywork applies myofascial release with Responsive Hold to the ocular motor muscles (with the utmost gentleness and applied stretch). Precise movement of eye muscles is needed to focus and accommodate. The cranial nerve only needs a slight torque to create muscle spasms, or the orbit pulls on the shape of one eye. *Bodywork for Sensory Integration* has been observed to assist self-correction of eye placement balance in the sockets in developmental optometric issues, post-concussion, head injuries, and general eye strain from overuse of electronics.
  - For inner ear mechanisms for optimal vestibular and auditory processing related to posture, balance, rhythm, and timing, the cranial bones surrounding and connecting

via fascia membranes need freedom from compression. The intracranial membranes need balanced tensegrity and absence of too great a tension for equality of dynamic movement between such structures as the cerebral hemispheres and cerebral ventricles.

Allison, Christine L., Helen Gabriel, Darrell Schlange, and Sara Fredrickson. "An optometric approach to patients with sensory integration dysfunction." *Optometry-Journal of the American Optometric Association* 78, no. 12 (2007): 644-651.

Dalton, Erik. Myoskeletal Alignment Techniques: Posture-Pain Performance. Freedom for Pain publisher. (2014).

Ghasemi, Cobra, Ali Amiri, Javad Sarrafzadeh, Mehdi Dadgoo, and Hasan Jafari. "Comparative study of muscle energy technique, craniosacral therapy, and sensorimotor training effects on postural control in patients with nonspecific chronic low back pain." *Journal of family medicine and primary care* 9, no. 2 (2020): 978.

Girsberger, Wanda, Ulricke Bänziger, Gerhard Lingg, Harald Lothaller, and Peter-Christian Endler. "Heart rate variability and the influence of craniosacral therapy on autonomous nervous system regulation in persons with subjective discomforts: a pilot study." *Journal of integrative medicine* 12, no. 3 (2014): 156-161.

Hack, Gary D, and Richard C. Hallgren. "Chronic headache relief after section of suboccipital muscle dural connections: a case report." *Headache: The Journal of Head and Face Pain* 44, no. 1 (2004): 84-89.

Hurst, Caroline MF, Sarah Van de Weyer, Claire Smith, and Paul M. Adler. "Improvements in performance following optometric vision therapy in a child with dyspraxia." *Ophthalmic and Physiological Optics* 26, no. 2 (2006): 199-210.

Kahkeshani, Kourosh, and Peter J. Ward. "Connection between the spinal dura mater and suboccipital musculature: evidence for the myodural bridge and a route for its dissection—a review." *Clinical Anatomy* 25, no. 4 (2012): 415-422.

Zheng, Nan, Beom Sun Chung, Yi-Lin Li, Tai-Yuan Liu, Lan-Xin Zhang, Yang-Yang Ge, Nan-Xing Wang et al. "The myodural bridge complex defined as a new functional structure." *Surgical and Radiologic Anatomy* 42 (2020): 143-153.

Dyspraxia

Praxis (Greek); /ˈpræk.sis/ "to do, to act"

Dyspraxia is a result of faulty sensory processing affecting fine and gross motor and organizational skills
Collin Dictionary of Medicine

Definition

Dyspraxia is a brain-based motor disorder affecting fine and gross motor skills, motor planning, and coordination. Dyspraxia of speech can be a presenting feature, though not everyone has speech involved. Motor planning dyspraxia is more challenging to recognize and, due to the complexity of the brain processing, requires skilled professional evaluation. Although cognitive skills can be affected, dyspraxia is not a disorder of intelligence. The opposite is true, which can add to the frustration and feelings of self-worth of the person dealing with it. The moment before practice begins. Based on "feeling" the actions that need to occur and generalize previously learned skills to carry out novel motor tasks. "Practice" over and over does not cure dyspraxia but instead teaches rote skills. Recovery from the effects of dyspraxia is a whole-brain activation through carefully constructed therapy activities rich in vestibular, proprioception, and tactile input with visual and auditory feedback. Moving from simple postures and balance challenges into more complex ones for all sub-skills of praxis to be mastered through a child-directed maturation process through active engagement in the therapy process.

Praxis is the moment before the expressed adaptive response to sensory input, representing the planning, organizing, and initiation (or lacking thereof). Skills of praxis (doing by executing movement abilities) emerge from the integration and functional response to sensory information of proprioception, kinesthesia, and vestibular. Visual motor and auditory perception add to the quality and depth of praxis competence. In addition, higher cortex brain centers contribute to the growing repertoire of motor skills matched with intrinsic motivation to engage in activities that development of motor memory and movement organization abilities. Children mature early in movement success to more complex physical abilities by learning and generalizing skills. The emerging science of human motivation likely plays a role in expressing praxis involving intrinsic and extrinsic factors. Subgroups of praxis skills have been identified to help address the components of mastering praxis skills: Motor planning, execution, organization, sequencing, constructional, and ideational.

When underlying sensory processing is inefficient, a condition known as "dyspraxia" can be the functional result. Collier first described this condition in the early 1900s. Occupational therapist and neuroscientist A. Jean Ayres offered the first scientific examination into the condition and formulated formalized testing and treatment (Sensory Integration and Praxis Test) in the 1970s. Other labels given to dyspraxia include: 'clumsy child syndrome,' minimal brain dysfunction, and versions of perceptual-motor dysfunction.

Several areas of development can be affected, which can persist into adulthood, especially without early childhood intervention. Dyspraxia has recently come to be called "Developmental Coordination Disorder (DCD)." subgroups of the condition include verbal apraxia, ideational dyspraxia, fine motor dyspraxia, and whole-body coordination disorder. Verbal apraxia is more readily recognized due to a lack of speech and language maturation, but body dyspraxia requires a more complex approach for differential assessment.

A breakdown or disorganization of praxis can interfere with success in life skills of play, learning, and work. A key element in the diagnosis is difficulties in motor performance in a novel, unpracticed task or lowered success at challenging movements or skills once mastered that had taken extraordinarily more time to achieve.

Dyspraxia, a subgroup of sensory processing dysfunction, may occur in isolation. Still, it frequently coexists with other conditions such as attention deficit and autism spectrum disorders, dyslexia, language disorders, or emotional and learning impairments. Current research supports the theory that developmental dyspraxia, or DCD, results from immaturity of neuron development rather than brain damage. Theoretical constructs suggest that the cerebellum plays a significant role in dyspraxia, and it must be mentioned that science is only now beginning to study the cerebellum seriously

As previously described, praxis skills of motor planning, ideational, sequential, constructional aptitude represents the brain and body's adaptive responses to multiple sensory, cognitive, and motivational center. Dyspraxia (motor type) can be generalized into body movements that are not smooth, executed well, awkward, or lacking timing between all body parts. Verbal and oral dyspraxia can be generalized by the tongue and the speech skills not forming (all) words optimally or moving food well inside the mouth for feeding skills and sound production.

Behavioral manifestation

Participation and mastery of routine tasks such as getting dressed each day promptly and adjusting clothing to look 'polished'; riding a bike safely; writing or typing with accuracy and legibility, enjoying dancing, or successful sports participation are but a few examples of skilled praxis. Time management and organization of space and things can be a co-occurring difficulty with long-term dyspraxia. There can be social-emotional ramifications for someone who is not organized or can't follow the sequential steps of an activity.

People with dyspraxia display difficulties in processing sensorimotor and cognitive information through problems in forming a goal or idea, planning and following steps of a sequence, or trying new unpracticed motor activity. Accident-prone, clumsy, and awkward describes their movements, or, conversely, being a couch potato and actively avoiding engagement in daily exercise or physical activity regularly, breaking toys or equipment, and lack of mastering fine motor or sports skills. Two coping strategies often adopted are excessive talking about things but not doing them or a demonstrated interest in something (riding a bike) but cannot get on, pedal, or balance and ride it.

The pre-school child

- Late milestones (rolling, sitting, standing, walking, and speaking)
- Less efficient than peers in running, jumping, hopping, catching, kicking balls
- Has difficulty in keeping friends; or judging how to behave in company
- Less than optimal spatial awareness, timing to avoid running into obstacles, bumps into things
- Has difficulty in walking up and down stairs
- Lesser organization and independence getting dressed
- Slow and hesitant in movements
- Appears not to be able to learn anything instinctively but must be taught skills
- Difficulty controlling oral structures, breath, and speech production skills
- Difficulty with control over chewing and swallowing during meals

The school age child

- Falls or trips frequently
- Movements are generally poorly organized, clumsy, uncoordinated
- Avoids gym class and playground games
- Performs better with one-on-one teaching than working independently on class assignments
- Has difficulty copying from the blackboard; difficulty forming letters and numbers
- Writes laboriously and immaturely
- Forming proper hand grips on pencils, utensils, and other tools
- Artwork is more immature than peers
- Difficulty recalling or following instructions, especially multi-step directions
- Keeping up with speed and legibility of daily writing assignments
- Needs to relearn movement steps with each attempt; needs to "talk out" movements (often seen in writing and drawing activities)
- Often anxious and easily distracted
- Speech impediment
- Disorganized in supplies, tasks, and physical appearance

Historical treatments in sensory integration

Ayres Sensory Integration® theory and intervention methods within the occupational therapy profession have included a well-researched and lengthy testing procedure. By design with a neurodevelopmental perspective, differential assessment of subtypes of sensory processing dysfunction could be identified and treatment individually tailored. Ayres' Sensory Integration and Praxis Test were comprehensive, reliable, and valid, which guided effective intervention planning and implementation.

There has been a long-standing need to revise Ayres' original testing methods for greater ease and time affordability while staying true to psychometrically strong, internationally appropriate, and accessible measurement tools. The Evaluation of Ayres Sensory Integration® (EASI) results from dedicated scientific clinicians now meet this demand for continuing the validity and reliability of sensory integration testing following Ayres' depth of investigation.

Ayres' sensory integration was uniquely designed for treating dyspraxia; she was the first clinician to recognize the underlying ramifications without apt treatment. Following the neuroscience of the time and adhering to the decades of discoveries, active participation changes the human brain. The art of doing helps brains develop and mature. There are effective strategies, including occupational and speech therapies, to manage Dyspraxia.

Sensory integration therapy is the professionally trained method of functional neurology that allows the child's whole brain becomes active. Unlike many practices that 'apply' procedures or dictate how a child moves through tasks, Ayres' fidelity of treatment is an artful design of following a child's lead into tasks of interest and tasks designed for therapeutic value. Occupation of playing, the key stage of development for a child, is to explore and master play capacities. When a therapy approach is designed for them instead of with them, rote skill teaching may result.

The Just Right Challenge – not too hard not too easy

Treatments were then child-directed designed to tap intrinsic motivation and scaffolded to facilitate complex motor challenges and higher-level adaptive responses. A sensory integration gym was a place rich with a plethora of movement activities for exposure to rich sources of vestibular, proprioception and tactile input. Auditory and visual stimulation were often included into therapeutic activities if identified as a need to improve processing those channels. Scaffolded for success means utilizing child's strengths and interests in the activity design, but hard enough to enhance weaker areas for the 'just right' challenge. Intervention based upon the subskills that are identified through testing to be weaker; using strengths to design therapeutic activities and exercises to remediate and strengthen the ineffectual areas.

New discoveries

Developmental Dyspraxia is now referred to as Developmental Coordination Disorder (DCD) by some sources. A subgroup of sensory integration dysfunction identified by the SIPT compares at "bilateral coordination" dysfunction. Theorized to be processing errors in vestibular, proprioception, and tactile input, bilateral coordination is manifested by errors in the timing left and right body sides with front-back and up-down synchronicity of movement. DCD is a generalized category as a functional diagnostic code rather than the differentiating between subskills of praxis, the latter identifies specific treatment needs and guides remediation.

Neuroimaging science has dramatically helped expand our understanding of brain region patterns in autism, attention deficit disorders, and dyslexia. Dyspraxia and DCD remain one of the least studied and understood neurodevelopmental disorders. However, science has yet to develop neuroimaging methods that can be carried out in anyway but an entirely still position. Typically, this is laying supine in a very tight canister of a machine. Studying movement and coordination dysfunction requires the child to be moving.

The cerebellum (Latin for "little brain") is separate from the cerebral cortex yet connected through vast highways that synthesize information received at the primary sensory homunculus. It has been theorized that as many as five sensorimotor regions synthesize that information and package it for functional use. The cerebellum has a role in virtually all physical movement as it monitors and regulates motor action, reaction, and feedback mechanisms. All without the need for conscious awareness. Not only are balance, postural control, and coordination controlled here, but cognition and speech synthesis are also a product of cerebellum processing.

Following the neurological theory that praxis skills express the synthesis of vestibular, proprioception, and tactile information, one must study the child's brain during active movement. Vestibular systems are believed to activate the cerebellar system. Linear acceleration is believed to be a task that raises awareness and starts deep postural and balance reactions within the cerebellum.

There is an acknowledged link between motor difficulties and brain features of DD (DCD), though few studies systemically explored the brains of children with these functional diagnoses. Based upon preliminary review of a few studies conducted to date, several brain areas are unquestionably linked to DCD: cerebellum, basal ganglia, parietal lobes, part of the frontal lobe (medial orbitofrontal cortex, and dorsolateral prefrontal cortex), though the evidence remains sparse for a thoroughly valid conclusion. In other words, the neural signature of DCD and Dyspraxia has yet to be obtained in brain imagery science.

Bodywork Observations

Treatment for the functional issues related to dyspraxia is one of the most challenging aspects of a sensory integration practice. Bodywork methods applied to this population takes into consideration that a whole-brain processing is involved between many brain centers, therefore a holistic approach is taken to address the individual needs of each child.

- Body considerations – fascia matrix as background and feedback sensation; peripheral sensory and motor pathways are sound in nerve conduction without structural resistance or compression
- Compression through cervical spine is very common and can compromise information flowing through the dorsal columns (main pathway for somatosensory information to brain)
    - Viscera organs – fascia suspension network of organs hangs off vertebral structures.
    - Rib cage mobility – increased ease of body rotation, and lack of lacks rotation
    - Compression into upper cervical structures, most likely retained from the birthing process.
- Assuming babies and small children self-correct mechanical distortions caused by extreme pressures of birthing, with or without intervention, is a false assumption Sutherland's own experiences proves that active compression into a brain can cause a variety of behavioral alterations, and these can resolve once the mechanical compression is relieved.
- Cranium consideration – cranial nerve structures and nerve pathways
    - The cerebellum can be affected structurally by inferior compression of occiput, lateral compressions of the temporal and parietal bones.
    - Segments of the occiput (not fully fused until age 8) can retain birthing compression of the spinal cord / brain stem at the foramen magnum level.
    - Frontal bone can be compressed posteriorly projecting to the sagittal suture, under which lies the sensory/motor/prefrontal cortexes.
    - Parietal lobe compression from the large squamous part of the parietal bones can compress perceptual and perceptual motor centers.
    - Temporal bones are vulnerable to lateral compressions into the cranium that can translate mechanical forces to the inner ear. Sensory systems believed to be involved in dyspraxia include vestibular processing, proprioception, tactile processing; with cognitive and motivational components.
- Intercranial membranes: Mechanical strains or compressions if spontaneous correction has not manifested can be absorbed and retained. Brain structures in close proximity to these membranes are involved in the sensory integration process. The thalamus,

cingulate gyrus, medulla and pons, the corpus collosum and brain stem peduncles are information crossing sites, and any other number of pathways of rapid transport of sensory and of motor information.

- Cranial base – Occiput: Medial compression of occiput condyles around the spinal cord (inferiorly) and the brain stem (superiorly). Occiput segments (4) are not fully fused until 8 years or older and this area (the cranial base) can retain mechanical compressions and tissue distortions from either the birthing process, in utero confinement, or acquired head injuries to the immature skull

Biotteau, Maëlle, Yves Chaix, Mélody Blais, Jessica Tallet, Patrice Péran, and Jean-Michel Albaret. "Neural signature of DCD: a critical review of MRI neuroimaging studies." *Frontiers in Neurology* 7 (2016): 227.

Christmas, Jill, and Rosaline Van de Weyer. *Hands on Dyspraxia: Developmental Coordination Disorder: Supporting Young People with Motor and Sensory Challenges*. Routledge, 2019.

Gibbs, John, Jeanette Appleton, and Richard Appleton. "Dyspraxia or developmental coordination disorder? Unravelling the enigma." *Archives of disease in childhood* 92, no. 6 (2007): 534-539.

Jegadeesan, T., & Nagalakshmi, P. (2020). Effect of sensory integration approach on children with dyspraxia. *Executive Editor*, *11*(12), 88.

Jegadeesan, T., and P. Nagalakshmi. "Effect of sensory integration approach on children with dyspraxia." *Executive Editor* 11, no. 12 (2020): 88.

Lane, Shelly J., Zoe Mailloux, Sarah Schoen, Anita Bundy, Teresa A. May-Benson, L. Diane Parham, Susanne Smith Roley, and Roseann C. Schaaf. "Neural foundations of ayres sensory integration®." *Brain sciences* 9, no. 7 (2019): 153.

Mailloux, Zoe, L. Diane Parham, Susanne Smith Roley, Laura Ruzzano, and Roseann C. Schaaf. "Introduction to the evaluation in ayres sensory integration®(EASI)." *The American Journal of Occupational Therapy* 72, no. 1 (2018): 7201195030p1-7201195030p7.

Smith Roley, Susanne, Zoe Mailloux, Heather Miller-Kuhaneck, and Tara J. Glennon. "Understanding Ayres' sensory integration." (2007). https://digitalcommons.sacredheart.edu/cgi/viewcontent.cgi?article=1017&context=ot_fac

Smith, Marlaine C. *Sensory integration: Theory and practice*. FA Davis, 2019.

Toe Walking

Definition

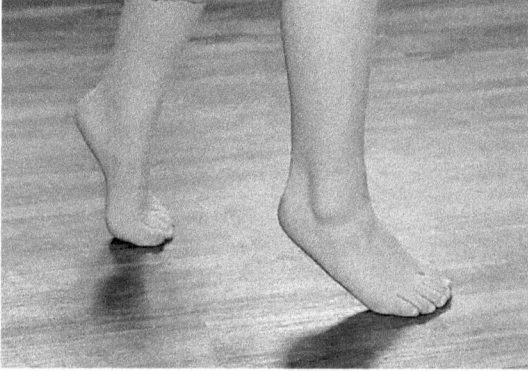

Toe walking is a gait pattern in which a child walks on the balls of their feet, with little to no contact between heels and the ground. It is a common condition to be treated in pediatric therapy clinics and can be a co-occurring issue with other neurodevelopmental conditions. The gastrocnemius and soleus muscles have historically been the anatomy of concern.

Behavioral manifestation

Although toe-walking is considered within the normal walking development of childhood, it becomes abnormal when persisting past two or three years of age. It is described as the child walking on the toes or balls of the foot without the heel or other parts of the foot coming in contact with the floor. Concurrent issues of ankle pronation with subtalar strain or collapse and an imbalance of rear-foot to forefoot dynamics are common. Active dorsiflexion is often missing, and posture and balance reactions can become problematic. Several known causes of toe walking include cerebral palsy, other neurological disorders, or muscular dystrophies. It is a common feature of autism spectrum disorder that occurs with other sensory-motor symptoms, including ataxia, hypotonia, and incoordination. Idiopathic toe walking is the term given when there is no discernable etiology.

Toe walking has also been theorized as a compensation gait pattern of advantage from weakness in other lower extremity muscles. Long-term consequences and complications from prolonged toe walking gait include secondary contractures and musculoskeletal deformity, the cost of prolonged therapies, and surgery with post-surgical management rehabilitation. Sensory disturbances with maladaptation to proprioception, tactile, and vibration input could also contribute.

Historical treatments for toe walking

- Physical therapy for stretching and progressive-resistance exercises of the leg and foot muscles
- Sensory-motor activities: marching in place, walking uphill and uneven surfaces, roller skating, ski boots, squatting exercises, horseback riding with feet in stirrups
- High top tennis shoes with rigid angle panels
- Leg braces, orthotics, splints, or serial casting
- Botox injections
- Surgery

New discoveries

Much of the current information on traditional theories and interventions of toe-walking is exclusive to the musculoskeletal system. However, manual therapies are not mentioned in addressing a "fluid model of human anatomy." It has been observed clinically that more than muscles and tendons are tight and restricted, created by the chronic posturing of feet and ankles. Fascia fields, blood vessels, interstitium, skin, and even soft tissue adhesions to bone contribute to the limits of lengthening of the gastrocnemius/soleus complex (allow for neutral ankle position in standing and walking).

A toxic internal environment with less-than-optimal liver drainage or health can hinder the detoxification pathway, leading to lymphatic stagnation systemically. Lymphatic anatomy is imperative as it is woven through the fascia matrix. Aggressive stretching or deep tissue massage may rupture engorged lymph nodes or push fluids in the wrong direction. Lymphatic fluids flow in distinct directions; protecting the lymphatic valves is imperative. The soft skeleton constructs the tissues that form, suspend, and drain the lymphatic fluid system. 15% of body wastes are pumped through lymphatic pathways, and the musculoskeletal system holds a vast field of veinous pathways. Chronic state of a lack of cleansing can lead to fibrosity of tissue, creating trigger points.

Wearing deep-pressure garments for athletic activities has long been proven to reduce lactic acid. These garments, made from compressive elastic fabric, aid muscle pumping for detoxification. (Therapists have recognized clinically that compression garments have reduced touch defensiveness).

Observations from clients receiving bodywork to address toe walking keep us vigilant for new ideas. Deep bands of fascia, such as the deep front or posterior lines, may also play a role in compensatory gait patterns. The dura mater tube around the spinal cord tightness can pull down through the sacral plexus. Psoas tightness can project restriction into the adjacent mesentery. If the iliopsoas muscles are tight and spastic from chronic spasms, then it is a safe bet that neighboring mesentery suspension of blood vessels could also be affected. Chronic gastrointestinal inflammation related to Crohn's Disease and the irritable bowel has co-occurred with some clients whose primary gait was toe walking.

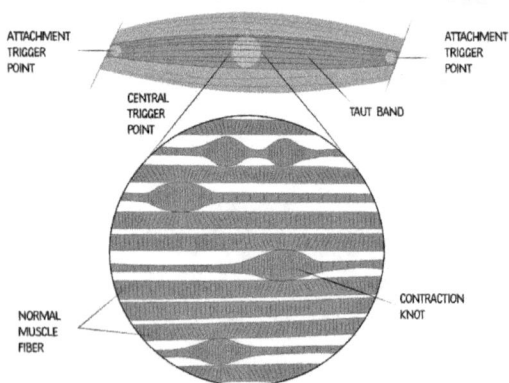

ATTACHMENT TRIGGER POINT

ATTACHMENT TRIGGER POINT

CENTRAL TRIGGER POINT

TAUT BAND

NORMAL MUSCLE FIBER

CONTRACTION KNOT

Bodywork for Sensory Integration has discovered with the gentle application of lymphatic drainage to entire muscle mass, as well as at periosteum and articular regions a consistent improvement in dorsiflexion and standing with full weight bearing is achieved. This has been especially true where historical aggressive stretching, foam rolling, and even dry needling has caused increased tension and tissue firmness.

Bodywork Observations

- Toe walkers may be having chronic muscle spasms (from lack of cellular and fluid cleansing)
- A full or modified Chikly sequence of lymphatic drainage massage routinely gains relaxation of gastrocnemius / soleus muscles, as well as other musculoskeletal tissues; allowing increased dorsiflexion and full foot contact with the floor
- Myofascial layers can become stuck to periosteum around bone (adhesions) causing lack of slide of muscles over bone in movement, especially in a stretch. With the addition of infrared light to soften and liquify adhesions where possible gains have been achieved in freeing length of muscle excursions
- Deep fascia trains, bands, and lines-of-pull have contributed to a whole-body treatment approach helping to reduce toe walking, typically in younger children

Most changes were observed in one to three sessions. Based upon consistent positive outcomes over several years, lymphatic drainage massage and other methods are valuable options to compare and contrast in both treatment and research.

Additional References

Further reading on the topics of anatomy

Adstrum, Sue, Gil Hedley, Robert Schleip, Carla Stecco, and Can A. Yucesoy. "Defining the fascial system." *Journal of bodywork and movement therapies* 21, no. 1 (2017): 173-177.

Armstrong, Colin. "The architecture and spatial organization of the living human body as revealed by intratissular endoscopy–an osteopathic perspective." *Journal of Bodywork and Movement Therapies* 24, no. 1 (2020): 138-146.

Baluk P., Fuxe J., Hashizume H., Romano T., Lashnits E., Butz S., et al. (2007). Functionally specialized junctions between endothelial cells of lymphatic vessels. J. Exp. Med. 204 2349–2362.

Benias, Petros C., Rebecca G. Wells, Bridget Sackey-Aboagye, Heather Klavan, Jason Reidy, Darren Buonocore, Markus Miranda et al. "Structure & distribution of unrecognized interstitium in human tissues." *Scientific reports* 8, no. 1 (2018): 1-8.

Bordoni, Bruno, David Lintonbon, and Bruno Morabito. "Meaning of the solid and liquid fascia to reconsider the model of biotensegrity." Cureus 10, no. 7 (2018).

Fathi, Amir H., Hooman Soltanian, and Alan A. Saber. "Surgical anatomy and morphologic variations of umbilical structures." *The American Surgeon* 78, no. 5 (2012): 540-544.

Glatte, Patrick, Sylvia J. Buchmann, Mido Max Hijazi, Ben Min-Woo Illigens, and Timo Siepmann. "Architecture of the cutaneous autonomic nervous system." *Frontiers in neurology* (2019): 970.

Guimberteau, Jean Claude, and Colin Armstrong. *Architecture of human living fascia: the extracellular matrix and cells revealed through endoscopy*. Handspring Publishing, 2015.

Hammond, George, Luis Yglesias, and James E. Davis. "The urachus, its anatomy and associated fasciae." *The Anatomical Record* 80, no. 3 (1941): 271-287.

Ingber, Donald E. "Tensegrity as the architecture of life." In *Proceedings of IASS Annual Symposia*, vol. 2018, no. 27, pp. 1-4. International Association for Shell and Spatial Structures (IASS), 2018.

Mathias, N., J. Li, and M. Sarkisian. "The human body: An instantaneously and dynamically self-equilibrating tensegrity structure." In *Advances in Engineering Materials, Structures and Systems: Innovations, Mechanics and Applications*, pp. 946-951. CRC Press, 2019.

Natale, Gianfranco, Guido Bocci, and Domenico Ribatti. "Scholars and scientists in the history of the lymphatic system." *Journal of anatomy* 231, no. 3 (2017): 417-429.

Petrova, Tatiana V., and Gou Young Koh. "Biological functions of lymphatic vessels." *Science* 369, no. 6500 (2020): eaax4063.

Pischinger, Alfred. *The extracellular matrix & ground regulation: Basis for holistic biological medicine.* North Atlantic Books, 2007.

Skobe, Mihaela, and Michael Detmar. "Structure, function, and molecular control of the skin lymphatic system." In *Jour Investigative Dermatology Symp Proceedings*, vol. 5, no. 1, pp. 14-19. Elsevier, 2000.

Stecco, Carla, and Robert Schleip. "A fascia and the fascial system." *Journal of bodywork and movement therapies* 20, no. 1 (2016): 139-140.

Ulvmar M. H., Mäkinen T. (2016). Heterogeneity in the lymphatic vascular system and its origin. Cardiovasc. Res. 111 310–321. 10.1093

Further reading on the topics of neurology

Armstrong, Jessica E., Ian Hutchinson, David G. Laing, and Anthony L. Jinks. "Facial electromyography: responses of children to odor and taste stimuli." *Chemical senses* 32, no. 6 (2007): 611-621.

Bjorefeldt, Andreas, Sebastian Illes, Henrik Zetterberg, and Eric Hanse. "Neuromodulation via the cerebrospinal fluid: insights from recent in vitro studies." *Frontiers in Neural Circuits* 12 (2018): 5.

Chang, Yi-Shin, Mathilde Gratiot, Julia P. Owen, Anne Brandes-Aitken, Shivani S. Desai, Susanna S. Hill, Anne B. Arnett, Julia Harris, Elysa J. Marco, and Pratik Mukherjee. "White matter microstructure is associated with auditory and tactile processing in children with and without sensory processing disorder." *Frontiers in neuroanatomy* (2016): 169.

Chang, Yi-Shin, Julia P. Owen, Shivani S. Desai, Susanna S. Hill, Anne B. Arnett, Julia Harris, Elysa J. Marco, and Pratik Mukherjee. "Autism and sensory processing disorders:

shared white matter disruption in sensory pathways but divergent connectivity in social-emotional pathways." *PloS one* 9, no. 7 (2014): e103038.

Farber, Shereen D. *Neurorehabilitation: A multisensory approach*. Saunders, 1982.

Lane, Shelly J., and Roseann C. Schaaf. "Examining the neuroscience evidence for sensory-driven neuroplasticity: Implications for sensory-based occupational therapy for children and adolescents." *The American journal of occupational therapy* 64, no. 3 (2010): 375-390.

Koller, William C. "Primitive reflexes in evaluation of the aging patient." *Clinical Gerontologist* 3, no. 2 (1984): 19-22.

Lane, Shelly J., Zoe Mailloux, Sarah Schoen, Anita Bundy, Teresa A. May-Benson, L. Diane Parham, Susanne Smith Roley, and Roseann C. Schaaf. "Neural foundations of Ayres Sensory Integration®." *Brain sciences* 9, no. 7 (2019): 153.

McIntosh, Daniel N., Lucy Jane Miller, Vivian Shyu, and Randi J. Hagerman. "Sensory-modulation disruption, electrodermal responses, and functional behaviors." *Developmental medicine and child neurology* 41, no. 9 (1999): 608-615.

Nagel, Maximilian, and Alexander T. Chesler. "PIEZO2 ion channels in proprioception." *Current Opinion in Neurobiology* 75 (2022): 102572.

Narayan, Anisha, Mikaela A. Rowe, Eva M. Palacios, Jamie Wren-Jarvis, Ioanna Bourla, Molly Gerdes, Annie Brandes-Aitken, Shivani S. Desai, Elysa J. Marco, and Pratik Mukherjee. "Altered cerebellar white matter in sensory processing dysfunction is associated with impaired multisensory integration and attention." *Frontiers in Psychology* 11 (2021): 618436.

Pryweller, Jennifer R., Kimberly B. Schauder, Adam W. Anderson, Jessica L. Heacock, Jennifer H. Foss-Feig, Cassandra R. Newsom, Whitney A. Loring, and Carissa J. Cascio. "White matter correlates of sensory processing in autism spectrum disorders." *NeuroImage: Clinical* 6 (2014): 379-387.

Owen, Julia P., Elysa J. Marco, Shivani Desai, Emily Fourie, Julia Harris, Susanna S. Hill, Anne B. Arnett, and Pratik Mukherjee. "Abnormal white matter microstructure in children with sensory processing disorders." *Neuroimage: clinical* 2 (2013): 844-853.

van Boxtel, Martin PJ, Hans Bosma, Jelle Jolles, and Fred W. Vreeling. "Prevalence of primitive reflexes and the relationship with cognitive change in healthy adults." *Journal of neurology* 253, no. 7 (2006): 935-941.

Zimmerman, Amanda, Ling Bai, and David D. Ginty. "The gentle touch receptors of mammalian skin." *Science* 346, no. 6212 (2014): 950-954.

Furthering reading on topics of sensory integration and processing

Ayres, J. "Sensory Integration and the Child. Los Angeles, Western Psychological Services, 1979.

Blanche, Erna Imperatore, and Gustavo Reinoso. "The use of clinical observations to evaluate proprioceptive and vestibular functions." *The American Occupational Therapy Association* (2008).

Bodison, Stefanie C., and L. Diane Parham. "Specific Sensory Techniques and Sensory Environmental Modifications for Children and Youth with Sensory Integration Difficulties: A Systematic Review." *The American Journal of Occupational Therapy* 72, no. 1 (2018): 1-7A.

Bundy, A., Lane, S., & Murray, E. (2002). Sensory integration -Theory and practice (2nd ed.) Philadelphia: F. A. Davis.

Critchley, Hugo D., and Sarah N. Garfinkel. "Interoception and emotion." *Current opinion in psychology* 17 (2017): 7-14.

Craig, Arthur D. *How Do You Feel?: An Interoceptive Moment with Your Neurobiological Self.* Princeton University Press, 2015.

Davies, Patricia L., and William J. Gavin. "Validating the diagnosis of sensory processing disorders using EEG technology." *The American Journal of Occupational Therapy* 61, no. 2 (2007): 176-189.

Farb, Norman, Jennifer Daubenmier, Cynthia J. Price, Tim Gard, Catherine Kerr, Barnaby D. Dunn, Anne Carolyn Klein, Martin P. Paulus, and Wolf E. Mehling. "Interoception, contemplative practice, and health." *Frontiers in psychology* 6 (2015): 763.

Gavin, William J., Alycia Dotseth, Kaylea K. Roush, Courtney A. Smith, Hayley D. Spain, and Patricia L. Davies. "Electroencephalography in children with and without sensory processing disorders during auditory perception." *The American Journal of Occupational Therapy* 65, no. 4 (2011): 370-377.

Hedman, Elliot, Sarah A. Schoen, Lucy J. Miller, and Rosalind Picard. "Wireless Measurement of Sympathetic Arousal During in-vivo Occupational Therapy Sessions." *Frontiers in integrative neuroscience* 14 (2020): 539875.

Khalsa, Sahib S., Ralph Adolphs, Oliver G. Cameron, Hugo D. Critchley, Paul W. Davenport, Justin S. Feinstein, Jamie D. Feusner et al. "Interoception and mental health: a roadmap." *Biological psychiatry: cognitive neuroscience and neuroimaging* 3, no. 6 (2018): 501-513.

Kuo, Arthur D. "An optimal state estimation model of sensory integration in human postural balance." Journal of neural engineering 2, no. 3 (2005): S235.

Lane, Shelly J., Zoe Mailloux, Sarah Schoen, Anita Bundy, Teresa A. May-Benson, L. Diane Parham, Susanne Smith Roley, and Roseann C. Schaaf. "Neural foundations of ayres sensory integration®." *Brain sciences* 9, no. 7 (2019): 153.

Mailloux, Zoe, L. Diane Parham, Susanne Smith Roley, Laura Ruzzano, and Roseann C. Schaaf. "Introduction to the evaluation in ayres sensory integration®(EASI)." *The American Journal of Occupational Therapy* 72, no. 1 (2018): 7201195030p1-7201195030p7.

May-Benson, T.A., Koomar, J.A., & Teasdale, A. (2009). Incidence of pre-, peri, and post-natal birth and developmental problems of children with sensory processing disorder and children with autism spectrum disorder. Frontiers in Integrative Neuroscience, 3, 31.

Mori, A. B. "Addressing sensory integration and sensory processing disorders across the lifespan: The role of occupational therapy." *American Occupational Therapy Association* (2015).

Parham, L. Diane. "The relationship of sensory integrative development to achievement in elementary students: Four-year longitudinal patterns." *The Occupational Therapy Journal of Research* 18, no. 3 (1998): 105-127.

Parham, L. Diane, Ellen S. Cohn, Susan Spitzer, Jane A. Koomar, Lucy Jane Miller, Janice P. Burke, Barbara Brett-Green et al. "Fidelity in sensory integration intervention research." *The American Journal of Occupational Therapy* 61, no. 2 (2007): 216-227.

Porges, Stephen W. *The polyvagal theory: Neurophysiological foundations of emotions, attachment, communication, and self-regulation (Norton Series on Interpersonal Neurobiology)*. WW Norton & Company, 2011.

Portwood, Madeleine. *Developmental Dyspraxia: Identification and Intervention A Manual for Parents and Professionals*. David Fulton Publishers, 2018.

Price, Cynthia J., and Carole Hooven. "Interoceptive awareness skills for emotion regulation: Theory and approach of mindful awareness in body-oriented therapy (MABT)." *Frontiers in psychology* 9 (2018): 798.

Reynolds, Stacey, and Shelly J. Lane. "Diagnostic validity of sensory over-responsivity: A review of the literature and case reports." *Journal of autism and developmental disorders* 38, no. 3 (2008): 516-529.

Smith Roley, S., Imperatore Blanche, E., & Schaaf, R. Sensory integration with diverse populations. Tuscon, Arizona: Therapy Skill Builders (2001).

Van Hulle, Carol A., Nicole L. Schmidt, and H. Hill Goldsmith. "Is sensory over-responsivity distinguishable from childhood behavior problems? A phenotypic and genetic analysis." *Journal of Child Psychology and Psychiatry* 53, no. 1 (2012): 64-72.

Zafeiriou, Dimitrios I. "Primitive reflexes and postural reactions in the neurodevelopmental examination." *Pediatric neurology* 31, no. 1 (2004): 1-8.

Furthering reading on topics of manual therapy for diagnostic groups

Au, Doreen WH, Hector WH Tsang, Paul PM Ling, Christie HT Leung, P. K. Ip, and W. M. Cheung. "Effects of acupressure on anxiety: a systematic review and meta-analysis." *Acupuncture in Medicine* 33, no. 5 (2015): 353-359.

Buffone, Francesca, Domenico Monacis, Andrea Gianmaria Tarantino, Fulvio Dal Farra, Andrea Bergna, Massimo Agosti, and Luca Vismara. "Osteopathic Treatment for Gastrointestinal Disorders in Term and Preterm Infants: A Systematic Review and Meta-Analysis." In *Healthcare*, vol. 10, no. 8, p. 1525. MDPI, 2022.

Cerritelli, Francesco, Gianfranco Pizzolorusso, Cinzia Renzetti, Vincenzo Cozzolino, Marianna D'Orazio, Mariacristina Lupacchini, Benedetta Marinelli et al. "A multicenter, randomized, controlled trial of osteopathic manipulative treatment on preterms." *PLoS One* 10, no. 5 (2015): e0127370.

Calk, P. "Best Practices for Oral Motor Stimulation to Improve Oral Feeding in Preterm Infants: A Systematic Review." *Ann Physiother Occup Ther* 2, no. 5 (2019).

Castejón-Castejón, M., M. A. Murcia-González, JL Martínez Gil, J. Todri, M. Suárez Rancel, O. Lena, and R. Chillón-Martínez. "Effectiveness of craniosacral therapy in the treatment of infantile colic. A randomized controlled trial." *Complementary Therapies in Medicine* 47 (2019): 102164.

Chen, Shu-Cheng, Branda Yee-Man Yu, Lorna Kwai-Ping Suen, Juan Yu, Fiona Yan-Yee Ho, Jun-Jun Yang, and Wing-Fai Yeung. "Massage therapy for the treatment of attention deficit/hyperactivity disorder (ADHD) in children and adolescents: a systematic review and meta-analysis." *Complementary Therapies in Medicine* 42 (2019): 389-399.

Field, Tiffany, Tracy Kilmer, Maria Hernandez-Reif, and Iris Burman. "Preschool children's sleep and wake behavior: effects of massage therapy." *Early Child Development and Care* 120, no. 1 (1996): 39-44.

Haller, Heidemarie, Gustav Dobos, and Holger Cramer. "The use and benefits of Craniosacral Therapy in primary health care: A prospective cohort study." *Complementary Therapies in Medicine* 58 (2021): 102702.

Honguten, Agsipa, Keerin Mekhora, Sopa Pichaiyongwongdee, and Sirikarn Somprasong. "Effects of lymphatic drainage therapy on autonomic nervous system responses in healthy subjects: A single blind randomized controlled trial." *Journal of Bodywork and Movement Therapies* 27 (2021): 169-175.

Lanaro, Diego, Nuria Ruffini, Andrea Manzotti, and Gianluca Lista. "Osteopathic manipulative treatment showed reduction of length of stay and costs in preterm infants: A systematic review and meta-analysis." *Medicine* 96, no. 12 (2017). The present systematic review showed the clinical effectiveness of OMT on the reduction of length of stay and costs in a large population of preterm infants.

Lane, S.J., Reynolds, S., & Thacker, L. (2010). Sensory over-responsivity and ADHD: differentiating using electrodermal responses, cortisol, and anxiety. Frontiers in Integrative Neuroscience, 4(8),1-14. doi:10.3389/fnint.2010.00008

Lu, Wei-Peng, Wen-Hui Tsai, Ling-Yi Lin, Rong-Bin Hong, and Yea-Shwu Hwang. "The beneficial effects of massage on motor development and sensory processing in young children with developmental delay: a randomized control trial study." *Developmental neurorehabilitation* 22, no. 7 (2019): 487-495.

Miller Lucy Jane, Nielsen DM, Schoen, SA (2012). Attention deficit hyperactivity disorder and sensory modulation disorder: A comparison of behavior and physiology. Research in Developmental Disabilities 33:804-818.

Miller, Lucy Jane, Joseph R. Coll, and Sarah A. Schoen. "A randomized controlled pilot study of the effectiveness of occupational therapy for children with sensory modulation disorder." *The American Journal of Occupational Therapy* 61, no. 2 (2007): 228-238.

Mur, E., J. Schmidseder, I. Egger, G. Bodner, G. Eibl, F. Hartig, K. P. Pfeiffer, and M. Herold. "Influence of reflex zone therapy of the feet on intestinal blood flow measured by color Doppler sonography." *Forschende Komplementarmedizin und klassische Naturheilkunde: Research in complementary and natural classical medicine* 8, no. 2 (2001): 86-89.

Piravej, Krisna, Preeda Tangtrongchitr, Parichawan Chandarasiri, Luksamee Paothong, and Saengaroon Sukprasong. "Effects of Thai traditional massage on autistic children's behavior." *The Journal of Alternative and Complementary Medicine* 15, no. 12 (2009): 1355-1361.

Pizzolorusso, Gianfranco, Patrizia Turi, Gina Barlafante, Francesco Cerritelli, Cinzia Renzetti, Vincenzo Cozzolino, Marianna D'orazio, Paola Fusilli, Fabrizio Carinci, and Carmine D'incecco. "Effect of osteopathic manipulative treatment on gastrointestinal function and length of stay of preterm infants: an exploratory study." *Chiropractic & manual therapies* 19, no. 1 (2011): 1-6

Pizzolorusso, Gianfranco, Francesco Cerritelli, Alessandro Accorsi, Chiara Lucci, Lucia Tubaldi, Jenny Lancellotti, Gina Barlafante, Cinzia Renzetti, Carmine D'Incecco, and Francesco Paolo Perri. "The effect of optimally timed osteopathic manipulative treatment on length of hospital stay in moderate and late preterm infants: results from a RCT." *Evidence-based complementary and alternative medicine* 2014 (2014).

Raith, Wolfgang, Peter B. Marschik, Constanze Sommer, Ute Maurer-Fellbaum, Claudia Amhofer, Alexander Avian, Elisabeth Löwenstein et al. "General Movements in preterm infants undergoing craniosacral therapy: a randomised controlled pilot-trial." *BMC Complementary and Alternative Medicine* 16 (2015): 1-9.

Raviv, Gil, Shai Shefi, Dalia Nizani, and Anat Achiron. "Effect of craniosacral therapy on lower urinary tract signs and symptoms in multiple sclerosis." *Complementary therapies in clinical practice* 15, no. 2 (2009): 72-75.

Rodrigues, Jorge Magalhães, Mariana Mestre, and Larry Ibarra Fredes. "Qigong in the treatment of children with autism spectrum disorder: A systematic review." *Journal of Integrative Medicine* 17, no. 4 (2019): 250-260.

Silva, Louisa MT, Mark Schalock, Robert Ayres, Carol Bunse, and Sarojini Budden. "Qigong massage treatment for sensory and self-regulation problems in young children with autism: A randomized controlled trial." *The American Journal of Occupational Therapy* 63, no. 4 (2009): 423-432.

Silva, Louisa MT, Mark Schalock, Jodi Garberg, and Cynthia Lammers Smith. "Qigong massage for motor skills in young children with cerebral palsy and Down syndrome." *The American Journal of Occupational Therapy* 66, no. 3 (2012): 348-355.

Song, Hyun Jin, Heejeong Son, Hyun-Ju Seo, Heeyoung Lee, Sun Mi Choi, and Sanghun Lee. "Effect of self-administered foot reflexology for symptom management in healthy persons: a systematic review and meta-analysis." *Complementary therapies in medicine* 23, no. 1 (2015): 79-89.

Steele, William, and Caelan Kuban. *Working with grieving and traumatized children and adolescents: Discovering what matters most through evidence-based, sensory interventions.* John Wiley & Sons, 2013.

Stub, Trine, Mona A. Kiil, Birgit Lie, Agnete E. Kristoffersen, Thomas Weiss, Jill Brook Hervik, and Frauke Musial. "Combining psychotherapy with craniosacral therapy for severe traumatized patients: A qualitative study from an outpatient clinic in Norway." *Complementary Therapies in Medicine* 49 (2020): 102320.

Waits, Alexander, You-Ren Tang, Hao-Min Cheng, Chen-Jei Tai, and Li-Yin Chien. "Acupressure effect on sleep quality: a systematic review and meta-analysis." *Sleep medicine reviews* 37 (2018): 24-34.

Watling, Renee, and Sarah Hauer. "Effectiveness of Ayres Sensory Integration® and sensory-based interventions for people with autism spectrum disorder: A systematic review." *The American Journal of Occupational Therapy* 69, no. 5 (2015): 6905180030p1-6905180030p12.

Wetzler, Gail, Melinda Roland, Sally Fryer-Dietz, and Dee Dettmann-Ahern. "Craniosacral therapy and visceral manipulation: new treatment intervention for concussion recovery." *Medical Acupuncture* 29 no. 4 (2017): 239-248.

Wójcik, Małgorzata, Inga Dziembowska, Paweł Izdebski, and Ewa Żekanowska. "Pilot randomized single-blind clinical trial, craniosacral therapy vs control on physiological reaction to math task in male athletes." *International Journal of Osteopathic Medicine* 32 (2019): 7-12.

Further reading on topics of autism spectrum disorders

Baj, Jacek, Wojciech Flieger, Michał Flieger, Alicja Forma, Elżbieta Sitarz, Katarzyna Skórzyńska-Dziduszko, Cezary Grochowski, Ryszard Maciejewski, and Hanna Karakuła-Juchnowicz. "Autism spectrum disorder: Trace elements imbalances and the pathogenesis and severity of autistic symptoms." *Neuroscience & Biobehavioral Reviews* 129 (2021): 117-132.

Bruchhage, Muriel MK, Maria-Pia Bucci, and Esther BE Becker. "Cerebellar involvement in autism and ADHD." *Handbook of clinical neurology* 155 (2018): 61-72.

Calaprice, Denise, Janice Tona, and Tanya K. Murphy. "Treatment of pediatric acute-onset neuropsychiatric disorder in a large survey population." *Journal of child and adolescent psychopharmacology* 28, no. 2 (2018): 92-103.

Case-Smith, Jane, and Marian Arbesman. "Evidence-based review of interventions for autism used in or of relevance to occupational therapy." *The American Journal of Occupational Therapy* 62, no. 4 (2008): 416-429.

Cowen, Virginia S. *Therapeutic Massage and Bodywork for Autism Spectrum Disorders: A Guide for Parents and Caregivers*. Singing Dragon, 2011.

Dadalko, Olga I., and Brittany G. Travers. "Evidence for brainstem contributions to autism spectrum disorders." *Frontiers in Integrative Neuroscience* 12 (2018): 47.

Exley, Christopher, and Elizabeth Clarkson. "Aluminium in human brain tissue from donors without neurodegenerative disease: A comparison with Alzheimer's disease, multiple sclerosis and autism." *Scientific reports* 10, no. 1 (2020): 1-7.

Fattorusso, Antonella, Lorenza Di Genova, Giovanni Battista Dell'Isola, Elisabetta Mencaroni, and Susanna Esposito. "Autism spectrum disorders and the gut microbiota." *Nutrients* 11, no. 3 (2019): 521.

Fetit, Rana, Robert F. Hillary, David J. Price, and Stephen M. Lawrie. "The neuropathology of autism: A systematic review of post-mortem studies of autism and related disorders." *Neuroscience & Biobehavioral Reviews* 129 (2021): 35-62.

Fiłon, Joanna, Jolanta Ustymowicz-Farbiszewska, and Elżbieta Krajewska-Kułak. "Analysis of lead, arsenic and calcium content in the hair of children with autism spectrum disorder." *BMC Public Health* 20, no. 1 (2020): 1-8.

Geier, David A., Janet K. Kern, Carolyn R. Garver, James B. Adams, Tapan Audhya, Robert Nataf, and Mark R. Geier. "Biomarkers of environmental toxicity and susceptibility in autism." *Journal of the Neurological Sciences* 280, no. 1-2 (2009): 101-108.

Kilroy, Emily, Lisa Aziz-Zadeh, and Sharon Cermak. "Ayres theories of autism and sensory integration revisited: What contemporary neuroscience has to say." *Brain sciences* 9, no. 3 (2019): 68.

Kratz, Susan Vaughan, Jane Kerr, and Lorraine Porter. "The use of CranioSacral therapy for Autism Spectrum Disorders: Benefits from the viewpoints of parents, clients, and therapists." *Journal of Bodywork and Movement Therapies* 21, no. 1 (2017): 19-29.

Lane, Alison E., Cynthia A. Molloy, and Somer L. Bishop. "Classification of Children With Autism Spectrum Disorder by Sensory Subtype: A Case for Sensory-Based Phenotypes." *Autism Research* 7, no. 3 (2014): 322-333.

Lane, Shelly J., Zoe Mailloux, Sarah Schoen, Anita Bundy, Teresa A. May-Benson, L. Diane Parham, Susanne Smith Roley, and Roseann C. Schaaf. "Neural foundations of ayres sensory integration®." *Brain sciences* 9, no. 7 (2019): 153.

Liao, Xiaoli, Yiting Liu, Xi Fu, and Yamin Li. "Postmortem studies of neuroinflammation in autism spectrum disorder: a systematic review." *Molecular Neurobiology* 57, no. 8 (2020): 3424-3438.

Mehra, Anshula, Geetakshi Arora, Manmohit Kaur, Hasandeep Singh, Balbir Singh, and Sarabjit Kaur. "Gut microbiota and Autism Spectrum Disorder: From pathogenesis to potential therapeutic perspectives." *Journal of Traditional and Complementary Medicine* (2022).

Mishra, Durga Prasad, and Anurupa Senapati. "Effectiveness of Combined approach of Craniosacral Therapy (CST) and Sensory-Integration Therapy (SIT) on reducing features in Children with Autism." *Indian Journal of Occupational Therapy* 47, no. 1 (2015): 3-8.

Mold, Matthew, Dorcas Umar, Andrew King, and Christopher Exley. "Aluminium in brain tissue in autism." *Journal of Trace Elements in Medicine and Biology* 46 (2018): 76-82.

Pfeiffer, Beth A., Kristie Koenig, Moya Kinnealey, Megan Sheppard, and Lorrie Henderson. "Effectiveness of sensory integration interventions in children with autism spectrum disorders: A pilot study." *The American journal of occupational therapy* 65, no. 1 (2011): 76-85.

Prizant, Barry M. *Uniquely human: A different way of seeing autism.* Simon and Schuster, 2015.

Thienemann, Margo, Tanya Murphy, James Leckman, Richard Shaw, Kyle Williams, Cynthia Kapphahn, Jennifer Frankovich et al. "Clinical management of pediatric acute-onset neuropsychiatric syndrome: part I—psychiatric and behavioral interventions." *Journal of child and adolescent psychopharmacology* 27, no. 7 (2017): 566-573.

Schaaf, Roseann C. "Interventions that address sensory dysfunction for individuals with autism spectrum disorders: Preliminary evidence for the superiority of sensory integration compared to other sensory approaches." In *Evidence-based practices and treatments for children with autism*, pp. 245-273. Springer, Boston, MA, 2011.

Sealey, L. A., B. W. Hughes, A. N. Sriskanda, J. R. Guest, A. D. Gibson, L. Johnson-Williams, D. G. Pace, and O. Bagasra. "Environmental factors in the development of autism spectrum disorders." *Environment international* 88 (2016): 288-298.

Schoen, Sarah A., Shelly J. Lane, Zoe Mailloux, Teresa May-Benson, L. Dianne Parham, Susanne Smith Roley, and Roseann C. Schaaf. "A systematic review of ayres sensory integration intervention for children with autism." *Autism Research* 12, no. 1 (2019): 6-19.

Sigra, Sofia, Eva Hesselmark, and Susanne Bejerot. "Treatment of PANDAS and PANS: a systematic review." *Neuroscience & Biobehavioral Reviews* 86 (2018): 51-65.

Thom, Robyn P., Christopher J. Keary, Michelle L. Palumbo, Caitlin T. Ravichandran, Jennifer E. Mullett, Eric P. Hazen, Ann M. Neumeyer, and Christopher J. McDougle. "Beyond the brain: a multi-system inflammatory subtype of autism spectrum disorder." *Psychopharmacology* 236, no. 10 (2019): 3045-3061.

Tad Wanveer, L. M. T. "Autism Spectrum Disorder: How CranioSacral Therapy Can Help." *Autism* 7, no. 07 (2007).

Yap, Chloe X., Anjali K. Henders, Gail A. Alvares, David LA Wood, Lutz Krause, Gene W. Tyson, Restuadi Restuadi et al. "Autism-related dietary preferences mediate autism-gut microbiome associations." *Cell* 184, no. 24 (2021): 5916-5931.

Further reading on topics of energy-based interventions

Ebrahimi, Mohammad, Sabokhi Sharifov, Maryam Salili, and Larysia Chernosova. "An introduction to impact of bio-resonance technology in genetics and epigenetics." In *Epigenetics Territory and Cancer*, pp. 495-513. Springer, Dordrecht, 2015.

Daish, Christian, Romane Blanchard, Kate Fox, Peter Pivonka, and Elena Pirogova. "The application of pulsed electromagnetic fields (PEMFs) for bone fracture repair: past and perspective findings." *Annals of biomedical engineering* 46, no. 4 (2018): 525-542.

Dartsch, Peter C. "Effects of the BICOM Optima Mobile Bioresonance Device on Cell Metabolism and Oxidative Burst of Inflammation-Mediating Cells." *Biomedical Journal of Scientific & Technical Research* 33, no. 2 (2021): 25616-25620.

Hennecke, Jürgen. *Bioresonance: A new view of medicine: Scientific principles and practical experience*. BoD–Books on Demand, 2012.

Herrmann, Eckart, and Michael Galle. "Retrospective surgery study of the therapeutic effectiveness of MORA bioresonance therapy with conventional therapy resistant patients suffering from allergies, pain and infection diseases." *European Journal of Integrative Medicine* 3, no. 3 (2011): e237-e244.

Islamov, B. I., R. M. Balabanova, V. A. Funtikov, Yu V. Gotovskii, and E. E. Meizerov. "Effect of bioresonance therapy on antioxidant system in lymphocytes in patients with rheumatoid arthritis." *Bulletin of experimental biology and medicine* 134, no. 3 (2002): 248-250.

Karakos, Periklis, Tripsiannis Grigorios, Konstantinidis Theodoros, and Lialiaris Theodoros. "The Impact of Bioresonance Therapy on Human Health." *Challenges in Disease and Health Research Vol. 10* (2021): 71-81.

Markov, Marko S. "Magnetic field therapy: a review." *Electromagnetic Biology and Medicine* 26, no. 1 (2007): 1-23.

Pihtili, Aylin, Michael Galle, Caglar Cuhadaroglu, Zeki Kilicaslan, Halim Issever, Feyza Erkan, Tulin Cagatay, and Ziya Gulbaran. "Evidence for the efficacy of a bioresonance method in smoking cessation: a pilot study." *Complementary Medicine Research* 21, no. 4 (2014): 239-245.

Resources for professional training

Both national and international education options are available for post-graduate pathways to develop competencies that shaped *Bodywork for Sensory Integration*. Practitioners may have local possibilities for training and mentoring. Training resources include, but are not limited to:

Ayres' Sensory Integration®

The long-awaited revision to formal training by the Collaborative Leadership in Ayres' Sensory Integration (CLASI) is now available for 21st century access. It aims to support the clinical leaders and provide evidenced-based practice with reliable and valid fidelity tools for use in clinical research on Ayres' Sensory Integration® intervention. Learn more about CLASI, their mission statement, board of directors, class schedule, and faculty list at:https://www.cl-asi.org

Barral Institute, USA – visceral and neural manipulation and other parallel trainings

Based upon anatomical and clinical discoveries of Jean Pierre Barral, DO, RPT healthcare practitioners can gain practical application of assisting the functional movements of visceral organs. Various and targeted manual therapies address specific organs, the motility within its confines, and the potential influence of such on the body as a whole. This work expanded into the missing body parts that may hold abnormal forces or distortions, and does so in extremely gentle approaches.

Barral continues to research and develop therapy techniques while maintaining a full clinical practice. Candidates in several European countries must now pass a rigorous test of his methods to earn a diploma in osteopathy, as this profession has adopted these techniques into standard curriculum. In the United States, where osteopathy has evolved away from manipulative interventions, other professionals with licenses to touch in non-invasive manners are invited to attend these classes. Learn more about both research and class schedules associated with this approach at:

https://www.barralinstitute.com/about/index.php

Chikly Health Institute

Bruno Chikly, MD, DO, LMT is a pioneer in contemporary lymphatic drainage method. Created from his aware-winning research on lymphatic system anatomy and other scientific discoveries, his methods teach how to detect and palpate specific rhythm, direction, depth, and quality of lymphatic fluid flow. Discrete tactile skills of listening and directing stagnant interstitial fluids out to the best alternate pathways have been taught internationally by Chikly ad his faculty for over three decades. Improvement to the parasympathetic and immune systems are consistently observed as students learn to work with self-correcting fluid evacuation in muscles and fascia, periosteum and organs, the brain and its meninges, sensory organs and blood vessels. Learn more about the history and class schedules associated with this approach at:

https://chiklyinstitute.com

Neurodevelopmental Therapy – multidisciplinary treatment for neuromuscular dysfunction

The Neurodevelopmental Therapy Association is a nonprofit organization dedicated for decades to the access and promotion of the original Bobath Approach. The NDT practice model operates through clinical education and research to guide practitioners through parallel advances in physical medicine, yet remains true to the original Bobath method of treatment for neuromuscular and movement disorders. The educational model has been a gold standard in physical, occupational, and speech therapy for decades for assessment and functional treatment. Learn more about the NDTA, class schedules, and faculty list at:

https://www.ndta.org/Organization?About-NDTA

Upledger Institute International – craniosacral therapy and other complementary approaches

An internationally recognized educational resource for the innovative and visionary clinical success of John E. Upledger, D.O., O.M.M. (1932-2012). The Institute continues this dedicated work through a large body of certified instructors, teaching in over 100 countries. The Institute is also historically associated with the promotion of visceral and lymphatic treatments as well as other complementary interventions. Upledger's philosophy was to give therapists many options for developing the art of manual therapy competencies. The original work, though, is the foundational craniosacral therapy discoveries made during his seven years of scientific research at Michigan State University and from his extensive clinical practice. His method stands alone in the understanding and clinical guidance of the body's hold of somato-emotional storage. He honed an approach to assist the therapist in navigating the client through recalled traumas or dramas with an artful use of verbal dialogue mixed with noninvasive therapeutic touch. Learn more about the extensive history of the UII, class schedules, and faculty list as well as a vast resource of research and clinical evidence at:

https://upledger.com/about/index.php

Index